Cognitive Training

Tilo Strobach • Julia Karbach
Editors

Cognitive Training

An Overview of Features and Applications

Second Edition

Springer

Editors
Tilo Strobach
Medical School Hamburg
Hamburg, Germany

Julia Karbach
University of Koblenz-Landau
Landau, Germany

ISBN 978-3-030-39294-9 ISBN 978-3-030-39292-5 (eBook)
https://doi.org/10.1007/978-3-030-39292-5

This Springer imprint is published by the registered company Springer Nature Switzerland AG
The registered company address is: Gewerbestrasse 11, 6330 Cham, Switzerland

Contents

About the Editors

Tilo Strobach finished his doctoral degree at Humboldt University Berlin. After that, he held postdoc positions at this university and at LMU Munich. Currently, he is full professor of cognitive psychology at the Medical School Hamburg. His research focuses on the analysis of cognitive plasticity as a result of training (e.g., video game, dual-task, working memory, and task switching training) and cognitive aging. Furthermore, he aims at specifying the cognitive processing architecture in situations that demand executive functions as well as the perception of complex objects.

Julia Karbach is a developmental cognitive psychologist holding a PhD in Psychology from Saarland University. Currently, she serves as full professor at the University of Koblenz-Landau, Germany. Her research interests include cognitive development across the lifespan, cognitive and neural plasticity, psychocardiology, and the prediction of academic achievement. Her work has been published in numerous leading peer-reviewed international journals.

Introduction

Tilo Strobach and Julia Karbach

Content

Abstract The area of cognitive training is a dynamically and fast growing research area that is also increasingly incorporated into scientific education. At the same time, it is characterized by an ongoing debate, particularly regarding the generalizability of training-induced performance gains. The present chapter provides an introduction into this research field and illustrates the framework of the second edition of *Cognitive Training: An Overview of Features and Applications*. This book includes 5 sections and 27 chapters providing comprehensive overviews of state-of-the art research in cognitive training. They focus on basic concepts and methodologies in cognitive training research, cognitive plasticity in different age groups and domains as well as on training in applied domains. The book addresses students and researchers on all academic levels as well as in applied contexts by outlining empirical findings and methodological approaches of cognitive training research in different populations, age groups, and cognitive domains.

Throughout the entire lifespan, individuals are required to adapt to the demands of changing contexts and dynamic social environments. The potential modifiability of a person's cognitive and neural system resulting from these adaptations has been referred to as cognitive and neural plasticity. One way to understand this plasticity is to apply training interventions and to measure the scope of their effects in order

T. Strobach (✉)
Medical School Hamburg, Department Psychology, Hamburg, Germany
e-mail: tilo.strobach@medicalschool-hamburg.de

J. Karbach
University of Koblenz-Landau, Landau, Germany

T. Strobach, J. Karbach (eds.), *Cognitive Training*,
https://doi.org/10.1007/978-3-030-39292-5_1

to identify the mechanisms underlying plastic changes in mind and brain (see Hertzog et al. 2008; Jaeggi et al. 2017; Lustig et al. 2009; Schubert et al. 2014 for reviews).

Over the last decades, the literature on cognitive training interventions has been growing rapidly, demonstrating that cognitive training is a timely issue of high academic as well as societal relevance. For instance, a literature search for "cognitive training" in the abstracting and indexing database PsycINFO on November 26, 2019, demonstrated a total of about 2.868 peer-reviewed contributions since 1966 while the same search in PubMed showed 2.471 peer-reviewed contributions since 1973. The search in PsycINFO demonstrated that more contributions on cognitive training were published between 2013 and 2019 than in all previous years together (1966–2012). This relation was even more extreme in the PubMed search where more contributions on cognitive training were published after 2016 than in all years before 2016. These impressive numbers raise the question why research on cognitive training became so popular in the second decade of this century.

Several factors may have contributed to this development. First – and this certainly influenced many psychological disciplines, including cognitive and experimental psychology – recent decades were characterized by tremendous technical advances. These technical advances also had a large impact on cognitive training research. For instance, they led to computerized experimental set-ups allowing to precisely assess changes in both behavior and neural processing; such precise assessments may be particularly relevant for demonstrating the sometimes rather small effect sizes in cognitive training studies. Further, technical advances also allow the efficient handling and sophisticated analyses of large longitudinal data sets that are very common in studies with extensive training protocols including numerous experimental sessions and groups of participants. With state-of-the art software, data cannot only be analyzed at the group level but training-induced changes can also be modeled at the individual and latent level.

Second, interest in cognitive training has been spurred tremendously as shown by findings that cognitive plasticity and neuronal plasticity are present up to very old age. Earlier accounts assumed that the brain is capable of significant plastic changes only early in life and becomes impervious to change afterwards (e.g., Wiesel and Hubel 1965), suggesting that basic processing capacities cannot be improved by training after early adulthood. However, recent work has clearly established that plasticity is not only present up to very old age (Bavelier et al. 2010; Buschkuehl et al. 2008; Green et al. 2014; Hertzog et al. 2008; Li et al. 2008) but that the mind and brain of older adults can be as plastic as in young adults (Karbach and Verhaeghen 2014, for a meta-analysis).

Another widely discussed issue in the field of cognitive training research is the "curse of specificity," that is, the transferability of training-induced performance gains to untrained tasks and abilities (Green and Bavelier 2008; Karbach and Unger 2014; Shipstead et al. 2012; Sala and Gobet 2017; Strobach et al. 2014). Early cognitive training studies showed that while individuals improved their performance on a task given appropriate training, little to no benefits of this training were seen on new tasks (even if they were seemingly similar to the trained task). Such task specific learning has been shown in nearly all fields of psychology from motor control, to problem-solving, reasoning, general cognition, and education (Ball and Sekuler 1982; Barnett and Ceci 2002; Detterman and Sternberg 1993; Fahle 2004; Schmidt and Bjork 1992). Nevertheless, recent work suggests that training can indeed produce relatively broad, generalizable effects. In fact, examples of general transfer effects are frequently reported in the literature, especially after cognitive training interventions focusing on basic processing capacities, such as working memory or executive functions (for meta-analyses see Au et al. 2015, 2016; Hindin and Zelinski 2012; Karbach and Verhaeghen 2014; Karr et al. 2014; Nguyen et al. 2019; Schwaighofer et al. 2015; Soveri et al. 2017). Moreover, playing video games of the "action video game" genre has been shown to improve a variety of cognitive skills (e.g., Li et al. 2009; Strobach et al. 2012; see Bediou et al. 2018; Toril et al. 2014, for meta-analyses). Aside from these cognitive trainings, there is also ample evidence for positive effects of musical training (Schellenberg 2004) and particularly physical training (see Bherer et al. 2013, for a review and Colcombe and Kramer 2003, Kramer and Colcombe 2018, for meta-analyses) on cognitive abilities across a wide range of ages. However, recent findings also showed the limits and specific conditions of benefits after cognitive training (e.g., Melby-Lervåg and Hulme 2013; Sala and Gobet 2017).

At this point in cognitive training research (more than 50 years after the first publications in the domain of "cognitive training" according to PsycINFO), we aimed at summarizing and reviewing the current state of findings of this first era of cognitive training research. From our perspective, this era is generally characterized as a rather heterogeneous phase in which (1) studies on cognitive research were published that included a variety of designs, methods, and training protocols which unsurprisingly yielded very mixed findings and (2) studies were often less theory driven and theoretical models describing the mechanisms underlying training and effects were mostly missing. Therefore, we aimed at integrating the state-of-the-art of different domains in the field of cognitive training research accompanied by theoretical models describing the mechanisms underlying training and transfer effects.

The fact that the area of cognitive training research is very dynamic is evident when we consider the advances that have occurred since we published the first edition of this book in 2016. In just a few years the field has evolved tremendously in terms of theoretical advances focusing on the underlying mechanisms of training effects and methodological advances and aims to improve methodological standards (e.g., Green et al. 2019). For instance, the field is now characterized by more rigorous research methods with adequate control and transfer conditions, advancing the understanding of cognitive and neural plasticity. Also, with the increase of empirical training studies, there is an increase in meta-analyses, providing an elaborated overview of these studies, which is reflected in the new and updated chapters in the second edition. Furthermore, this field increased in diversity and shows a substantial amount of research in areas that have been under-investigated when the first edition of this book was published (e.g., meta-cognitive training and training of socio-cognitive processes).

We also acknowledge the fact that the last years have shown a change from merely analyzing between-group differences towards focusing on individual differences in training-induced performance gains (e.g. Guye et al. 2017; Jaeggi et al. 2014; Karbach et al. 2017; Könen and Karbach 2015; Lövdén et al. 2012). Many studies shave confirmed patterns of distinct compensation and magnification effects after different types of training, suggesting that individuals respond differently to the same training intervention depending on their age, baseline ability, motivation, personality, genetic predisposition etc. These findings have let many researchers to argue that interventions following the "one size fits all" principle may not be very promising and that interventions tailored to specific needs (on the level of the individual or homogeneous groups of individuals) may be far more promising in order to maximize training-related gains (e.g., Karbach and Unger 2014; Kliegel and Bürki 2012).

As a consequence of these advances in the research on cognitive training, the second edition of this book on cognitive training has a new structure including five sections. The first section covers basic concepts, theory, and methodological issues from a very general perspective (i.e., relevant for different populations, age groups, and cognitive domains). Hence, Taatgen (this volume) presents and elaborates on general theoretical models of training and transfer effects. In addition, Wenger and Kühn (this volume) focus on explaining training and transfer effects by referring to the plasticity of neurons and neuronal networks as a consequence of cognitive training. Researchers who investigate these effects can draw on a well-established methodology for the evaluation of psychological interventions. Doing so, they face the equally well-established long list of critical issues, reducing the validity of findings in studies on cognitive training. Therefore, Schmiedek (this volume) discusses the most common and relevant issues as well as possible methodological solutions while Cochrane and Green (this volume) elaborate on the latest methodological developments in the training field and Könen and Auerswald (this volume) provide statistical solutions to analyze training results on a latent level. Katz et al. (this volume) present the state-of-the-art regarding individual differences in the effectiveness of cognitive training and the role of motivational processes. In addition, Byrne et al. (this volume) provide in introduction into noninvasive brain stimulation techniques to modulate outcomes of cognitive training effects.

Cognitive training is relevant throughout the entire lifespan. Thus, the second section of this book elaborates on the cognitive and neural plasticity in different age groups from a developmental perspective. Since effective cognitive skills are key to learning, socialization, and success to a wide range of real-world outcomes, Rueda et al. (this volume) present the great body of literature on the extent to which cognitive skills can be enhanced through training interventions during childhood and adolescence. Furthermore, probably the most prominent way of applying cognitive training is to use it as a tool against age-related decline in cognitive abilities. Guye et al. (this volume) illustrate promising avenues in this domain.

After these general perspectives on theory, methodology, and age groups of cognitive training, the third section provides details regarding specific cognitive domains targeted during training. Several prominent types of domain-specific training focused on memory training. Therefore, training and transfer effects are reviewed in the domain of working memory (Könen et al., this volume), episodic memory (Wenger et al., this volume), prospective memory (Umanath et al., this volume), and executive functions (Karbach and Kray, this volume).

Similar to the third section, the fourth one is structured by the type of training. However, in contrast to the theoretically well-defined training domains presented in section three, the chapters of this section are structured by more general characteristics. The trainings discussed in these chapters tap different cognitive domains (multidomain training). For instance, video game training – more specifically "action video games" – is characterized by complex visual displays, fast-paced speed, and motivational elements. Therefore, Bediou et al. (this volume) discuss the effects of playing these games on perception and attentional control, while Strobach and Schubert (this volume) focus on potential influences of the action video game playing on executive functions. The following chapters cover the effects of mindfulness training (Verhaeghen this volume), music training (Swaminathan and Schellenberg, this volume), physical training (Bherer and Pothier, this volume), meta-cognitive training (Schaeffner et al., this volume), commercial brain training (Strobach and Kupferberg, this volume), and socio-cognitive processes (Thompson and Steinbeis, this volume).

The focus of the fifth book section is on the applied perspective. Promising ways to apply cognitive training in the educational context are discussed by Johann and Karbach (this volume). DeVries and Geurts (this volume) review findings on cognitive training in children with neurodevelopmental disorders. Focusing on cognitive training as a tool against age-related decline in cognitive abilities, Falkenstein and Gajewski (this volume) summarize training-related neurophysiological changes in older adults and relate them to a discussion of data from EEG training studies with elderly workers. Also with a focus on older adults, Boller et al. (this volume) present different types of cognitive training and show their training and transfer effects in patients with mild cognitive impairment (MCI).

While the previous sections largely focus on the past findings in cognitive training research with a strong theoretical perspective, the final section draws conclusions for future research. That is, Colzato and Hommel (this volume) discuss future developments in this area. For instance, they emphasize the need to develop more specific theories guiding cognitive training programs. With this emphasis, they conclude the theoretical perspective of this book and pave the way for future studies on the effects of cognitive training.

The area of cognitive training is a dynamically and fast growing research area that is increasingly incorporated into scientific teaching and education. The sections of this book provide comprehensive overviews of state-of-the art research in cognitive training. They address students and researchers of all academic levels (i.e., from undergraduates to professors) as well as professionals in applied contexts (e.g., teachers, clinicians, etc.) by outlining empirical findings and methodological approaches of cognitive training research in different populations, age groups, and cognitive domains. We hope that this volume not only serves to summarize the current state of research but also inspires new, exiting, well-designed, and informative studies in this fast-growing scientific field. One of the largest potentials in this area of research lies in the fact that it is very multidisciplinary, integrating research from cognitive, neuropsychological, developmental, educational, and medical science on a theoretical, methodological, and applied level. We believe that this potential may be used in future studies to uncover the cognitive and neural mechanisms underlying training-induced performance benefits and to design adaptive, individually tailored training interventions that can be applied in various contexts, including scientific, educational, and clinical settings.

References

Au, J., Sheehan, E., Tsai, N., Duncan, G. J., Buschkuehl, M., & Jaeggi, S. M. (2015). Improving fluid intelligence with training on working memory: A meta-analysis. *Psychonomic Bulletin & Review, 22*, 366–377.

Au, J., Buschkuehl, M., Duncan, G. J., & Jaeggi, S. M. (2016). There is no convincing evidence that working memory training is NOT effective: A reply to Melby-Lervåg and Hulme (2015). *Psychonomic Bulletin & Review, 23*, 331–337.

Ball, K., & Sekuler, R. (1982). A specific and enduring improvement in visual motion discrimination. *Science, 218*, 697–698.

Barnett, S. M., & Ceci, S. J. (2002). When and where do we apply what we learn?: A taxonomy for far transfer. *Psychological Bulletin, 128*, 612–637.

Bavelier, D., Levi, D. M., Li, R. W., Dan, Y., & Hensch, T. K. (2010). Removing brakes on adult brain plasticity: From molecular to behavioral interventions. *Journal of Neuroscience, 30*, 14964–14971.

Bediou, B., Adams, D. M., Mayer, R. E., Tipton, E., Green, C. S., & Bavelier, D. (2018). Meta-analysis of action video game impact on perceptual, attentional, and cognitive skills. *Psychological Bulletin, 144*, 77–110.

Bherer, L., Erickson, K. I., & Liu-Ambrose, T. (2013). A review of the effects of physical activity and exercise on cognitive and brain functions in older adults. *Journal of Aging Research, 2013*, 657508.

Buschkuehl, M., Jaeggi, S. M., Hutchison, S., Perrig-Chiello, P., Däpp, C., Müller, M., Breil, F., Hoppeler, H., & Perrig, W. J. (2008). Impact of working memory training on memory performance in old-old adults. *Psychology and Aging, 23*, 743–753.

Colcombe, S., & Kramer, A. F. (2003). Fitness effects on the cognitive function of older adults a meta-analytic study. *Psychological Science, 14*, 125–130.

Detterman, D. K., & Sternberg, R. J. (1993). *Transfer on trial: Intelligence, cognition, and instruction*. Westport: Ablex Publishing.

Fahle, M. (2004). Perceptual learning: A case for early selection. *Journal of Vision, 4*, 879–890.

Green, C. S., & Bavelier, D. (2008). Exercising your brain: A review of human brain plasticity and training-induced learning. *Psychology and Aging, 23*, 692–701.

Green, C. S., Strobach, T., & Schubert, T. (2014). On methodological standards in training and transfer experiments. *Psychological Research, 78*, 756–772.

Green, C. S., Bavelier, D., Kramer, A. F., Vinogradov, S., Ansorge, U., Ball, K. K., ... & Facoetti, A. (2019). Improving methodological standards in behavioral interventions for cognitive enhancement. *Journal of Cognitive Enhancement, 3*, 2–29.

Guye, S., De Simoni, C., & von Bastian, C. C. (2017). Do individual differences predict change in cognitive training performance? A latent growth curve modeling approach. *Journal of Cognitive Enhancement, 1*, 374–393.

Hertzog, C., Kramer, A. F., Wilson, R. S., & Lindenberger, U. (2008). Enrichment effects on adult cognitive development can the functional capacity of older adults be preserved and enhanced? *Psychological Science in the Public Interest, 9*, 1–65.

Hindin, S. B., & Zelinski, E. M. (2012). Extended practice and aerobic exercise interventions benefit untrained cognitive outcomes in older adults: A meta-analysis. *Journal of the American Geriatrics Society, 60*, 136–141.

Jaeggi, S. M., Buschkuehl, M., Shah, P., & Jonides, J. (2014). The role of individual differences in cognitive training and transfer. *Memory & Cognition, 42*, 464–480.

Jaeggi, S., Karbach, J., & Strobach, T. (2017). Enhancing brain and cognition through cognitive training. *Journal of Cognitive Enhancement, 4*, 353–357.

Karbach, J., & Unger, K. (2014). Executive control training from middle childhood to adolescence. *Frontiers in Psychology, 5*, 390.

Karbach, J., & Verhaeghen, P. (2014). Making working memory work a meta-analysis of executive-control and working memory training in older adults. *Psychological Science, 25*, 2027–2037.

Karbach, J., Könen, T., & Spengler, M. (2017). Who benefits the most? Individual differences in the transfer of executive control training across the lifespan. *Journal of Cognitive Enhancement, 1*, 394–405.

Karr, J. E., Areshenkoff, C. N., Rast, P., & Garcia-Barrera, M. A. (2014). An empirical comparison of the therapeutic benefits of physical exercise and cognitive training on the executive functions of older adults: A meta-analysis of controlled trials. *Neuropsychology, 28*, 829–845.

Kliegel, M., & Bürki, C. (2012). Memory training interventions require a tailor-made approach: Commentary on McDaniel and Bugg. *Journal of Applied Research in Memory and Cognition, 1*, 58–60.

Könen, T., & Karbach, J. (2015). The benefits of looking at intraindividual dynamics in cognitive training data. *Frontiers in Psychology, 6*, 615.

Kramer, A. F., & Colcombe, S. (2018). Fitness effects on the cognitive function of older adults: A meta-analytic study—Revisited. *Perspectives on Psychological Science, 13*, 213–217.

Li, S. C., Schmiedek, F., Huxhold, O., Röcke, C., Smith, J., & Lindenberger, U. (2008). Working memory plasticity in old age: Practice gain, transfer, and maintenance. *Psychology and Aging, 23*, 731–742.

Li, R., Polat, U., Makous, W., & Bavelier, D. (2009). Enhancing the contrast sensitivity function through action video game training. *Nature Neuroscience, 12*, 549–551.

Lövdén, M., Brehmer, Y., Li, S. C., & Lindenberger, U. (2012). Training-induced compensation versus magnification of individual differences in memory performance. *Frontiers in Human Neuroscience, 6*, 141.

Lustig, C., Shah, P., Seidler, R., & Reuter-Lorenz, P. A. (2009). Aging, training, and the brain: A review and future directions. *Neuropsychology Review, 19*, 504–522.

Melby-Lervåg, M., & Hulme, C. (2013). Is working memory training effective? A meta-analytic review. *Developmental Psychology, 49*, 270–291.

Nguyen, L., Murphy, K., & Andrews, G. (2019). Immediate and long-term efficacy of executive functions cognitive training in older adults: A systematic review and meta-analysis. *Psychological Bulletin, 145*, 698–733.

Sala, G., & Gobet, F. (2017). Does far transfer exist? Negative evidence from chess, music, and working memory training. *Current Directions in Psychological Science, 26*, 515–520.

Schellenberg, E. G. (2004). Music lessons enhance IQ. *Psychological Science, 15*, 511–514.

Schmidt, R. A., & Bjork, R. A. (1992). New conceptualizations of practice: Common principles in three paradigms suggest new concepts for training. *Psychological Science, 3*, 207–217.

Schubert, T., Strobach, T., & Karbach, J. (2014). New directions in cognitive training: On methods, transfer, and application. *Psychological Research, 78*, 749–755.

Schwaighofer, M., Fischer, F., & Bühner, M. (2015). Does working memory training transfer? A meta-analysis including training conditions as moderators. *Educational Psychologist, 50*, 138–166.

Shipstead, Z., Redick, T. S., & Engle, R. W. (2012). Is working memory training effective? *Psychological Bulletin, 138*, 628–654.

Soveri, A., Antfolk, J., Karlsson, L., Salo, B., & Laine, M. (2017). Working memory training revisited: A multi-level meta-analysis of n-back training studies. *Psychonomic Bulletin & Review, 24*, 1077–1096.

Strobach, T., Frensch, P. A., & Schubert, T. (2012). Video game practice optimizes executive control skills in dual-task and task switching situations. *Acta Psychologica, 140*, 13–24.

Strobach, T., Salminen, T., Karbach, J., & Schubert, T. (2014). Practice-related optimization and transfer of executive functions: A general review and a specific realization of their mechanisms in dual tasks. *Psychological Research, 78*, 836–851.

Toril, P., Reales, J. M., & Ballesteros, S. (2014). Video game training enhances cognition of older adults: A meta-analytic study. *Psychology and Aging, 29*, 706–716.

Wiesel, T. N., & Hubel, D. H. (1965). Comparison of the effects of unilateral and bilateral eye closure on cortical unit responses in kittens. *Journal of Neurophysiology, 28*, 1029–1040.

Part I
Basic Concepts and Methodology

Methods and Designs

Florian Schmiedek

Contents

Abstract Cognitive training research faces a number of methodological challenges. Some of these are general to evaluation studies of behavioral interventions, like selection effects that confound the comparison of treatment and control groups with preexisting differences in participants' characteristics. Some challenges are also specific to cognitive training research, like the difficulty to tell improvements in general cognitive abilities from improvements in rather task-specific skills. Here, an overview of the most important challenges is provided along an established typology of different kinds of validity (statistical conclusion, internal, external, and construct validity) that serve as the central criteria for evaluating intervention studies. Besides standard approaches to ensure validity, like using randomized assignment to experimental conditions, emphasis is put on design elements that can help to raise the construct validity of the treatment (like adding active control groups) and of the outcome measures (like using latent factors based on measurement models). These considerations regarding study design are complemented with an overview of data-analytical approaches based on structural equation modeling, which have a number of advantages in comparison to the still predominant approaches based on analysis of variance.

F. Schmiedek (✉)
DIPF | Leibniz Institute for Research and Information in Education, Frankfurt am Main, Germany
e-mail: schmiedek@dipf.de

T. Strobach, J. Karbach (eds.), *Cognitive Training*,
https://doi.org/10.1007/978-3-030-39292-5_2

Introduction

Researchers who aim to investigate the effectiveness of cognitive trainings can draw on the well-established methodology for the evaluation of behavioral interventions in psychology and education (Murnane and Willett 2011; Shadish et al. 2002). Doing so, they face a long list of potential issues that can be characterized as threats to different types of the validity of findings. Here, the most common and relevant threats, as well as possible methodological approaches and study design elements to reduce or rule out these threats in the context of cognitive training studies, will be discussed.

The commonly preferred design for investigating cognitive training interventions is one with random assignment of a sample of participants to training and control groups with pre- and posttest assessments of a selection of tasks chosen to represent one or more cognitive abilities that the training might potentially improve. Significantly larger average improvements on such outcome measures in the training than in a control group are taken as evidence that the training benefits cognition. Such a design indeed clears out a number of potential issues. Certain problems that arise when evaluating cognitive trainings, however, require solutions that go beyond, or modify, commonly used of-the-shelf study design elements. For example, the inclusion of no-treatment control groups for ruling out threats to internal validity and the use of single tasks as outcome measures of transfer effects are associated with certain deficits. In the following, methodological problems and challenges will be discussed along the established typology of statistical conclusion validity, internal and external validity, as well as construct validity (Shadish et al. 2002).

Statistical Conclusion Validity

Statistical conclusion validity refers to whether the association between the treatment and the outcome can be reliably demonstrated. Such demonstration is based on inferential statistics, which can provide evidence that observed differences between experimental groups in posttest scores, or in pretest-to-posttest changes, are unlikely to be due to sampling error (i.e., one group having higher scores simply by chance). Given that existing training studies mostly have relatively small sample sizes (with experimental groups of more than 30–40 participants being rare exceptions), the statistical power to do so often is low, and the findings are in danger of being difficult to replicate and being unduly influenced by outliers and violations of statistical assumptions.

Furthermore, and in light of recent discussions about the replicability of findings and deficient scientific standards in psychological research (e.g., Maxwell et al. 2015), there is the problem that low power might increase researchers' propensity to lapse into fishing-for-effect strategies. Given that (a) the researchers' desired hypothesis often will be that a training has a positive effect, (b) that training studies are resource-intensive, and (c) that the nonregistered analysis of data allows for a number of choices of how exactly to be conducted (Fiedler 2011), it has to be considered a danger that such choices (like choosing subsamples or subsets of outcome tasks) are made post hoc in favor of "finding" significant effects and thereby invalidate the results of inferential test statistics. In combination with publication biases that favor statistically significant over nonsignificant results, such practices in a field with typically low power could lead to a distorted picture of training effectiveness, even in meta-analyses. A general skepticism should therefore be in place regarding all findings that have not been replicated by independent research groups. Regarding the danger of fishing-for-effects practices, preregistration of training studies, including the specific hypotheses and details of data preparation and analysis, is a possible solution, which is well established in the context of clinical trials and gaining acceptance, support, and utilization in science in general (Nosek et al. 2018). In general, effort should be invested to increase statistical power and precision of effect size estimates. Besides large enough sample sizes, this also includes ensuring high reliability of outcome measures and of treatment implementation.

As an alternative to null hypothesis significance testing, which still dominates most of the cognitive training research, the use of a Bayesian inference framework should also be considered (Wagenmakers et al. 2018). A dedicated implementation of such a framework would require the use of knowledge and expectations regarding the distribution of effect sizes as priors in the analyses. Even without consent to such a fully Bayesian perspective, however, the use of Bayes factors offers a useful and sensible alternative to null hypothesis significance testing (Dienes 2016). Particularly when it is not clear whether a training program has any notable effect, and therefore the null hypothesis of no effect is a viable alternative, Bayes factors have the advantage that they allow quantifying evidence for the null hypothesis as well as for the hypothesis of an effect being present. When studies have sufficient statistical power, such analyses can result in strong and conclusive evidence for the null hypothesis, and thereby allow for a sobering acceptance of a certain training not producing the desired effects – something null hypothesis testing cannot provide (see von Bastian et al. 2020, for an evaluation of working memory training studies using Bayes factors).

Internal Validity

Internal validity, that is, a study's ability to unambiguously demonstrate that the treatment has a causal effect on the outcome(s), deserves getting a strong weight when judging the quality of intervention studies. It involves ruling out alternative explanations for within-group changes (including practice effects, maturation, or statistical regression to the mean from pretest to posttest) and/or between-group differences (e.g., systematic selection effects into the treatment condition). Common reactions to these problems are requests to (a) use a control group that allows to estimate the size of the effects due to alternative explanations and to (b) randomly assign participants into the different groups. While intact random assignment assures that the mean differences between groups can be unbiased estimates of the *average causal effect* of the treatment (Holland 1986), several cautionary notes are at place regarding this "gold standard" of intervention studies.

First, the unbiasedness of the estimate refers to the expected value. This does not rule out that single studies (particularly if sample sizes are small) have groups that are not well comparable regarding baseline ability or other person characteristics that might interact with the effectiveness of the training. Therefore, the amount of trust in effect size estimates should only be high for studies with large samples or for replicated (meta-analytic) findings. For single studies with smaller samples, matching techniques based on pretest scores can help to reduce random differences between groups that have an effect on estimates of training effects.

Second, the benefits of randomization get lost if the assignment is not "intact," that is, if participants do not participate in the conditions they are assigned to or do not show up for the posttest. Such lack of treatment integrity or test participation can be associated with selection effects that turn an experiment into a quasi-experiment – with all the potential problems of confounding variables that can affect the estimate of outcome differences. In such cases of originally randomized, but later on nonintact experiments, instrumental variable estimation (using the randomized assignment as an instrument for the realized treatment variable) can be used to still get unbiased estimates of the causal effect of the treatment for the subpopulation of participants who comply with the treatment assignment (Angrist et al. 1996). Instrumental variable estimation requires larger samples, however, than those available in many cognitive training studies.

Third, formal analysis of causal inference based on randomized treatment assignment (Holland 1986) shows that the interpretation of mean group differences as average causal effects is only valid if participants do not interact with each other in ways that make individual outcomes dependent on whether or not particular other participants are assigned to the treatment or the control condition. While this is unlikely to pose a problem if training is applied individually, it could be an issue that has received too little attention in studies with group-based interventions – where interactions among participants might, for example, influence motivation. In such cases, a viable solution is to conduct a cluster-randomized experiment and randomize whole groups of participants into the experimental conditions. If groups systematically differ in outcome levels before the training, however, the power of such a study can be considerably lower than it would be if the same number of participants would be assigned individually to experimental conditions. To achieve sufficient power, often much larger total sample sizes and a careful choice of covariates at the different levels of analysis (i.e., individuals and groups) will be necessary (Raudenbush et al. 2007).

Whenever treatment assignment cannot be random, due to practical or ethical considerations, or when randomization breaks down during the course of the study, careful investigation of potential selection effects is required. This necessitates the availability of an as-complete-as-possible battery of potential confounding variables at pretest. If analyses of such variables indicate group differences, findings cannot unambiguously be attributed to the treatment. Attempts to remedy such group differences with statistical control techniques are associated with strong conceptual (i.e., exhaustiveness of the available information regarding selection effects and correctness of the assumed causal model) and statistical assumptions (e.g., linearity of the relation with the outcome) and should therefore be regarded with great caution. An alternative to regression-based control techniques is post hoc matching and subsample selection based on propensity score analyses (Guo and Fraser 2014). This requires sample sizes that are typically not available in cognitive training research, however. Beneficial alternative design approaches for dealing with situations in which randomization is not possible, or likely to not stay intact, are available, like regression discontinuity designs or instrumental variable approaches (Murnane and Willett 2011), but have received little attention in cognitive training research so far.

Construct Validity

While the demonstration of causal effects of the treatment undoubtedly is a necessity when evaluating cognitive trainings, a strong focus on internal validity and randomization should not distract from equally important aspects of construct validity. Addressing the question of whether the investigated variables really represent the theoretical constructs of interest, construct validity is relevant for both, the treatment as well as the outcome measures.

Regarding the treatment, high internal validity does only assure that one or more aspects that differentiate the treatment from the control condition causally influence the outcome. It does not tell which aspect of the treatment it is, however. Given the complexity of many cognitive training programs and the potential involvement of cognitive processes as well as processes related to motivation, self-concept, test anxiety, and other psychological variables in producing improvements in performance, the comparison to so-called *no-contact control conditions* typically cannot exclude a number of potential alternative explanations of why an effect has occurred. In the extreme case, being in a no-contact control condition and still having to redo the assessment of outcome variables at posttest is so demotivating that performance in the control group declines from pre- to posttest. Such a pattern has been observed in several cognitive training studies and renders the interpretation of significant interactions of groups (training vs. control) and occasions (pretest vs. posttest) as indicating improved cognitive ability very difficult to entertain (Redick 2015). As from a basic science perspective, the main interest is in effects that represent plastic changes of the cognitive system; "active" control conditions therefore need to be designed, which are able to produce the same nonfocal effects, but do not contain the cognitive training ingredient of interest. This is a great challenge, however, given the number and complexity of cognitive mechanisms that potentially are involved in processing of, for example, working memory tasks and that can be affected by training (Von Bastian and Oberauer 2014; Könen et al., this volume). For many of these mechanisms, like the use of certain strategies, practice-related improvements are possible, but would have to be considered exploitations of existing behavioral flexibility, rather than extensions of the range of such behavioral flexibility (Lövdén et al. 2010). If motivational effects are partly due to the joy of being challenged by complex tasks, it also will be difficult to invent tasks of comparably joyful complexity but little demand on working memory. In addition to inventive and meticulous creation of control conditions, it is therefore necessary to assess participants' expectations, task-related motivation, and noncognitive outcomes, before, during, and after the intervention (see also Cochrane and Green, Katz et al., this volume).

Regarding the outcome variables, construct validity needs to be discussed in light of the issue of transfer distance and the distinction between skills and abilities. When the desired outcome of a training is the improvement of a specific skill or the acquisition of a strategy tailored to support performing a particular kind of task, the assessment of outcomes is relatively straightforward – it suffices to measure the trained task itself reliably at pre- and posttest. As the goal of cognitive trainings typically is to improve an underlying broad ability, like fluid intelligence or episodic memory, demonstrating improvements on the practiced tasks is not sufficient, however, as those confound potential changes in ability with performance improvements due to the acquisition of task-specific skills or strategies. It is therefore common practice to employ transfer tasks that represent the target ability but are different from the trained tasks. The question of how different such transfer tasks are from the trained ones is often answered using arguments of face validity and classifications as "near" and "far" that are open to criticism and difficult to compare across studies. What seems far transfer to one researcher might be considered near transfer by another one. Particularly if only single tasks are used as outcome measure for a cognitive ability, it is difficult to rule out alternative explanations that explain improvements with a task-specific *skill*, rather than with improvements in the underlying *ability* (see, e.g., Hayes et al. 2015, or Moody 2009).

The likelihood of such potential alternative explanations can be reduced if the abilities that a training is thought to improve are operationalized with several heterogeneous tasks that all have little overlap with the trained tasks and are dissimilar from each other in terms of paradigm and task content. The analysis of effects can then be conducted on the shared variance of these tasks, preferably using confirmatory factor models. This allows to analyze transfer at the level of latent factors that represent the breadth of the ability construct, replacing the arbitrary classification of "near vs. far" with one that defines "narrow" or "broad" abilities by referring to well-established structural models of cognitive abilities (Noack et al. 2009). If transfer effects can be shown for such latent factors, this renders task-specific explanations less likely.

External Validity

External validity encompasses the generalizability of a study's results to other samples, as well as to other contexts, variations of the intervention's setting, and different outcome variables. As few training studies are based on samples that are representative for broad populations, mostly little is known regarding generalizability to different samples. Furthermore, as findings for certain training programs are only rarely replicated by independent research groups, we only have very limited evidence so far regarding the impact of variations of the context, setting, and of the exact implementation of cognitive trainings. As one rare exception, the Cogmed working memory training (http://www.cogmed.com/) has been evaluated in a number of studies by different research groups and with diverse samples. This has resulted in a pattern of failed and successful replications of effects that has been reviewed as providing little support for the claims that have been raised for the program (Shipstead et al. 2012a, b).

Similarly, generalizations of effects for certain transfer tasks to real-life cognitive outcomes, like everyday competencies and educational or occupational achievement, are not warranted, unless shown with direct measures of these outcomes. Even if transfer tasks are known to have strong predictive validity for certain outcomes, this does not ensure that *changes* in transfer task performance show equally strong relations to *changes* in the outcomes (Rode et al. 2014). Finally, relatively little is known about maintenance and long-term effects of cognitive trainings. Here, the combination of training interventions and longitudinal studies would be desirable. In sum, there is a need for studies that reach beyond the typically used convenience samples and laboratory-based short-term outcomes, as well as beyond research groups' common practice of investigating their own pet training programs – to explore the scope, long-term effects, and boundary conditions of cognitive trainings in a systematic way.

Types of Studies

Trying to optimize the different kinds of validity often leads to conflicts because limited resources prohibit maximization of all aspects simultaneously. Furthermore, certain decisions regarding research design may need be to made against the background of direct conflicts among validity aspects. Maximizing statistical conclusion validity by running an experiment in strictly controlled laboratory conditions, for example, may reduce external validity. Balancing the different kinds of validity when planning studies requires to acknowledge that intervention studies may serve quite different purposes. Green et al. (2019) differentiate *feasibility studies*, *mechanistic studies*, *efficacy studies*, and *effectiveness studies* and discuss important differences between these regarding the study methodology, some of which shall be briefly summarized here (see also Cochrane and Green, this volume).

Feasibility studies serve to probe, for example, the viability of new approaches, the practicality of technological innovations, or the applicability of a training program to a certain population. They are typically implemented before moving to one or more of the other kinds of studies. In feasibility studies, the samples may be small in size, but carefully drawn from the target population to, for example, identify potential implementation problems early on. Control groups may often not be necessary, as the focus is not on demonstrating a causal effect yet. Outcome variables may also be more varied and include aspects like compliance rates or subjective ratings of aspects of the training program.

Mechanistic studies test specific hypotheses deducted from a theoretical framework with the aim of identifying the causally mediating mechanisms and moderating factors underlying training-related performance improvements. As such, they provide the basic research fundamentals on which interventions with applied aims can be built. Furthermore, cognitive intervention studies may also serve to answer general questions about cognitive development and the range of its malleability, as for example in the testing-the-limits paradigm (Lindenberger and Baltes 1995), without the goal of generating available training programs. Trying to confirm or explore specific mechanisms of training-related cognitive changes, mechanistic studies will often require different kinds of training and control conditions (to generate the appropriate experimental contrasts) than efficacy and effectiveness studies, which are rather interested in the combined effect of all cognitive change processes involved. Similarly, the outcome variables of mechanistic studies may rather serve to identify a specific cognitive process than to demonstrate broad transfer effects of practical relevance.

Efficacy studies aim at establishing a causal effect of an intervention in comparison to some placebo or other standard control conditions and at thereby answering the question *"Does the paradigm produce the anticipated outcome in the exact and carefully controlled population of interest when the paradigm is used precisely as intended by the researchers?"* (Green et al. 2019, p. 6). Here, ensuring internal validity is of critical importance, as is construct validity of treatment and outcomes and the consideration of sufficient statistical power.

Finally, effectiveness studies aim at evaluating the outcomes of an intervention when implemented in real-world settings. Because such deployment and scaling up of interventions typically is associated with less control over the sampling of participants and fidelity of the dosage and quality of the intervention; the weighting of prime criteria shifts from internal validity to external validity. Control conditions typically will be the "business-as-usual" that is present without an intervention and a relatively stronger focus will lie on evaluating real-life outcome criteria, unwanted side effects, and long-term maintenance of training gains (Green et al. 2019).

Data Analysis

The standard data-analytical approach to the pretest–posttest control-group design in most studies still is a repeated measures ANOVA with *group* (training vs. control) as a between- and *occasion* (pretest vs. posttest) as a within-subject factor, and with a significant interaction of the two factors taken as evidence that observed larger improvements in the training than in the control group indicate a reliable effect of treatment. If there is interest in individual differences in training effects (Katz et al., this volume), either subgroups or interactions of the within-factor with covariates are analyzed. This approach comes with a number of limitations, however.

First, the associated statistical assumptions of sphericity and homogeneity of (co)variances across groups might not be met. For example, when a follow-up occasion (months or years after training) is added, sphericity is unlikely to hold across the unequally spaced time intervals. When the training increases individual differences in performance more than the control condition, homogeneity of variances might not be provided. Second, participants with missing data on the posttest occasions have to be deleted listwise (i.e., they are completely removed from the analysis). Third, analyses have to be conducted on a single-task level. This means that unreliability of transfer tasks can bias results and that, if several transfer tasks for the same ability are available, analyses have to be conducted either one by one or on some composite score. Fourth, when comparability of experimental groups is not ensured by randomized assignment to conditions, the prominent use of ANCOVA, using the pretest as a covariate to adjust for potential pretreatment group differences in the outcome, can be associated with further problems. Regarding causal inference, controlling for pretest scores will only lead to an unbiased estimate of the causal effect of the treatment if the pretest (plus other observed confounders entered as additional covariates) can be assumed to sufficiently control for all confounding that is due to unmeasured variables (Kim and Steiner in press). If this assumption cannot be made with confidence, but instead the assumptions that unmeasured confounders do influence pretest and posttest scores to the same degree (i.e., that confounding variables are time-invariant trait-like characteristics of the participants) and that the pretest does not influence the treatment assignment are likely to hold, then the use of analyses based on gain scores may be preferable over ANCOVA (Kim and Steiner in press).

The first three potential problems mentioned above can be cleared out by basing analyses on a structural equation modeling framework and using latent change score models (McArdle 2009; see also Könen and Auerswald, this volume). Provided large enough samples, multigroup extensions of these models (Fig. 1) allow testing all the general hypotheses typically addressed with repeated measures ANOVA – and more – while having several advantages: First, assumptions of sphericity and homogeneity of (co)variances are not necessary, as (co)variances are allowed to vary across groups and/or occasions. Second, parameter estimation based on full information maximum likelihood allows for missing data. If there are participants who took part in the pretest but dropped out from the study and did not participate in the posttest, their pretest score can still be included in the analysis and help to reduce bias of effect size estimates due to selective dropout (Schafer and Graham 2002). Third, change can be analyzed using latent factors. This has the advantage that effects can be investigated with factors that (a) capture what is common to a set of tasks that measure the same underlying cognitive ability and (b) are free of measurement error. This provides estimates of training effects that are not biased by unreliability of tasks. It also allows investigating individual differences in change in a way that is superior to the use of individual difference scores, which are known to often lack reliability. For example, the latent change score factor for a cognitive outcome could be predicted by individual differences in motivation, be used to

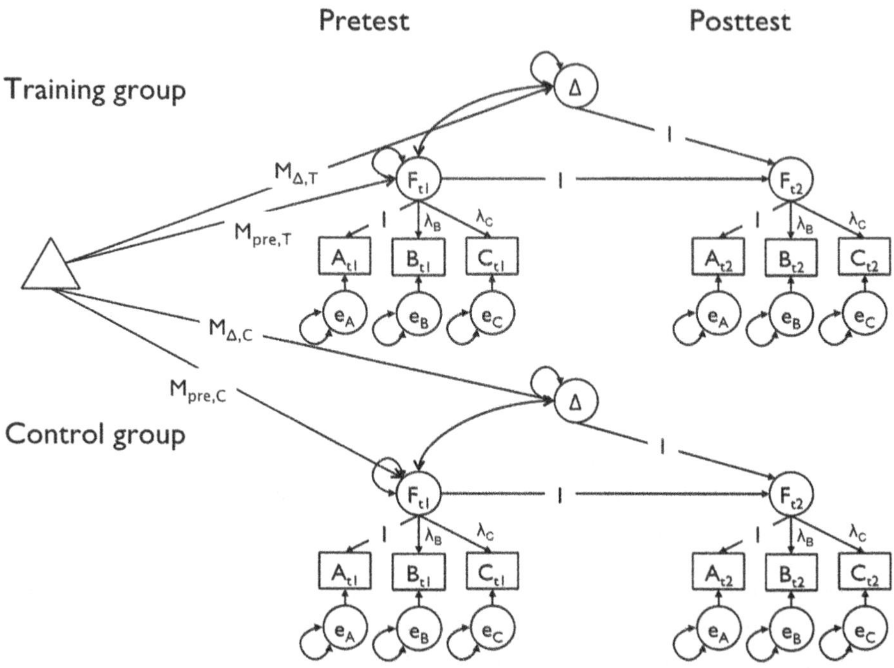

Fig. 1 Two-group latent change score model for pretest–posttest changes in a cognitive training study. Changes are operationalized as the latent difference (Δ) between latent factors at pretest (F_{t1}) and posttest (F_{t2}). These factors capture the common variance of a set of indicator tasks (A, B, and C). Ideally, factor loadings (λ), variances of the residual terms (e), and task intercepts (not shown) are constrained to be equal across groups and occasions (i.e., strict measurement invariance). Based on this model, hypotheses regarding group differences in pretest mean levels (M_{Pre}) and mean changes from pre- to posttest (M_{Δ}) can be investigated, as well as hypotheses regarding the variance and covariance of individual differences in pretest levels and changes (double-headed curved arrows on latent factors)

predict other outcomes (e.g., wellbeing), or be correlated with latent changes in other trained or transfer tasks (e.g., McArdle and Prindle 2008).

Regarding the fourth potential problem of potentially biased estimates in experiments with nonrandom assignment to conditions, latent change score models also allow for a choice between both general options – either analyzing (latent) gain scores or conducting ANCOVA-like adjustments for pretest scores – depending on which assumptions are thought to be more likely to hold.

Furthermore, these models can be extended using the full repertoire of options available in advanced structural equation models. These include multilevel analysis (e.g., to account for the clustering of participants in school classes), latent class analysis (e.g., to explore the presence of different patterns of improvements on a set of tasks), item response models (e.g., to model training-related changes at the level of responses to single items), and more.

Besides a lack of awareness of these advantages, three requirements of latent change score models might explain why they have been used relatively little in cognitive training research so far (Noack et al. 2014). First, these models typically require larger sample sizes than those available in many training studies. When analyzed in a multigroup model with parameter constraints across groups, however, it may be sufficient to have smaller sample sizes in each group than those typically requested for structural equation modeling with single groups. Second, the models require measurement models for the outcome variables of the training. As argued above, operationalizing outcomes as latent variables with heterogeneous task indicators also has conceptual advantages. If only single tasks are available, it still might be feasible to create a latent factor using parallel versions of the task (e.g., based on odd and even trials) as indicator variables. Third, these measurement models need to be invariant across groups and occasions to allow for unequivocal interpretation of mean changes and individual differences therein at the latent factor level (Vandenberg and Lance 2000; see also Könen and Auerswald, this volume). This includes equal loadings, intercepts, and preferably also residual variances of indicator variables. While substantial deviations from measurement invariance can prohibit latent change score analyses, they at the same time can be highly informative, as they can indicate the presence of task-specific effects.

Summary and Outlook

The field of cognitive training research is likely to stay active, due to the demands from societies with growing populations of older adults and attempts to improve the fundamentals of successful education and lifelong learning. As reviewed along the different validity types, this research faces a list of challenges, to which still more could be added (for other methodological reviews and recently discussed issues, see Boot and Simons 2012; Green et al. 2014; Strobach and Schubert 2012; Shipstead et al. 2012a, b; Tidwell et al. 2013). At the same time, awareness of the methodological issues seems to be increasing so that there is a reason to be optimistic that evaluation criteria for commercial training programs (like preregistration of studies) will be established, methodological standards regarding research design will rise, and available advanced statistical methods and new technological developments (like ambulatory assessment methods to assess outcomes in real-life contexts) will be used. Together with basic experimental and neuroscience research on the mechanisms underlying plastic changes in cognition (Wenger and Kühn, this volume), this should lead to better understanding of whether, how, and under which conditions different cognitive training interventions produce desirable effects.

References

Angrist, J. D., Imbens, G. W., & Rubin, D. B. (1996). Identification of causal effects using instrumental variables. *Journal of the American Statistical Association, 91*, 444–455.
Boot, W. R., & Simons, D. J. (2012). Advances in video game methods and reporting practices (but still room for improvement): A commentary on Strobach, Frensch, and Schubert (2012). *Acta Psychologica, 141*, 276–277.
Dienes, Z. (2016). How Bayes factors change scientific practice. *Journal of Mathematical Psychology, 72*, 78–89.
Fiedler, K. (2011). Voodoo correlations are everywhere – not only in neuroscience. *Perspectives on Psychological Science, 6*, 163–171.
Green, C. S., Strobach, T., & Schubert, T. (2014). On methodological standards in training and transfer experiments. *Psychological Research, 78*, 756–772.
Green, C. S., Bavelier, D., Kramer, A. F., Vinogradov, S., Ansorge, U., Ball, K. K., ... & Facoetti, A. (2019). Improving methodological standards in behavioral interventions for cognitive enhancement. *Journal of Cognitive Enhancement, 3*, 2–29.
Guo, S., & Fraser, W. M. (2014). *Propensity Score Analysis: Statistical Methods and Applications* (2nd ed.). Thousand Oaks: Sage Publications, Inc..
Hayes, T. R., Petrov, A. A., & Sederberg, P. B. (2015). Do we really become smarter when our fluid-intelligence test scores improve? *Intelligence, 48*, 1–15.
Holland, P. W. (1986). Statistics and causal inference. *Journal of the American Statistical Association, 81*, 945–960.
Kim, Y., & Steiner, P. M. (in press). Gain scores revisited: A graphical models perspective. *Sociological Methods & Research*. https://doi.org/10.1177/0049124119826155.
Lindenberger, U., & Baltes, P. B. (1995). Testing-the-limits and experimental simulation: Two methods to explicate the role of learning in development. *Human Development, 38*, 349–360.
Lövdén, M., Bäckman, L., Lindenberger, U., Schaefer, S., & Schmiedek, F. (2010). A theoretical framework for the study of adult cognitive plasticity. *Psychological Bulletin, 136*, 659–676.
Maxwell, S. E., Lau, M. Y., & Howard, G. S. (2015). Is psychology suffering from a replication crisis? What does "failure to replicate" really mean? *American Psychologist, 6*, 487–498.
McArdle, J. J. (2009). Latent variable modelling of differences and changes with longitudinal data. *Annual Review of Psychology, 60*, 577–605.
McArdle, J. J., & Prindle, J. J. (2008). A latent change score analysis of a randomized clinical trial in reasoning training. *Psychology and Aging, 23*, 702–719.
Moody, D. E. (2009). Can intelligence be increased by training on a task of working memory? *Intelligence, 37*, 327–328.
Murnane, R. J., & Willett, J. B. (2011). *Methods matter: Improving causal inference in educational and social science research*. Oxford: Oxford University Press.
Noack, H., Lövdén, M., Schmiedek, F., & Lindenberger, U. (2009). Cognitive plasticity in adulthood and old age: Gauging the generality of cognitive intervention effects. *Restorative Neurology and Neuroscience, 27*, 435–453.
Noack, H., Lövdén, M., & Schmiedek, F. (2014). On the validity and generality of transfer effects in cognitive training research. *Psychological Research, 78*, 773–789.
Nosek, B. A., Ebersole, C. R., DeHaven, A. C., & Mellor, D. T. (2018). The preregistration revolution. *PNAS, 115*, 2600–2606.
Raudenbush, S. W., Martinez, A., & Spybrook, J. (2007). Strategies for improving precision in group-randomized experiments. *Educational Evaluation and Policy Analysis, 29*, 5–29.
Redick, T. S. (2015). Working memory training and interpreting interactions in intelligence interventions. *Intelligence, 50*, 14–20.
Rode, C., Robson, R., Purviance, A., Geary, D. C., & Mayr, U. (2014). Is working memory training effective? A study in a school setting. *PLoS One, 9*, e104796.
Schafer, J. L., & Graham, J. W. (2002). Missing data: Our view of the state of the art. *Psychological Methods, 7*, 147–177.

Shadish, W. R., Cook, T. D., & Campbell, D. T. (2002). *Experimental and quasi-experimental designs for causal inference*. Boston: Houghton Mifflin.

Shipstead, Z., Hicks, K. L., & Engle, R. W. (2012a). Cogmed working memory training: Does the evidence support the claims? *Journal of Applied Research in Memory and Cognition, 1*, 185–193.

Shipstead, Z., Redick, T. S., & Engle, R. W. (2012b). Is working memory training effective? *Psychological Bulletin, 138*, 628–654.

Strobach, T., & Schubert, T. (2012). Video game experience and optimized cognitive control skills–On false positives and false negatives: Reply to Boot and Simons (2012). *Acta Psychologica, 141*, 278–280.

Tidwell, J. W., Dougherty, M. R., Chrabaszcz, J. R., Thomas, R. P., & Mendoza, J. L. (2013). What counts as evidence for working memory training? Problems with correlated gains and dichotomization. *Psychonomic Bulletin & Review, 21*, 620–628.

Vandenberg, R. J., & Lance, C. E. (2000). A review and synthesis of the measurement invariance literature: Suggestions, practices, and recommendations for organizational research. *Organizational Research Methods, 3*, 4–70.

Von Bastian, C. C., & Oberauer, K. (2014). Effects and mechanisms of working memory training: A review. *Psychological Research, 78*, 803–820.

von Bastian, C. C., Guye, S., & de Simoni, C. (2020). How strong is the evidence for the effectiveness of working memory training. In J. M. Novick, M. F. Bunting, M. R. Dougherty, & R. W. Engle (Eds.), *Cognitive and working memory training: Perspectives from psychology, neuroscience, and human development* (pp. 58–78). New York: Oxford University Press.

Wagenmakers, E. J., Marsman, M., Jamil, T., Ly, A., Verhagen, J., Love, J., ... & Matzke, D. (2018). Bayesian inference for psychology. Part I: Theoretical advantages and practical ramifications. *Psychonomic Bulletin & Review, 25*, 35–57.

New Directions in Training Designs

Aaron Cochrane and C. Shawn Green

Contents

Abstract Cognitive training is a rapidly expanding domain, both in terms of academic research and commercial enterprise. Accompanying this expansion is a continuing evolution of training design that is driven by advances on various fronts. Foundational learning principles such as spacing and interleaving have always, and continue to, inform the design of training for cognitive improvements, yet advances are constantly made in how to best instantiate these principles in training paradigms. Improvements in hardware have allowed for training to be increasingly immersive (e.g., using virtual reality) and to include multifaceted measurements and dynamics (e.g., using wearable technology and biofeedback). Further, improved training algorithms and gamification have been hallmarks of advances in training software. Alongside the development of these tools, researchers have also increasingly established cognitive training as a more coherent field through an emerging consensus regarding the appropriate methods (e.g., control group selection and tasks to test generalization) for different possible studies of training-related benefits. Hardware, software, and methodological developments have

A. Cochrane (✉) · C. S. Green
Department of Psychology, University of Wisconsin-Madison, Madison, WI, USA
e-mail: akcochrane@wisc.edu; cshawn.green@wisc.edu

T. Strobach, J. Karbach (eds.), *Cognitive Training*,
https://doi.org/10.1007/978-3-030-39292-5_3

quickly made cognitive training an established field, yet many questions remain. Future studies should address the extent and type of generalization induced by training paradigms while taking into account the many possible patterns of improvements from training. Patterns of benefits vary across training types as well as individuals, and understanding individual differences in training benefits will help advance the field. As the field of cognitive training matures, the upcoming years are set to see a proliferation of innovation in training design.

Introduction

Cognitive training has existed, in something like its current form, for only a few decades. It is therefore not surprising that, like many fledgling domains, the field continues to be rife with rapid change and advancement. This is especially true given the fact that, unlike many other areas of psychology, many questions in the cognitive training sphere are not of purely academic or theoretical nature. Instead, the potential for the commercialization of cognitive training has frequently pushed current practices as well (although not always with methodology to demonstrate efficacy to match – see below). Concurrently, advancements in computer hardware as well as training software have facilitated research and applications of training in increasingly diverse and ecologically valid contexts. Here we focus on recent advances (e.g., improvements in hardware and software capabilities), endemic challenges (e.g., as related to methods for controlling for expectation effects or how to best translate from broad principles of effective learning to specific instantiations in cognitive training paradigms), and future directions in the field of cognitive training.

Cognitive Training: Built upon Foundational Principles of Learning and Neuroplasticity

Although the field of cognitive training continues to develop, in most cases these improvements are situated squarely within the existing work in the learning sciences. For instance, one of the best single predictors of the extent to which a new skill will be learned is time on task (e.g., the "total time hypothesis," Ebbinghaus 1913). Simply put, the more time that individuals spend on a given task, the more they will learn. It is thus not surprising that this appears to be the case in perceptual and cognitive training as well (Jaeggi et al. 2008; Stafford and Dewar 2014; Stepankova et al. 2014), with some recent work truly pushing the envelope in terms of length of training (Schmiedek et al. 2010). Next, while the total amount of time spent learning is clearly important, not all time is equally well spent. One of the most replicated findings in the learning literature is that learning is more efficient (i.e., in terms of improvement per unit time) when training sessions are distributed rather than massed in time (Baddeley and Longman 1978). While this general finding is likely due to multiple mechanisms working in concert (e.g., decay of irrelevant learning, homeostatic regulation associated with sleep, etc.), it nonetheless indicates a clear design recommendation for cognitive training: many shorter training sessions are better than fewer longer training sessions. Indeed, the potential importance of both total training time and distribution of practice can be seen in comparing the results of two similar studies utilizing video game training – one that employed 50 total hours of training with each training session generally lasting around 1 hour (Green et al. 2010), and which produced generally positive results, and a second that employed up to 40 fewer hours of training and sessions that lasted up to four times as long, and which produced largely null results (Van Ravenzwaaij et al. 2014).

Another principle of effective learning common across domains is that of adaptivity of the to-be-learned material. In many cases this adaptivity takes the form of increasing difficulty as learner ability increases. That is, as a participant becomes proficient at completing training tasks, those tasks should become more difficult – thus keeping the participant at the edge of what they are able to handle (Deveau et al. 2015; Vygotsky 1981). Feedback during learning is also key. While a full discussion of the topic requires more nuance than is possible here, generally speaking learning is more effective when learners are provided with immediate and informative feedback related to their performance (Seitz and Dinse 2007). Finally, many other principles of effective learning find their empirical roots, at least partially, in the study of neuroplasticity (see also Wenger and Kühn, this volume). For instance, elegant basic science work has delineated the importance of various neuromodulatory systems in activating neuroplastic brain states (e.g., the cholinergic system via the nucleus basalis (Kilgard et al. 1998), and the dopaminergic system via the ventral tegmental area (Bao et al. 2001)). This has, in turn, served to strongly underscore the importance of designing training paradigms so as to induce a certain degree of physiological arousal and to make proper use of reward in order to maximize the potential efficacy of the training (Green and Bavelier 2010).

Other core principles that are foundational to the field of cognitive training focus not on the learning of the training tasks themselves, but on the extent to which the learning that occurs generalizes to untrained tasks (Schmidt and Bjork 1992). In essentially all areas of learning there exists a tension between learning that is highly specific to the trained paradigm and learning that transfers to untrained contexts and situations. A host of core learning task characteristics are known to increase the degree to which learning generalizes. Interestingly, most of these characteristics simultaneously decrease the overall rate of improvement. The goal of most cognitive training paradigms is to maximize the extent to which the learning generalizes broadly, and relevant principles of learning might therefore fall under the category of what have been dubbed *desirable difficulties* (Schmidt and Bjork 1992). For example, increases in both overall training heterogeneity and the extent to which training tasks are intermixed improve the generality of learning. Generalization tends to be increased when training is not homogeneous, but instead includes variation (Deveau et al. 2015; Dunlosky et al. 2013; Xiao et al. 2008); note though that effects may vary across populations of interest, see (Karbach and Kray 2009).

Yet, while the principles above have clearly been influential in the development of the paradigms employed in the cognitive training literature, as we will see later in the chapter, (1) it is not always clear how to best instantiate the principles in practice (e.g., how to engender motivation) and (2) these principles can interact in multiple, and sometimes unexpected ways.

Advances in Hardware for Cognitive Training

Before considering the training paradigms themselves, it is worth briefly considering changes in available hardware, as this represents the first bottleneck of training design. Over the past decade portable technology such as tablets have become increasingly common in cognitive training interventions (e.g., Ge et al. 2018; Oei and Patterson 2013; Shin et al. 2016; Wang et al. 2016). Tablets are relatively inexpensive, easy to use across a wide range of age groups, can be readily available for participants to train at their convenience, and can provide continuous updates of data for researchers. They can also be easily paired with wearable technology able to track heart rate, physical activity, and an increasing number of other variables (Piwek et al. 2016). These benefits though are accompanied by a loss in control over the administration of training and, as such, compliance with training regimens may be impossible to perfectly ensure. Even compliant learners may not adhere strictly to training instructions, and many sources of unwanted variance may be completely out of the control of training designers (e.g., screen viewing distance, device volume, and distracting environments). Although improvements in online psychological studies have addressed and mitigated some issues regarding experimental control, there will inevitably be some compromises when training is completed outside of controlled settings (Yung et al. 2015). The use of tablets, cell phones, or other portable devices thus involves accepting a tradeoff between the amount of data that is collected and the variability in the data.

Virtual reality (VR) headsets are another recently-developed type of hardware that has the potential for cognitive training applications. By using VR headsets, training programs can be more aligned with the field of view, depth, and actions of naturalistic settings. While cognitive training research utilizing VR is in its infancy, there have been some attempts to adapt typical monitor-based tasks to 3-dimensional virtual reality (Nyquist 2019; Nyquist et al. 2016). The immersion and ecological validity promised by VR could have the potential to improve many cognitive training paradigms. Barriers to effective deployment of virtual reality training continue to exist, however. Powerful computers are necessary for rendering virtual environments, and even the best computers for VR cannot yet compete with the spatial and temporal resolutions available on high-end monitors. And even as this technology improves, challenges will remain with respect to the human experience of VR. One clear example is nausea; the subtle mismatches between perceptual-motor predictions and simulated realities in VR can compound into debilitating "simulator sickness" (Allen et al. 2016; Kim et al. 2018).

Like virtual reality, wearable technology is increasingly available and likely to play a major role in future studies of cognitive training. Combined effects of physical training and cognitive training have promised greater improvements than either in isolation (Hertzog et al. 2008). Furthermore, even when implementing cognitive training with minimal physical demands, physiological measurements may nonetheless be informative to researchers regarding mediators or moderators of training outcomes. As examples, physical activity and sleep are each linked to neuroplasticity (Atienza et al. 2004; Bavelier et al. 2010; Tononi and Cirelli 2003). For each of these factors measurement with wearable technologies is simple. Even technology formerly relegated to research such as electroencephalography (EEG) is now available in portable formats and has been used in biofeedback-based training paradigms (Shin et al. 2016). As with EEG, increased interest in transcranial direct current stimulation (tDCS) has led to studies of efficacy of tDCS in concert with behavioral cognitive training (Martin et al. 2014; Martin et al. 2013).

Given the possibilities afforded to cognitive training by advances in hardware, the face of training is rapidly changing. Training in the future will likely be designed to be more immersive (e.g., virtual reality or always-available tablets), will integrate a more diverse set of measurements (e.g., heart rate and sleep tracking), many of which can be fed back directly into adaptive training algorithms, and may utilize methods to put the brain in a more plastic state (Hensch 2004; Seitz and Dinse 2007).

Advances in Software for Cognitive Training

One hardware issue not discussed above is the simple increase in computational power that comes with each passing year. This aspect in turn allows ever more complex training algorithms to be implemented (Deveau et al. 2015). Classic training algorithms in perceptual and cognitive fields have relied on unidimensional measures (e.g., correct/incorrect) aggregated across many training trials to determine performance, which then allowed adjustment of difficulty. In contrast, modern training in educational domains has developed more nuanced methods for understanding performance and correspondingly adapting difficulty (Liu et al. 2019; Ritter et al. 2007). In the latter case, interleaved training of various target skills is a straightforward implementation of another well-established principle of learning (e.g., Schmidt and Bjork 1992). The ability to track performance in each of the target skills, and provide on-the-fly adjustment of training demands in order to balance new content with refreshing old content, is a much more difficult task from a training perspective (Zhang et al. 2019). Indeed, in cognitive training research, targets of training are often homogeneous (e.g., only working memory), or trained processes are simply interleaved in a balanced design. This represents an opportunity for cognitive training research to improve as the field matures; while improved assessments and algorithms are increasingly possible, the efficacy of competing assessments and algorithms is still poorly understood. As with educational apps and intelligent tutoring systems, cognitive training can include many principles from basic learning research. These include interleaving, spacing, and adapting training as learners progress through a program. Additionally, personalization of training is a valuable ability facilitated by sensitive on-the-fly assessments of ability.

Possibly the most obvious design trend in cognitive training has been so-called "gamification" (Jaeggi et al. 2011; Squire 2003). Off-the-shelf recreational video games themselves have been used frequently in the cognitive training domain (for a review see Bediou et al. 2018; see also Bediou, Bavelier, and Green, Strobach and Schubert, this volume). These games provide natural instantiations for many of the learning principles discussed earlier and thus are an obvious source material from which designers may develop more dedicated forms of training (Deveau et al. 2015; Gentile and Gentile 2008; Nyquist et al. 2016). For instance, well-designed games produce both external and internal motivation to play, leading to a great deal of time on task. Video games also induce a great deal of physiological arousal and activation of the neural reward systems, which together create a brain state that is capable of efficient learning. Video games often involve a variety of tasks, types of decisions, and varying load on different attention and memory systems. As such, these games conform to the principle of variety and interleaving of learning. By frequently changing the demands placed on players, fast-paced video games are able to produce benefits in overlapping domains (e.g., attention to a wide visual field of view), while avoiding specificity in learning and maintaining adaptive difficulty that supports efficient learning (Deveau et al. 2015).

The increase in gamification has been supported by improved software for developing games or game-like environments. This is in stark contrast to game production in the past which required a set of highly skilled programmers and designers. Ease of game production does not necessarily mean high-quality games, however, and gamification does not directly imply that cognitive training would have the benefits of video games. Gamification should add rewards, engagement, arousal, and/or variety to cognitive training in order to introduce any benefits above, and should go beyond simply training on a cognitive task (Deveau et al. 2015). As noted early however, this may be easier said than done. While creating games has become easier, designing engaging, enjoyable, and effective training games remains challenging. In one test of motivational game-like features in cognitive training of children, Katz et al. (2014) found that none of the motivational game-like features that were implemented produced improvements on training-task learning. There may be various reasons for this outcome, including a highly stimulating base training (i.e., before adding motivational features), distracting effects of features such as points or levels (i.e., that the motivating features took attention away from the critical to-be-learned skills), or an insufficient timescale to detect differences (3 days of training). However, with limited tests of generalization, it may also have been the case that process-level benefits differed between training groups, and these differences were not apparent in the training data. Indeed, as discussed above, a classic finding is that *desirable difficulties* in learning may inhibit initial learning while boosting generalization (Schmidt and Bjork 1992).

Advances in Methods for Studying the Impact of Cognitive Training

There are clearly many outstanding questions regarding the most appropriate and efficacious interventions for given contexts and populations. Yet, many of the deepest questions in the field today concern studies' structural choices and assumptions and best-practice methodologies (see also Könen and Auerswald, Schmiedek, this volume). As an example, while Boot and colleagues have argued that training results are interpretable only if both intervention and control groups improve from pretest to posttest (Boot et al. 2011), Green and colleagues argue that these test–retest effects are theoretically unnecessary, and in fact, reduce the power to observe training-related benefits (Green et al. 2014). As an important step toward establishing a common methodological framework for diverse training paradigms and populations, over 50 leading researchers in the field recently collaborated in the publication of a consensus regarding methodological standards (Green et al. 2019). This section will briefly discuss the four dimensions of relevant methodological issues: control group choice, blinding, randomization, and tests of generalization.

Control Groups

Studies in experimental psychology are only as good as the contrasts utilized, and cognitive training is no exception. In order to demonstrate effectiveness of a training paradigm, and to identify the relevant processes undergoing change, appropriate experimental controls must be implemented. Control group selection in cognitive training is far from simple, and depending on the questions that are being posed, experimenters may choose to maximize the perceptual similarity of the control training with that completed by the experimental training group, to induce similar expectations and/or affective states, to match levels of engagement and interest, or to implement training grounded in alternative hypotheses regarding mechanism or efficacy (Green et al. 2014). The choice of active control is necessarily linked to the specific aims of a study, and there is no one-size-fits-all approach. Such study-specific control design poses difficulty for comparison of results across studies, however, which in turn hinders the ability for the field to move forward. Simply put, because the effects of interest in the field are usually a difference of differences (i.e., changes from pretest to posttest in the experimental group as compared to the pretest to posttest changes in the control group), massive differences in the characteristics of the control group make it difficult-to-impossible to effectively compare and contrast the impact of the experimental training paradigms. Thus, in order to ensure one-to-one comparisons of training effect sizes across studies with varying active control groups, it has recently been suggested that studies should implement no-contact controls in addition to their active control groups (Green et al. 2019). These business-as-usual comparison groups allow for clear qualitative and quantitative matching between effects of varying training regimes and will facilitate future work (Colzato and Hommel, this volume).

Blinding: Managing and Measuring Expectations

Expectation effects refer to changes in studies' outcomes in response to beliefs regarding the purpose or hypothesis of the studies. One well-known example is the placebo effect, in which positive beliefs regarding the efficacy of an intervention lead to beneficial outcomes even in the absence of the proposed mechanism of benefit (e.g., an inert sugar pill producing a similar analgesic effect as acetaminophen). The reduction of these expectation effects is largely accomplished through effective blinding, or ensuring that learners (and experimenters) are unaware of the expectations regarding their condition. For example, in a pain study, participants could be assigned to receive one of the two outwardly identical pills – one of which is a sugar pill, the other being a true analgesic. Because the participants will not know which of the two pills they are receiving, the expected benefit should be matched across groups, and thus any differences in outcome could not be attributed to expectations alone. In the cognitive training domain, it is not possible to produce two outwardly identical paradigms, where one is "inert" (like the sugar pill) and one is "active" (like the true analgesic). The outward appearance of a behavioral training platform is, after all, intractably linked to the extent to which the training is inert or active. As such, the best that can be done in the domain of cognitive training is to devise control experiences that appear plausible as interventions (Green et al. 2019). This is not necessarily trivial. Indeed, it is not even clear how to best measure the success of such attempted blinding (e.g., how to determine what expectations participants in the various groups hold). Advances in this area will therefore be critical for the field going forward.

We note that while minimizing expectation effects is necessary for demonstrating that any experimental training has true efficacy, expectations themselves may be used for the benefit of training once such a demonstration has been made. By intentionally creating expectations and maximizing their influence through conditioning, these expectations may become tools for increasing the effectiveness of training regimens (Green et al. 2019). Benefits of utilizing expectations may be especially pronounced in young populations due to the possibility of compounding long-term effects of small early-life benefits and attitudes (Stanovich 1986). Even if early benefits are "only" placebo effects (e.g., not true improvements in core cognitive processes), these benefits may still have very real positive downstream effects.

Randomization: Ensuring Interpretability of Results

Conventional wisdom in behavioral research is that study participants should be randomly assigned to experimental groups. However, truly random assignment is liable to create inter-group variation at pretest that reduces the interpretability of postintervention results. Given that the intentions of randomization and of group comparisons are each to reduce noise and clarify study-specific differences in behavior (i.e., learning), targeted efforts to match groups' performance on pretests will increase the interpretability of statistical tests of change from pretest to posttest (Green et al. 2014). Several methods exist to establish this masking, ranging from stratified or grouped random sampling (i.e., randomizing group membership after categorizing by other measures such as age or cognitive performance) to condition-difference minimization (i.e., assigning each new participant to whichever condition minimizes the between-condition pretest differences).

But What Is Learned? The Use of Pretest and Posttest Batteries

The target of cognitive training is often a specific process or set of processes. In order to test for changes to this target, or even to detect baseline individual differences, a variety of tasks loading on the target process can be used. By identifying the common component underlying, for example, complex span working memory tasks, individual variation and possible training-related benefits can be better identified (Engle et al. 1999; Green et al. 2014). Null results are likewise strengthened by process-level tests of generalization. By testing generalization to processes that are explicitly not expected to benefit from training, the contrast between null effects and nonnull effects can be used to clarify mechanisms of learning and falsify competing hypotheses. That is, if the mechanism of improvement was simply an increased effort on all tasks, all tests of generalization should benefit uniformly; to the extent that there are some null effects, any nonnull effects are more interpretable.

Despite the benefits of large numbers of pretest and posttest tasks, there are clear limitations. With continued testing fatigue will eventually diminish the quality of behavioral data. Fatigue is especially problematic in lower-functioning populations such as young children or older adults. While normally-functioning young adults may be expected to complete several hours of testing with a uniformly minimal decrement in performance, lower-functioning populations are likely to have a wider variance in their susceptibility to fatigue. In these populations patterns of performance may be shaped by participants' differential abilities to maintain attention and vigilance throughout demanding tasks. Training-related benefits may then be obscured or confounded by individual differences in the ability to complete long task batteries. As such, the size and scope of pretest and posttest batteries should be as large as feasible given the resources, context of training, and population of interest.

Frontiers: Questions and Practices for the Field

Benefits of Training: General or Specific?

All cognitive training is, justifiably, subject to scrutiny regarding the degree to which benefits observed within the training environment also extend to other behaviors. Robust improvements on trained tasks are often accompanied by little or no benefit to untrained tasks. This fact is far from unique to cognitive training; in areas as disparate as math education and visual contrast sensitivity training, learning can be surprisingly specific to the trained task. The lack of generalized benefits observed after using some common "brain training" apps has led to increased scrutiny of cognitive training from the popular press as well as the United States government, with a highly publicized rebuke and fine of one company occurring in 2016 (Federal Trade Commission 2016).

Even in tightly controlled studies, the generalization of cognitive benefits is sometimes not observed. However, we caution against interpretations of absences of generalization as "failures." Rather, specificity of a given training paradigm provides important information about the limiting cases in which cognitive training may or may not be efficacious. This may be relevant, for example, when matching interventions to appropriate populations. As discussed above, in young populations it may be the intention of training to improve scores on (and, ideally, the lifelong downstream consequences of) these specific cognitive abilities (see de Vries, Kenworthy, Dovis, and Geurtz, Johann and Karbach, Rueda et al., this volume).

Re-framing our understanding of generalization or specificity is only a small part of the larger problem: evidence regarding efficacy of training paradigms has been sparse. This problem is exacerbated by varying methodologies in training which make cross-study comparisons problematic at best; only by developing an aggregated estimate of efficacy can the understanding of generalization be advanced. Attempts have been made at aggregation, often with conflicting results (Au et al. 2015; Melby-Lervåg and Hulme 2013). The inter-study variation that causes these divergences is a key motivation for the push toward methodological consensuses mentioned above. Understanding generalization as a function of training design necessitates more data using common methods.

Multiple Forms of Generalization

There is also ambiguity regarding the expected mechanisms of generalization. While "transfer of learning" has typically been understood as immediate benefits observed in untrained contexts or tasks, there are a variety of ways in which initial training can benefit later performance (Barnett and Ceci 2002). Generalization of learning may also cause multiplicative benefits rather than additive benefits to generalized performance, leading to patterns of transfer that appear as learning to learn rather than immediate improvements. That is, even if performance in a test of generalization is not immediately benefited, performance may improve faster on tests of generalization than they would have prior to training (Kattner et al. 2017).

Delayed benefits in training generalization are a largely under-explored area, yet these effects are mechanistically aligned with the theoretical basis of cognitive training. If the targets of training are core cognitive abilities, it is possible that the benefits of these enhanced abilities would not be evident immediately on novel tasks due to task-specific factors (e.g., idiosyncratic interference from prior experience). Indeed, at different points in learning, separate processes may be constraining performance (Ackerman and Cianciolo 2000). This underscores the need to understand learning and generalization as time-evolving processes; the changes and generalized benefits of learning may be evident at some times and obscured at other times by other limiting processes (Rebok et al. 2014).

A different delayed training benefit may occur due to enhancement of cognitive abilities associated with more rapid learning in novel contexts. The locus of this change could be one of the various possibilities (e.g., faster speed of processing and improved perceptual template; (Bejjanki et al. 2014)). In this case of learning to learn, improvements on tests of generalization would be delayed due to the mechanism of generalization causing a divergence in performance with increased experience on a test of generalization. That is, if training causes an improvement in learning ability, there is little reason to believe that immediate benefits would be observed on novel tasks, but benefits should quickly become apparent with time. This is likely the case, for example, in cognitive benefits observed from action video game playing (Green et al. 2010).

Yet another cause for delayed generalization effects of cognitive training concerns the developmental timescales on which benefits are supposed to emerge. Early in the lifespan, interventions may have downstream effects due to trained children's ability to succeed in early school years, leading to an improved ability to use school resources themselves for improvement (Stanovich 1986). This is, for example, one theoretical motivation behind many early-childhood interventions outside the purely cognitive domain (e.g., Head Start, Ludwig and Phillips 2008). Later in life, too, interventions may have long-lasting effects by mitigating the downward trajectory of cognitive decline (Hertzog et al. 2008; Rebok et al. 2014; Willis et al. 2006).

In each of these cases of delayed generalization effects, the training should be designed appropriately for the observation of training-related benefits. That is, if there are very few observations of potential generalization (such as low trial numbers in cognitive assessments), there would inevitably be insufficient evidence to determine the presence or absence of delayed generalization effects. Likewise, if long-term developmental trajectories may be influenced by training, then assessments on the appropriate timescale must be implemented.

Alongside appropriate training design, evidence regarding generalization should also be considered using methods that allow for detection of delayed effects and dissociation between immediate and delayed generalization. In the case of learning to learn, in particular, it is important to understand the time course of performance on generalization tasks. In this case the mechanism of generalization manifests as a difference in performance that may be evident only after some a priori indeterminate amount of task experience. It is important, then, to approach generalization as a dynamically unfolding process in which training-related benefits may cause a divergence in performance between trained and untrained individuals (Bray and Dziak 2018). Each time (e.g., trial within a task) is therefore an important point at which generalization may be occurring, and generalization performance can be quantitatively modeled as a time-dependent process. By utilizing this by-trial modeling of performance, four possible outcomes can be dissociated: (1) immediate generalization, (2) delayed generalization (e.g., learning to learn), (3) lack of generalization, or (4) both (1) and (2). In the absence of time-dependent models of generalization, superficially unrelated factors such as generalization-task number of observations may obscure the effects of training (Kattner et al. 2017).

Variance in Outcomes: Individual Differences in Training Benefits

In an insightful analysis of learning data from several classic studies, Heathcote et al. (2000) noted that a canonical power-law function of learning did not exist in any individual learner, but the power function was instead an artifact of averaging performance across individuals. A similar possibility has the potential for reducing the accuracy of inferences regarding the efficacy of cognitive training. That is, group-level estimates of training efficacy may obscure individual-level changes in cognitive abilities (Bürki et al. 2014). Certain factors, such as genetics, attitudes toward training, or compliance may even mediate positive effects of training on cognition (Colzato et al. 2014; Jaeggi et al. 2014). Further, group-level estimates of change may hide the possibility that some learners actually perform worse at posttest than at pretest. This pattern is obviously not desirable, but it is a very important addition to the field's understanding of training design and efficacy. That is, in real-world applications, training should ideally benefit each learner. While ubiquitous success is an unlikely outcome, it is possible that the time spent training takes away from the time spent on other beneficial activities (e.g., rehabilitation exercises or classroom exercises). If certain populations are unresponsive to training and are better served by "business-as-usual," then the main effects of training vs. control groups can hide this mechanistic nuance. Thus, as far as what is feasible, researchers should consider individual trajectories of improvement, and should develop tools for identifying individuals who do not benefit from the training intervention. This will be an important aspect of adaptivity algorithms in future applied training contexts. As with any intervention that should be stopped when a lack of efficacy has been demonstrated in a certain patient (e.g., administration of medication), cognitive training must not algorithmically "keep trying" when an individual is not responsive to the intervention.

The power of individual-level data is also an important feature of understanding the results of training. While statistical power to detect the effect of an intervention is often understood in terms of the number of participants in a study, the features of the study itself also influence the power to detect any training-related effects. That is, there is a clear resource allocation trade-off between studying few people trained following the best practices, and studying many people trained using practices with less likelihood to detect any effect. In fact, depending on the timescale on which plasticity in target processes would change, it is possible that training programs of different lengths (e.g., 3 days vs. 25 days) would not simply be quantitatively different in their power to detect training-related benefits, but also be qualitatively different in the types of benefits able to be induced in that timescale. Quantitative reviews of various training studies may exacerbate the problem. That is, if studies in a meta-analysis are weighted according to the number of participants, then studies that have emphasized the participant number over training integrity would be more influential in drawing conclusions. Even if other variables are statistically controlled for (e.g., time training per session, number of different training tasks, or number of sessions), there is little way to know whether the target processes of various studies are qualitatively similar enough to justify quantitative aggregation. Nonetheless, to the degree that methods such as the total time and spacing are qualitatively similar across participants and studies, hierarchical and meta-analytic statistical models provide the ability to simultaneously estimate both individual-level and aggregate parameter estimates that can indicate the efficacy of training paradigms.

The Next Generation of Training Design: Integrated, Informed, and More Powerful than Ever

The direction of cognitive training design is toward increasingly engaging, available, and well-informed programs. Recent consensus statements from scientists in the field provide guidelines for theoretically understanding, and methodologically implementing, studies for the advancement of the field (Green et al. 2019; Max Planck Institute for Human Development and Stanford Center on Longevity 2014). These statements encourage healthy skepticism regarding the results of any single program or study, but they also encourage innovation through the recognition that studies and paradigms have widely differing intentions and populations. Advances may be attempted through the use of novel hardware, software, or even cognitive targets of training, and even null results add to the community's understanding of training mechanisms and efficacy (Green et al. 2014).

References

Ackerman, P. L., & Cianciolo, A. T. (2000). Cognitive, perceptual-speed, and psychomotor determinants of individual differences during skill acquisition. *Journal of Experimental Psychology: Applied, 6*, 259–290.

Allen, B., Hanley, T., Rokers, B., & Green, C. S. (2016). Visual 3D motion acuity predicts discomfort in 3D stereoscopic environments. *Entertainment Computing, 13*, 1–9.

Atienza, M., Cantero, J. L., & Stickgold, R. (2004). Posttraining sleep enhances automaticity in perceptual discrimination. *Journal of Cognitive Neuroscience, 16*, 53–64.

Au, J., Sheehan, E., Tsai, N., Duncan, G. J., Buschkuehl, M., & Jaeggi, S. M. (2015). Improving fluid intelligence with training on working memory: A meta-analysis. *Psychonomic Bulletin & Review, 22*, 366–377.

Baddeley, A. D., & Longman, D. J. A. (1978). The influence of length and frequency of training session on the rate of learning to type. *Ergonomics, 21*, 627–635.

Bao, S., Chan, V. T., & Merzenich, M. M. (2001). Cortical remodelling induced by activity of ventral tegmental dopamine neurons. *Nature, 412*, 79–83.

Barnett, S. M., & Ceci, S. J. (2002). When and where do we apply what we learn?: A taxonomy for far transfer. *Psychological Bulletin, 128*, 612–637.

Bavelier, D., Levi, D. M., Li, R. W., Dan, Y., & Hensch, T. K. (2010). Removing brakes on adult brain plasticity: From molecular to behavioral interventions. *Journal of Neuroscience, 30*, 14964–14971.

Bediou, B., Adams, D. M., Mayer, R. E., Tipton, E., Green, C. S., & Bavelier, D. (2018). Meta-analysis of action video game impact on perceptual, attentional, and cognitive skills. *Psychological Bulletin, 144*, 77–110.

Bejjanki, V. R., Zhang, R., Li, R., Pouget, A., Green, C. S., Lu, Z. L., & Bavelier, D. (2014). Action video game play facilitates the development of better perceptual templates. *Proceedings of the National Academy of Sciences, 111*, 16961–16966.

Boot, W. R., Blakely, D. P., & Simons, D. J. (2011). Do action video games improve perception and cognition? *Frontiers in Psychology, 2*, 226.

Bray, B. C., & Dziak, J. J. (2018). Commentary on latent class, latent profile, and latent transition analysis for characterizing individual differences in learning. *Learning and Individual Differences, 66*, 105–110.

Bürki, C. N., Ludwig, C., Chicherio, C., & de Ribaupierre, A. (2014). Individual differences in cognitive plasticity: An investigation of training curves in younger and older adults. *Psychological Research, 78*, 821–835.

Colzato, L. S., van den Wildenberg, W. P. M., & Hommel, B. (2014). Cognitive control and the COMT Val^{158}Met polymorphism: Genetic modulation of videogame training and transfer to task-switching efficiency. *Psychological Research, 78*, 670–678.

Deveau, J., Jaeggi, S. M., Zordan, V., Phung, C., & Seitz, A. R. (2015). How to build better memory training games. *Frontiers in Systems Neuroscience, 8*, 243.

Dunlosky, J., Rawson, K. A., Marsh, E. J., Nathan, M. J., & Willingham, D. T. (2013). Improving students' learning with effective learning techniques: Promising directions from cognitive and educational psychology. *Psychological Science in the Public Interest, 14*, 4–58.

Ebbinghaus, H. (1913). *Memory: A contribution to experimental psychology*. University Microfilms. Available at: https://books.google.com/books?id=oRSMDF6y3l8C

Engle, R. W., Tuholski, S. W., Laughlin, J., & Conway, A. R. A. (1999). Working memory, short-term memory and general fluid intelligence: A latent variable model approach. *Journal of Experimental Psychology: General, 128*, 309–331.

Federal Trade Commission. (2016). *Lumosity to pay $2 million to settle FTC deceptive advertising charges for Its "Brain Training" program*. Available at: https://www.ftc.gov/system/files/documents/cases/160105lumoslabsstip.pdf

Ge, S., Zhu, Z., Wu, B., & McConnell, E. S. (2018). Technology-based cognitive training and rehabilitation interventions for individuals with mild cognitive impairment: A systematic review. *BMC Geriatrics, 18*, 213.

Gentile, D. A., & Gentile, J. R. (2008). Violent video games as exemplary teachers: A conceptual analysis. *Journal of Youth and Adolescence, 37*, 127–141.

Green, C. S., Pouget, A., & Bavelier, D. (2010). Improved probabilistic inference as a general learning mechanism with action video games. *Current Biology, 20*, 1573–1579.

Green, C. S., Strobach, T., & Schubert, T. (2014). On methodological standards in training and transfer experiments. *Psychological Research, 78*, 756–772.

Green, C. S., Bavelier, D., Kramer, A. F., Vinogradov, S., Ansorge, U., Ball, K. K., ... & Facoetti, A. (2019). Improving methodological standards in behavioral interventions for cognitive enhancement. *Journal of Cognitive Enhancement, 3*, 2–29.

Heathcote, A., Brown, S., & Mewhort, D. J. (2000). The power law repealed: The case for an exponential law of practice. *Psychonomic Bulletin and Review, 7*, 185–207.

Hensch, T. K. (2004). Critical period regulation. *Annual Review of Neuroscience, 27*, 549–579.

Hertzog, C., Kramer, A. F., Wilson, R. S., & Lindenberger, U. (2008). Enrichment effects on adult cognitive development: Can the functional capacity of older adults be preserved and enhanced? *Psychological Science in the Public Interest, 9*, 1–65.

Jaeggi, S. M., Buschkuehl, M., Jonides, J., & Perrig, W. J. (2008). Improving fluid intelligence with training on working memory. *Proceedings of the National Academy of Sciences, 105*, 6829–6833.

Jaeggi, S. M., Buschkuehl, M., Jonides, J., & Shah, P. (2011). Short-and long-term benefits of cognitive training. *Proceedings of the National Academy of Sciences (USA), 108*, 10081–10086.

Jaeggi, S. M., Buschkuehl, M., Shah, P., & Jonides, J. (2014). The role of individual differences in cognitive training and transfer. *Memory and Cognition, 42*, 464–480.

Karbach, J., & Kray, J. (2009). How useful is executive control training? Age differences in near and far transfer of task-switching training. *Developmental Science, 12*, 978–990.

Kattner, F., Cochrane, A., Cox, C. R., Gorman, T. E., & Green, C. S. (2017). Perceptual learning generalization from sequential perceptual training as a change in learning rate. *Current Biology, 27*, 840–846.

Katz, B., Jaeggi, S., Buschkuehl, M., Stegman, A., & Shah, P. (2014). Differential effect of motivational features on training improvements in school-based cognitive training. *Frontiers in Human Neuroscience, 8*, 242.

Kilgard, M. P., & Merzenich, M. M. (1998). Cortical map reorganization enabled by nucleus basalis activity. *Science, 279*, 1714–1718.

Kim, H. K., Park, J., Choi, Y., & Choe, M. (2018). Virtual reality sickness questionnaire (VRSQ): Motion sickness measurement index in a virtual reality environment. *Applied Ergonomics, 69*, 66–73.

Liu, R., Stamper, J., Davenport, J., Crossley, S., McNamara, D., Nzinga, K., & Sherin, B. (2019). Learning linkages: Integrating data streams of multiple modalities and timescales. *Journal of Computer Assisted Learning, 35*, 99–109.

Ludwig, J., & Phillips, D. A. (2008). Long-term effects of head start on low-income children. *Annals of the New York Academy of Sciences, 1136*, 257–268.

Martin, D. M., Liu, R., Alonzo, A., Green, M., Player, M. J., Sachdev, P., & Loo, C. K. (2013). Can transcranial direct current stimulation enhance outcomes from cognitive training? A randomized controlled trial in healthy participants. *International Journal of Neuropsychopharmacology, 16*, 1927–1936.

Martin, D. M., Liu, R., Alonzo, A., Green, M., & Loo, C. K. (2014). Use of transcranial direct current stimulation (tDCS) to enhance cognitive training: Effect of timing of stimulation. *Experimental Brain Research, 232*, 3345–3351.

Max Planck Institute for Human Development and Stanford Center on Longevity. (2014). *A consensus on the brain training industry from the scientific community (full) – Stanford Center on Longevity*. Available at: http://longevity.stanford.edu/a-consensus-on-the-brain-training-industry-from-the-scientific-community-2/. Accessed 29 July 2019.

Melby-Lervåg, M., & Hulme, C. (2013). Is working memory training effective? A meta-analytic review. *Developmental Psychology, 49*, 270–291.

Nyquist, J.B. (2019). NeuroTrainer labs. *NeuroTrainer labs*. Available at: https://neurotrainer.com. Accessed 27 July 2019.

Nyquist, J. B., Lappin, J. S., Zhang, R., & Tadin, D. (2016). Perceptual training yields rapid improvements in visually impaired youth. *Scientific Reports, 6*, 37431.

Oei, A. C., & Patterson, M. D. (2013). Enhancing cognition with video games: A multiple game training study J. J. Geng (ed). *PLoS One, 8*, e58546.

Piwek, L., Ellis, D. A., Andrews, S., & Joinson, A. (2016). The rise of consumer health wearables: Promises and barriers. *PLoS Medicine, 13*, e1001953.

Rebok, G. W., Ball, K., Guey, L. T., Jones, R. N., Kim, H. Y., King, J. W., ... & Willis, S. L. (2014). Ten-year effects of the advanced cognitive training for independent and vital elderly cognitive training trial on cognition and everyday functioning in older adults. *Journal of the American Geriatrics Society, 62*, 16–24.

Ritter, S., Anderson, J. R., Koedinger, K. R., & Corbett, A. (2007). Cognitive tutor: Applied research in mathematics education. *Psychonomic Bulletin and Review, 14*, 249–255.

Schmidt, R. A., & Bjork, R. A. (1992). New conceptualizations of practice: Common principles in three paradigms suggest new concepts for training. *Psychological Science, 3*, 207–217.

Schmiedek, F., Lövdén, M., & Lindenberger, U. (2010). Hundred days of cognitive training enhance broad cognitive abilities in adulthood: Findings from the COGITO study. *Frontiers in aging neuroscience, 2*, 27.

Seitz, A., & Dinse, H. R. (2007). A common framework for perceptual learning. *Current Opinion in Neurobiology, 17*, 148–153.

Shin, M. S., Jeon, H., Kim, M., Hwang, T., Oh, S. J., Hwangbo, M., & Kim, K. J. (2016). Effects of smart-tablet-based neurofeedback training on cognitive function in children with attention problems. *Journal of Child Neurology, 31*, 750–760.

Squire, K. (2003). Video games in education. *Int. J. Intell. Games and Simulation, 2*, 49–62.

Stafford, T., & Dewar, M. (2014). Tracing the trajectory of skill learning with a very large sample of online game players. *Psychological Science, 25*, 511–518.

Stanovich, K. E. (1986). Matthew effects in reading: Some consequences of individual differences in the acquisition of literacy. *Reading Research Quarterly, 21*, 360–407.

Stepankova, H., Lukavsky, J., Buschkuehl, M., Kopecek, M., Ripova, D., & Jaeggi, S. M. (2014). The malleability of working memory and visuospatial skills: A randomized controlled study in older adults. *Developmental Psychology, 50*, 1049–1059.

Tononi, G., & Cirelli, C. (2003). Sleep and synaptic homeostasis: A hypothesis. *Brain Research Bulletin, 62*, 143–150.

Van Ravenzwaaij, D., Boekel, W., Forstmann, B. U., Ratcliff, R., & Wagenmakers, E.-J. (2014). Action video games do not improve the speed of information processing in simple perceptual tasks. *Journal of Experimental Psychology: General, 143*, 1794–1805.

Vygotsky, L. S. (1981). *Mind in society: The development of higher psychological processes Nachdr*. Cambridge, MA: Harvard Univ. Press.

Wang, Z., Zhang, H., Wang, J., Wang, X., Yu, X., & Wang, H. (2016). Tablet-based multi-domain cognitive training increases the parahippocampus volume in patients with amnestic mild cognitive impairment. *Alzheimer's & Dementia, 12*, P509–P510.

Willis, S. L., Tennstedt, S. L., Marsiske, M., Ball, K., Elias, J., Koepke, K. M., ... & Wright, E. (2006). Long-term effects of cognitive training on everyday functional outcomes in older adults. *JAMA, 296*, 2805–2814.

Xiao, L. Q., Zhang, J. Y., Wang, R., Klein, S. A., Levi, D. M., & Yu, C. (2008). Complete transfer of perceptual learning across retinal locations enabled by double training. *Current Biology, 18*, 1922–1926.

Yung, A., Cardoso-Leite, P., Dale, G., Bavelier, D., & Green, C. S. (2015). Methods to test visual attention online. *Journal of Visualized Experiments, 96*, 52470.

Zhang, P., Zhao, Y., Dosher, B. A., & Lu, Z.-L. (2019). Evaluating the performance of the staircase and quick change detection methods in measuring perceptual learning. *Journal of Vision, 19*, 14.

Theoretical Models of Training and Transfer Effects

Niels A. Taatgen

Contents

Abstract Models of cognitive transfer are typically based on the theory of identical elements: knowledge from one task can only be used by another task if the elements of knowledge are identical. However, this leaves many open questions regarding the nature of the knowledge and the mechanisms of transfer. The central idea presented in this chapter is that the elements of knowledge can be identified at several levels of abstraction, and that knowledge can transfer at a very low level. Moreover, the mechanism of transfer is that general knowledge is a byproduct of learning. The PRIMs (primitive information processing element) theory offers a process model of transfer. In this chapter we discuss the basis of PRIMs and show how it can predict phenomena around brain training, cognitive development, and learning from instructions at different levels of abstraction.

Introduction

Despite the large increase in interest and research on cognitive training, there is very little theory that can explain the effectiveness or lack thereof of cognitive training. There are two reasons for this. The first is the prevailing idea that cognitive training

N. A. Taatgen (✉)
University of Groningen, Groningen, Netherlands
e-mail: n.a.taatgen@rug.nl

T. Strobach, J. Karbach (eds.), *Cognitive Training*,
https://doi.org/10.1007/978-3-030-39292-5_4

is similar to training muscles and, therefore, requires little explanation. The second is that most detailed theories of cognitive training assume that what is learned in training is highly specific, which implies that general cognitive training is not really possible.

The muscle analogy of cognitive or "brain training" is quite pervasive. In particular, the term "brain training" suggests that it involves a physiological system that needs to become better or stronger (or has to be trained to prevent it from becoming weaker). It is also consistent with a tradition of viewing cognition as a collaboration between a set of cognitive functions or systems. For example, if we consider working memory as a system with a certain capacity, then the logical result of training working memory is the expansion of that capacity. Following the analogy leads to the idea that the various cognitive systems make up the muscles of the mind. However, there are several reasons why the muscle analogy may not be the most appropriate. First of all, the brain is not a muscle, nor anything like a muscle. Moreover, if brain training is like muscle training, why are the results so inconsistent? Reports of unsuccessful training are as common as successful reports, even without considering the publication bias that favors success over null results. Perhaps the successful training or testing regimens find some right combination, and the unsuccessful ones do not (see also Guye et al., and Umanath et al., this volume).

What makes humans such a successful species is not the strength of muscles but the capability to fit in almost every niche in nature. In other words, humans are almost infinitely adaptable to different circumstances. Therefore cognitive training can better be viewed as acquisition and prioritizing of cognitive skills. For example, working memory training may not literally increase our capacity but instead expand our cognitive strategies for maintaining information for relative short periods of time. This was certainly the case with subject SF in Chase and Ericsson's digit span training where SF managed to expand his digit span from an average 7 to around 80 after 44 weeks of practice (Chase and Ericsson 1982). However, SF was not able to use this skill for anything else but digits. This brings us to the second reason why there is little theory about cognitive training, which is the strong belief that skills are seldom transferrable. The origin of this idea stems from Thorndike who proposed the *identical elements* theory of transfer (Thorndike and Woodworth 1901). Only when the knowledge components are identical between two skills can there be transfer. Although Thorndike produced some evidence for lack of transfer between certain tasks, his methodology would not be considered as convincing when viewed in the context of today's standards. Singley and Anderson (1989) introduced a modern version of this theory by specifying the *production rule* as the element that has to be identical between tasks in order to produce transfer. Production rules are a form of knowledge representation that specifies how to achieve goals that take multiple actions. For example, to represent the task of doing multi-column addition, production rules are needed that specify the different steps in performing that task: focusing on a particular column, retrieving arithmetic facts from memory, writing down an answer underneath a column, remembering that there is a carry, and then handling that carry in the next column. To properly sequence these production rules, each of them has several conditions that need to be satisfied before they can be carried out. For example, having just finished a column is a condition for moving to the next column. The modern production rule assumes that it coordinates the exchange of information between more specialized cognitive modules (Anderson 2007). For example, a production rule may route information from visual perception to the memory retrieval system, route information from the memory system to the working memory, or an item from working memory to the motor system.

Typical procedural representations are quite specific for a particular task: the production rules for multi-column addition cannot be used in multi-column subtraction, despite some similarity between the two. This is why the Singley and Anderson theory, following Thorndike, predicts that transfer is a limited phenomenon. Empirical research followed this tradition with several studies that showed limited transfer through the use of analogy (e.g., Gick and Holyoak 1980). However, the existing methods are insufficient to explain result of many recent training studies. This means we need a new model paradigm to explain general effects of cognitive training.

The PRIMs Theory

If the effects of cognitive training can be explained by neither the muscle analogy nor the transfer of knowledge between tasks, what alternatives are left? Fortunately, there is another possible solution. The assumption of this solution is that when people learn *specific* cognitive tasks, the byproduct of the learning process consists of *general* cognitive skills. The general skills can be reused for different tasks without the need of explicit transfer between tasks. Moreover, the two tasks that share general skills can be quite different: they just share the same patterns of routing information through the cognitive system.

To implement a system along those lines, a more fine-grained representation than production rules is needed. Production rules typically carry out multiple smaller steps, only some of which are specific to the task. The PRIMs theory (Taatgen 2013b) breaks up production rules into these basic elements of information processing (PRIM = primitive information processing element) and separates the task-specific from the task-general steps. The PRIMs software, including documentation, articles, examples and tutorials, can be downloaded from https://www.ai.rug.nl/~niels/prims/index.html.

What is called a rule in most production systems is an *operator* in PRIMs. Although it typically takes a single cycle to carry out a standard rule, an operator in PRIMs can take many more cycles, depending on the complexity of the operator, and the experience the system has with that operator. Operators themselves are carried out by one or more PRIM rules. PRIM rules operate at a smaller scale than standard production rules.

For example, in a model of multi-column addition, a production rule might take the two values in the column that has just been attended and retrieve an addition fact from memory to calculate the sum. Such a rule would be useless in multi-column subtraction or multiplication because the goal is different (and production rules a linked to the goal) and because a different kind of arithmetic fact has to be retrieved. However, part of the information exchange is the same: two elements of information from the visual system have to be transferred to the memory retrieval system along with the information that we are trying to retrieve the sum of these two. We only do this when certain conditions are met: in this case the condition that we are attending a column with two numbers with nothing underneath it.

In PRIMs, this production rule would be represented by an operator that specifies the individual information processing steps. Before any learning has occurred, each of these steps is carried out by a separate PRIM rule (the production rule that initiates a memory retrieval in multi-column addition consists of six of these steps and, therefore, needs six PRIM rules initially). However, once a particular sequence of two steps is used often enough, the learning mechanism called production compilation (Taatgen and Anderson 2002) combines them into a single new PRIM rule. This means that after some repetitions, the initial six PRIM rules have been combined into three composite PRIM rules that each carry out two basic steps. But this learning process continues when composite rules compile into larger composite rules that carry out four elementary steps, that then compile into a PRIM rule that carries out the whole operator at once. Figure 1 illustrates the learning sequence.

The PRIM rule at the bottom-right of Fig. 1 is a composite rule that carries out a relative complex pattern of information exchange but is still independent of the particular task. This PRIM rule, or any of the other PRIM rules higher in the tree, can be reused for other tasks that need the same pattern of information exchange.

Multi-column addition consists of six operators that deal with retrieving arithmetic facts, writing down the answer, moving from one column to the next, and dealing with carries. Multi-column multiplication needs eight operators to do the multiplication part, and another six to add up the results. The latter six are, of course, identical to multi-column addition. However, there is also partial overlap between the multiplication part and multi-column addition because the procedure also deals with columns, carries, and the retrieval of arithmetic facts but in a slightly different way.

Figure 2 shows a graphical representation of the two models. The six red nodes represent the six operators for multi-column addition. The bottom-right red node (retrieve addition fact) represents the operator we discussed in detail: it has three condition PRIMs (the string of three grey nodes) and three action PRIMs (the string of three white nodes). When we subsequently add the multi-column multiplication task, we only need an additional five operators instead of 14. We, of course, save the six operators that implement the multi-column addition subtask in multi-column multiplication. In addition, three more operators from multi-column addition can be directly reused for multi-column multiplication. In the figure these are the red nodes that are pointed to by an arrow from central multi-column multiplication node, for instance, the "write answer" operator that writes the result of an arithmetic retrieval underneath a column. But even operators that are new can benefit from prior knowl-

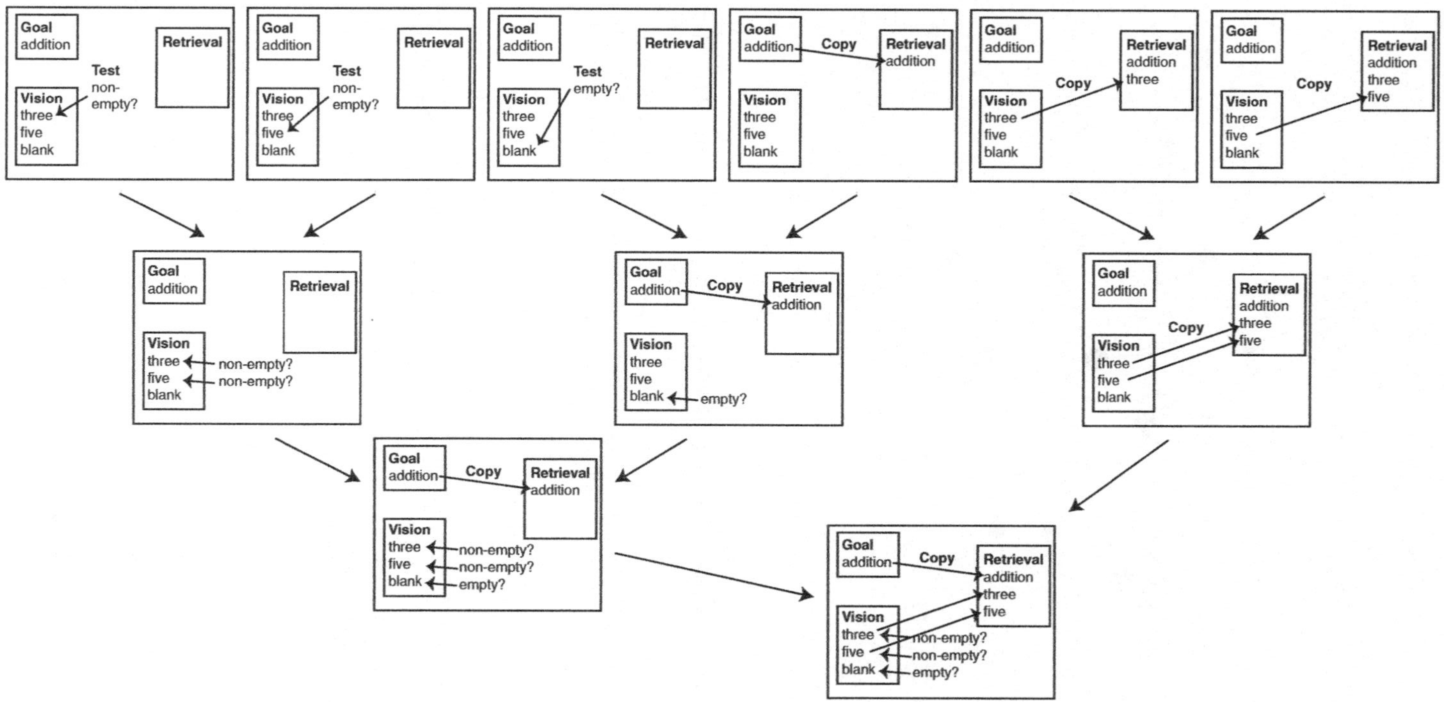

Fig. 1 Illustration of how a composite PRIM rule in multi-column addition (bottom right box) is learned from six elementary PRIM rules (top row). The three smaller boxes in each larger box represent the interface to the goal representation, to the long-term memory system ("retrieval") and to the visual system. Rules carry out checks and transformations between these systems. Rules in the top row just carry out a single check or transformation but after learning rules are composed that can do several of these at once

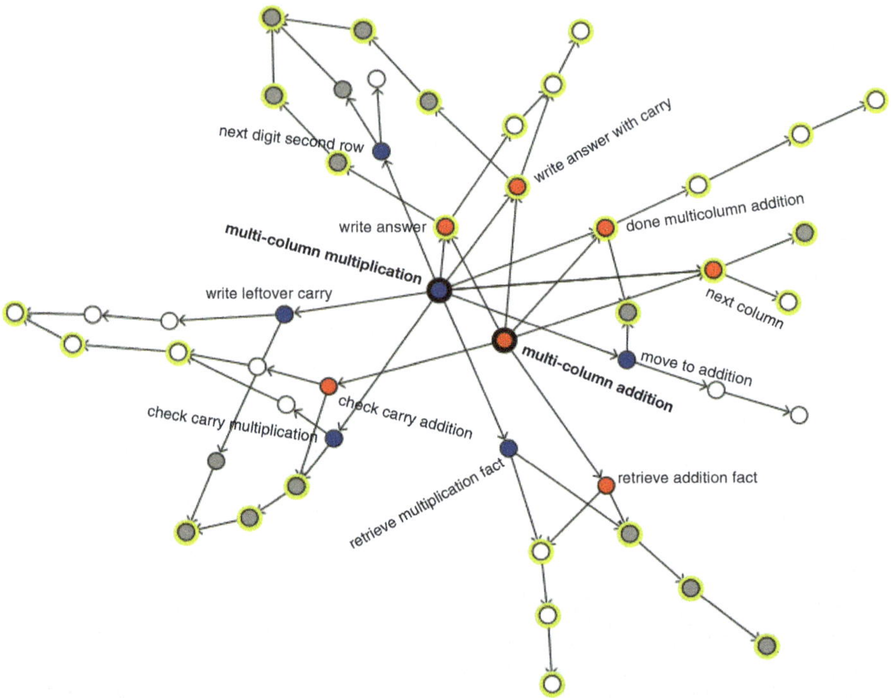

Fig. 2 Illustration of the operators for multi-column addition (red nodes), multi-column multiplication (blue nodes), and the primitive steps (PRIMs) connected to each of these operators (white and gray, gray nodes are conditions, while white nodes are actions). PRIMs that are used by both tasks, and therefore produce transfer, are indicated by a yellow halo. The two colored central nodes (multi-column multiplication and multi-column addition) are not operators but just nodes that connect to all the operators

edge: the "retrieve multiplication fact" operator has the same PRIMs as the "retrieve addition fact" operator. This means that if the model has learned the composite rule shown in Fig. 1, it can also use this rule to carry out the "retrieve multiplication fact" operator. The only reason we need a separate operator in this case is that one operator sets the value in the goal to "addition" while the other sets it to "multiplication." Partial overlap is also possible: the "check carry addition" and "check carry multiplication" operators differ by one PRIM but share the others.

Figure 2 already shows that learning multi-column multiplication is much easier after first learning multi-column addition. Only five new operators have to be learned but also the rules learned in multi-column addition can be reused. As a consequence, this model predicts that it is easier to learn multi-column multiplication after mastering multi-column addition. Moreover, it predicts that learning multi-column addition is harder than learning multi-column multiplication (assuming multi-column addition is learned first). But PRIMs does more than allow us to analyze the structure of the knowledge. Because it is based on the powerful ACT-R cognitive architecture (Anderson 2007), it can simulate how humans carry out these tasks and make predictions about reaction times, errors, learning curves, and precise characteristics of transfer.

Higher-Level Transfer

Transfer in the multi-column addition and multiplication model plays out at two levels of abstraction. The first level is that operators share PRIMs, and, therefore, training on one operator benefits the other. But the model also reuses whole operators, such as "write-answer." PRIMs supports transfer at an even higher level, namely the level of skills. The level of skill is situated between operators and tasks: a skill consists of several operators, and several skills are combined to perform a task.

The level of skills is necessary if we want to explain how people can perform new tasks without any learning. In many psychological experiments, subjects are given an instruction for a task they have never done before but that they are nevertheless able to perform without problem just based on that instruction. An explanation for this "one-shot learning" is that people already have the required skills to perform the new task but that they just need to combine these skills in a novel way, similar to how language produces new meaning by combining words in novel ways.

A skill consists of a collection of operators that have a number of variables that need to be instantiated when a skill is used as a part of a task. For example, in the multi-column model, the operator that retrieves facts from memory has a variable that determines the type, which is addition or multiplication. Variables can also be used to link skills together.

Explaining task performance in terms of skills allows us to take prior experience and knowledge into account, and this can help us understand certain cases of suboptimal performance. As a simple example, interference in the Stroop tasks is due to prior learning because children that cannot yet read (but can name colors) will not show the Stroop effect. Here we will elaborate an example by Hoekstra et al. (2019) in which they built a model of the Attentional Blink task.

In the Attentional Blink (AB) task (Raymond et al. 1992), subjects are presented with a rapid stream of characters (typically 100 ms/character), most of which are distractors (e.g., digits), but two are targets (e.g., letters). The task is to report the targets at the end of the stream. The typical AB effect (Fig. 3) is that accuracy on the second target is strongly affected if it is 200–400 ms after the first but not if it is only 100 ms or more than 600 ms after the first. There are many explanations for the AB, ranging from memory limitations, attentional limitations, and issues with control. However, there are many circumstances that attenuate the AB, for example, distraction, instructions, and certain types of training. An alternative explanation is, therefore, that the AB is due to the choice of skills to perform the task. To demonstrate this, Hoekstra et al. have decomposed the AB task into three skills that they have taken out of other models: target detection (taken from a visual search task model), memory consolidation, and memory retrieval (taken from a complex working memory task model). By only instantiating the variables in these skills differently, they were able build a model that showed the AB effect.

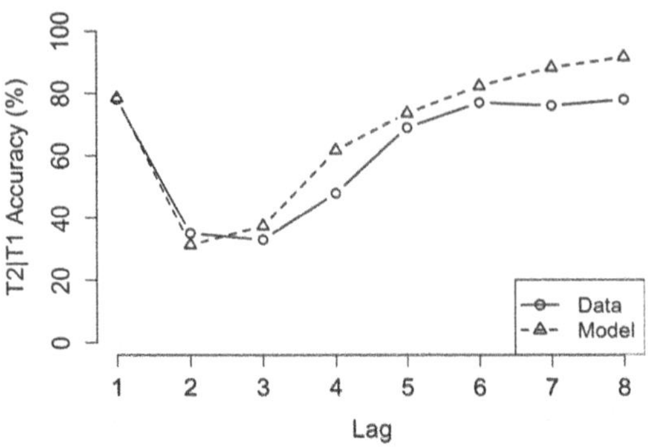

Fig. 3 AB-model fit for the accuracy of the second target by Lag. (Lag is the distance between the two targets in hundreds of ms.) (Data are from Raymond et al. (1992))

The AB effect in the model is produced by the consolidation strategy: if the first target has been detected and there is no second target yet, the model consolidates the target as a single element in memory and can, therefore, miss later targets as long as it is consolidating (which takes 200 ms on average). This is consistent with the instruction that states that (independent) targets have to be reported. However, it is also possible to use a different consolidation strategy that tries to put to be memorized items into chunks as has been observed in simple working memory tasks. If a chunk consolidation strategy is used, the model does not exhibit an AB because it will wait for the second target before it starts consolidation. An example from the literature that is consistent with this chunking idea is a study by Ferlazzo et al. (2007) in which they gave subjects the instruction to report the *syllable* in the stream. This alternative instruction eliminated the AB, contrary to a condition with the standard "report the two letters" instruction. According to the model, the syllable instruction prompts the right consolidation strategy that avoids the blink.

PRIMs' Current Scope

The multi-column addition and multiplication example is a good illustration of the essence of PRIMs, but does not give us clear testable hypotheses. However, PRIMs has been successful in modeling several phenomena related to transfer and cognitive training.

Reuse of Skills in Text Editing

A first example is an experiment involving text editors that Singley and Anderson (1985) used to support their identical productions theory. In that experiment, subjects with no prior computer experience were trained on using three different text editors. Two of these were very similar in use (ED and EDT, both so-called line editors) while the third was different (Emacs, a screen-based editor). During the six-day experiment, subjects switched editors once, twice, or never at all, depending on which of the five conditions of the experiment they were in. This design allowed Singley and Anderson to determine how much of the knowledge of one editor could be used for another editor. The amount of transfer between ED and EDT turned out to be very high; around 95% of the knowledge for one editor could be used for the other editor. The Singley and Anderson model based on identical productions was able to capture this to a large extent (85%). However, the experimental data also showed decent transfer between ED/EDT and Emacs, around 60%, of which the identical productions model could only explain 33%. The PRIMs model (Taatgen 2013b) that was based on roughly the same productions (translated into PRIMs operators) was able to fit the data much more precisely (predicting around 90% transfer between ED and EDT, and 63% between ED/EDT and Emacs). The reason is that PRIMs predicts transfer even if productions are not completely but only partially identical. This meant that a large part of the knowledge gained by training on ED and EDT that was not directly applicable to Emacs could still be reused.

Training Skills in Brain Training

In the two examples that we have examined, arithmetic and text editing, the overlap between the tasks was quite substantial and directly determined the amount of transfer between tasks. Another possibility is that the overlap between tasks is relatively small but critical in determining a difference in performance. This is the case when there are multiple strategies to perform a task, and prior training leads to the selection of a better strategy than the one that would normally be preferred.

An example of this is a model of an experiment by Karbach and Kray (2009 see also Karbach and Kray, this volume). In that experiment, subjects were trained on a particular variation of task switching. The training was effective in improving performance on several tasks, among which the Stroop task and a complex working memory task. All three of these tasks are considered to be tasks that measure *cognitive control*, but they are also tasks that have different strategies. We can broadly categorize the strategies as either *proactive* or *reactive* (Braver 2012). A reactive strategy means that behavior is driven by the stimuli in the experiment. For example, in the Stroop task a reactive strategy involves waiting for the stimulus, attending it, and then naming the color of the letters. During the attending step the identity of the word has the opportunity to interfere with the color of the letters, producing the Stroop effect. However, a proactive strategy can reduce this effect. Proper preparation before the stimulus appears can change the "default" attention step to be replaced by an attending step that focuses on just the color of the letters, strongly reducing the interference. Proactive strategies are, therefore, characterized by task-related processing or preparation that is not directly cued by the stimuli. Proactive strategies tend to be more complex than reactive strategies in the sense they need more operators when modeled, and also more vulnerable to mental distraction, because there is no external cue that triggers the proactive step.

In most task-switching paradigms, subjects also have a choice to proactively prepare for the next stimulus or to wait for the stimulus and then decide what to do (De Jong 1995). However, in the Karbach and Kray experiment subjects were forced to be proactive because there was no external cue from which the current task could be deduced. As a consequence, whether by accident or design, task switching effectively trained a proactive strategy. In the PRIMs model, the proactive strategy consisted of an operator that initiated task preparation before the stimulus appeared (Taatgen 2013b). A subsequent operator would react to this preparation after the stimulus appeared. In the case of the Stroop task, the subsequent operator would focus attention on the color of the stimulus, overruling the default operator that would attend all attributes of the stimulus. In the case of task switching, the preparing operator would adjust the task goals before the stimulus after which the subsequent operator immediately carry out the task goal. Training on task switching was effective because the proactive operators were trained and, therefore, became more efficient to use. After training the choice for a proactive strategy on the Stroop task became more attractive because that strategy could use the same operators.

The model can also explain the improvement on the complex working memory task. In complex working memory tasks, subjects typically have to remember a sequence of items but between the presentation of these items they have to perform another task. In order to be successful on the memory part of the task, it is necessary to perform maintenance rehearsal, but due to the continuous nature of the task as a whole there is no natural moment to do this. Subjects, therefore, need to force themselves to do rehearsal at moments that new stimuli also demand their attention. Therefore, in a complex working memory task, a reactive strategy is to not rehearse at all, or only in brief moments that there is no stimulus, whereas a proactive strategy tries to insert a rehearsal even in the presence of stimuli that demand a response.

If the model of training in the Karbach and Kray experiment is correct, this has repercussions on the effectiveness of brain training. It predicts that brain training is only effective if the cognitive skill that is trained is useful for the tasks that the subjects are tested on, and also effective if the skill in question is not one that subjects would normally have preferred anyway.

Diminishing Return in Expertise

If we reject the notion of the muscle analogy to cognitive training, we should also question cognitive training regimens in which the same task is repeated very often even if that task is gradually increased in difficulty. It is generally assumed that cognitive training adheres to the laws of diminishing returns. For example, there is some evidence that if children learn chess, this has a positive impact on cognitive performance in other areas. However, a chess grandmaster probably has much less benefit of the more advanced strategies in chess (Doll and Mayr 1987).

Frensch (1991) demonstrated this in a study in which subjects repeatedly solved a set of six equations. Some subjects received the training in a blocked paradigm, which means they were trained on equation 1 a number of times, then equation 2 a number of times, etc. A second group would have to solve the equations in a fixed order: equation 1 first, then equation 2, equation 3, etc., and after the last equation back to equation 1. The third group would have to solve the equations in random order. After a certain amount of training in one of the three conditions, all conditions switched to the fixed order version of the task. Frensch found that if the switch took place after a modest amount of training (25 × 6 equations), subjects all performed equally well after the switch to the fixed order. However, if the switch appeared after a large amount of training (75 × 6 equations), the pattern was different: subjects who were trained in the fixed condition now performed much better than those trained in the other conditions. In particular, in the blocked condition performance after short training was identical to performance after long training. In other words, the training between 25 and 75 blocks was only helpful for solving the equations in a particular order.

A PRIMs model of this task (Taatgen 2013a) shows a large overlap in operators that are needed to solve the equations. They generally consist of operators that substitute variables by values and operators that do basic arithmetic. Because of this overlap, the model will rapidly become more efficient at solving the particular equations. The difference between the conditions is mainly in terms of task control. In the fixed order condition, control mainly involves anticipating and moving to the next equation. This operation is less frequent than the operators that solve equations and is, therefore, learned more slowly. Early transfer is, therefore, characterized by transfer in solving the equations. Late transfer is characterized by control, which is different for each of the conditions.

Apart from the specific prediction the model makes for this experiment, generally longer training on the same task leads to rules that are able to handle longer chains of PRIMs. The probability that such a long, specialized chain can be transferred to another task becomes smaller (see also Newell and Rosenbloom 1981, for a similar argument).

Stages or Phases in Development

There is an ongoing debate in developmental psychology on the status of stages or phases in development (Piaget 1952). Although few now believe children progress from one stage to another across the board, a process that could explain a sudden progression in different areas of development is the acquisition of general cognitive skills that are useful for many different things. For example, van Rijn et al. (2002) built a cognitive model that described the progression through various stages of the balance beam task. In that task, children have to determine which side of a balance goes down, taking into account both weight and distance from the middle of the balance. In order to reach the higher stages of performance, children have to be able to integrate the two dimensions (weight and distance) in this task. Van Rijn et al. needed to assume in their model that this multi-dimensional integration skill was one that the child discovered elsewhere and could then apply it in the balance beam task.

To explore this idea, Gittelson and Taatgen (2014) reimplemented three of the stages of the van Rijn model in PRIMs along with three models of decision making of increasing complexity that follow the heuristics of Gigerenzer and Goldstein (1996): the recognition heuristic, the take-the-best-heuristic, and the weighted averages heuristic. These heuristics are used to make choices between two options, for example, which of two cities is larger. The take-the-best heuristic prescribes that you take the most important attribute of each of the cities (e.g., does the city have an airport) and base your decision on that attribute. If you cannot make a choice on the basis of the most important attribute, you move to the second most important attribute (e.g., does the city have a premier league soccer team). This heuristic is similar to stage 2 in the balance beam task where children first look at weight and only if the weights are equal pay attention to distance.

The models showed considerable transfer, not just *vertical transfer*, in the sense that it is easier to learn the second stage of the balance beam task once the first stage is mastered, but also *horizontal transfer*, in the sense that mastery of the take-the-best heuristic makes it easier to learn the second stage of the balance beam task. This means that discovery of the take-the-best heuristic also facilitates moving from stage 1 to stage 2 in the balance beam task, and maybe other tasks as well. Therefore discovery of a "stage 2" strategy may trigger advances in several different tasks, giving the impression of an across-the-board stage wise developmental transition.

Conclusions

The central idea of PRIMs is that general cognitive strategies are learned as a byproduct of task-specific learning. This principle can explain classical transfer effects, the effects of brain training, the limitations of expertise, and, potentially, aspects of cognitive development. It provides superior explanations for transfer data of text editing because the more fine-grained representation of the skill was able to predict transfer between line-based and screen-based editors.

The explanation for brain training is that it primes strategies that are more proactive, and therefore lead to better performance on tasks that also benefit from proactive strategies. However, it also predicts that this form of brain training has a limited scope, and that the benefits do no persist with long-term training. However, if brain training would teach a variety of generally useful skills, it might be beneficial for those individuals that do not already possess these skills.

PRIMs can offer explanations of transfer at different levels of abstraction. The effects of transfer between text editors and brain training are at a relatively low level in which elementary procedure skills are transferred. In contrast, the explanation for the attentional blink is based on the choice of a wrong combination of cognitive skills.

Acknowledgments This research was supported by the European Research Council Starting Grant MULTITASK 283597.

References

Anderson, J. R. (2007). *How can the human mind occur in the physical universe?* New York: Oxford university press.

Braver, T. S. (2012). The variable nature of cognitive control: A dual mechanisms framework. *Trends in Cognitive Sciences, 16*, 106–113.

Chase, W. G., & Ericsson, K. A. (1982). Skill and working memory. In G. H. Bower (Ed.), *The psychology of learning and motivation* (pp. 1–58). New York: Academic Press.

De Jong, R. (1995). The role of preparation in overlapping-task performance. *Quarterly Journal of Experimental Psychology Section A: Human Experimental Psychology, 48*, 2–25.

Doll, J., & Mayr, U. (1987). Intelligenz und schachleistung—Eine untersuchung an schachexperten. [intelligence and success in chess playing: An examination of chess experts.]. *Psychologische Beitrage, 29*, 270–289.

Ferlazzo, F., Lucido, S., Di Nocera, F., Fagioli, S., & Sdoia, S. (2007). Switching between goals mediates the attentional blink effect. *Experimental Psychology, 54*, 89–98.

Frensch, P. A. (1991). Transfer of composed knowledge in a multistep serial task. *Journal of Experimental Psychology: Learning, Memory, and Cognition, 17*, 997–1016.

Gick, M. L., & Holyoak, K. J. (1980). Analogical problem solving. *Cognitive Psychology, 12*, 306–355.

Gigerenzer, G., & Goldstein, D. G. (1996). Reasoning the fast and frugal way: Models of bounded rationality. *Psychological Review, 103*, 650–669.

Gittelson, L., & Taatgen, N. A. (2014). Transferring primitive elements of skill within and between tasks. In P. Bello, M. Guarini, M. McShane, & B. Scassellati (Eds.), *Proceedings of the 36th annual meeting of the cognitive science society* (pp. 529–534). Quebec city.

Hoekstra, C., Martens, S., & Taatgen, N. A. (2019). A skill-based approach to modeling the attentional blink. In *Proceedings of the international conference on cognitive modelling*. Canada: Montreal.

Karbach, J., & Kray, J. (2009). How useful is executive control training? Age differences in near and far transfer of task-switching training. *Developmental Science, 12*, 978–990.

Newell, A., & Rosenbloom, P. S. (1981). Mechanisms of skill acquisition and the law of practice. In J. R. Anderson (Ed.), *Cognitive skills and their acquisition* (pp. 1–56). Hillsdale: Erlbaum.

Piaget, J. (1952). *The origins of intelligence in children*. New York: International University Press.

Raymond, J. E., Shapiro, K. L., & Arnell, K. M. (1992). Temporary suppression of visual processing in an RSVP task: An attentional blink? *Journal of Experimental Psychology: Human Perception and Performance, 18*, 849–860.

Singley, M. K., & Anderson, J. R. (1985). The transfer of text-editing skill. *Journal of Man-Machine Studies, 22*, 403–423.

Singley, M. K., & Anderson, J. R. (1989). *The transfer of cognitive skill*. Cambridge, MA: Harvard University Press.

Taatgen, N. A. (2013a). Diminishing return in transfer: A PRIM model of the frensch (1991) arithmetic experiment. In *Proceedings of the international conference on cognitive modeling* (pp. 29–34). Ottawa: Carleton University.

Taatgen, N. A. (2013b). The nature and transfer of cognitive skills. *Psychological Review, 120*, 439–471.
Taatgen, N. A., & Anderson, J. R. (2002). Why do children learn to say "broke"? A model of learning the past tense without feedback. *Cognition, 86*, 123–155.
Thorndike, E. L., & Woodworth, R. S. (1901). The influence of improvement in one mental function upon the efficiency of other functions. *Psychological Review, 8*, 247–261.
van Rijn, H., van Someren, M., & van der Maas, H. (2002). Modeling developmental transitions on the balance scale task. *Cognitive Science, 107*, 1–31.

Statistical Modeling of Latent Change

Tanja Könen and Max Auerswald

Contents

Abstract The focus of this chapter is on a selected class of statistical models: latent change models. They are especially eligible for typical applications in cognitive training research with two or three groups (e.g., training, active control, passive control) and two or three time points (pretest, posttest, follow-up). Latent variable models have a long tradition in cognitive science because they can separate task-, paradigm-, and ability-specific variance in performance tasks. Latent change modeling allows to study latent means, latent intraindividual mean changes, and interindividual differences in both. This chapter addresses how the effectiveness of training programs can be evaluated with latent change models and typical misunderstandings in this context. Statistical power considerations and measurement invariance across experimental groups and time points are discussed. The benefits and risks of analyzing predictors and correlates of latent change variables are particularly relevant for cognitive training research. They provide valuable correlative information about possible mechanisms moderating training outcomes (e.g., compensation or magnification effects) but are no causal test of these mechanisms. Taken together, latent change modeling does not only allow testing whether a cognitive training works on average, but also studying interindividual differences in training outcomes.

T. Könen (✉)
Department of Psychology, University of Koblenz-Landau, Landau, Germany
e-mail: koenen@uni-landau.de

M. Auerswald
Department of Psychology, Ulm University, Ulm, Germany
e-mail: max.auerswald@uni-ulm.de

T. Strobach, J. Karbach (eds.), *Cognitive Training*,
https://doi.org/10.1007/978-3-030-39292-5_5

Latent Variable Models in Cognitive Science

Latent variable models have a long tradition in cognitive science (e.g., Hertzog and Schaie 1986; Sternberg 1978; see also Cochrane and Green, this volume) and offer characteristics which are particularly useful for studying cognitive performance. They allow not only to differentiate true score and error variance of a construct, but also to separate important sources of variance in cognitive tasks. For example, working memory updating tasks (Salthouse et al. 1991, see also Könen et al., this volume) require the continuous updating of the status of multiple stimuli (e.g., of spatial movements of multiple objects or of simple calculations with multiple numbers) before the final results must be recalled. Variance in this task performance can thus be attributed to task-specific (stimuli types), paradigm-specific (continuous updating), and ability-specific effects (simultaneous storage and processing). For almost all types of research questions, it is informative to know whether an effect of interest is valid on the ability level (e.g., working memory as system for simultaneous storage and processing, Baddeley and Hitch 1994), or is based on a specific mechanism which is captured by selected task paradigms (e.g., updating), or is task specific (e.g., updating of letters). Because a latent variable is equivalent to whatever is common among its indicators (and *not* a combination of its indicators; Rhemtulla et al. 2019), using different established task paradigms from more than one domain (spatial, numerical, verbal) and/or modality (e.g., visual, acoustic) as indicators for a latent variable allows for inferences on a cognitive ability level. For example, a latent variable with diverse working memory tasks (different paradigms and domains/modalities) as indicators captures simultaneous storage and processing as it is their central common requirement (Fig. 1). In this case, paradigm- and

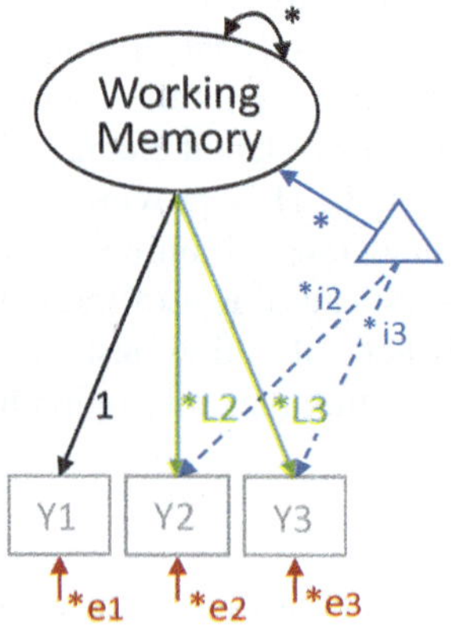

Fig. 1 Confirmatory factor model of a cognitive ability, for example, working memory. The circle represents a latent variable, squares represent observed variables, asterisks represent estimated parameters, and the triangle represents mean- and intercept-information (dashed lines are intercepts). For model identification, the first factor loading is fixed to one. Factor loadings (L2, L3), intercepts (i2, i3), and error terms (e1, e2, e3) are estimated. Observed indicator variables (Y1–Y3) could be, for example, a spatial updating, numerical n-back, and verbal complex span task (see Wilhelm et al. 2013, for task descriptions)

task-specific variances are considered an indicator-specific measurement error and are thus separated from the latent ability variance.

Using tasks of the same paradigm but from different domains and/or modalities as indicators of a latent variable supports inferences about the central mechanism assessed by this paradigm. For example, updating is the central common requirement of spatial, numerical, and verbal updating tasks. This demonstrates how latent variable modeling supports testing effects on the level of interest in cognitive psychology. More general introductions highlight that many psychological constructs are inherently latent (i.e., not directly observable; Borsboom 2008, for details) and should be represented accordingly in statistical analyses.

The cognitive training literature could profit from an increased application of latent variable models. As Noack et al. (2014) argue, if training programs aim at improving a cognitive ability, then this ability should be theoretically defined and represented as latent. Its indicators should be multiple heterogeneous transfer tasks (i.e., non-trained tasks), which are sampled from the theoretically determined task space (Little et al. 1999). This strengthens claims of ability improvements as it rules out task-specific effects as alternative explanation for performance improvements, such as the development and automatization of task-specific strategies. Controversies in the cognitive training literature about the presence (Au et al. 2015, 2016; Karbach and Verhaeghen 2014) or absence (Melby-Lervåg and Hulme 2016; for details see Guye et al., this volume; Könen et al., this volume) of far transfer effects (i.e., improvements in cognitive functions other than the trained one/s) could also be addressed and possibly solved on the latent ability level.

In this chapter, we focus on a selected class of statistical models for analyzing latent change: latent change models. Latent change models are a particularly useful framework for cognitive training studies because they are especially eligible for typical applications with two or three groups (e.g., training, active control, passive control) and two or three time points (pretest, posttest, if applicable follow-up). Hence, they have been increasingly applied in the training literature over the last decade (e.g., McArdle and Prindle 2008; Schmiedek et al. 2010, 2014; Zelinski et al. 2014). Below, we present an introduction to latent change modeling and important concepts (e.g., measurement invariance) and further discuss possible practical challenges and limitations.

Introduction to Latent Change Modeling

Latent change models (McArdle and Hamagami 2001; for an overview see McArdle 2009) are also called latent change score models, latent difference (score) models, and latent true change models. They can be estimated as *multiple-group latent change models* and allow analyzing latent variables and latent changes in these variables across both time points and groups. Latent change models utilize a set of fixed coefficients (fixed to 1) to define a later measurement occasion (Fig. 2: $f[2]$) as the sum of an earlier occasion ($f[1]$) and the difference ($\Delta f_{[2-1]}$) between both: $f[2] = f[1]$

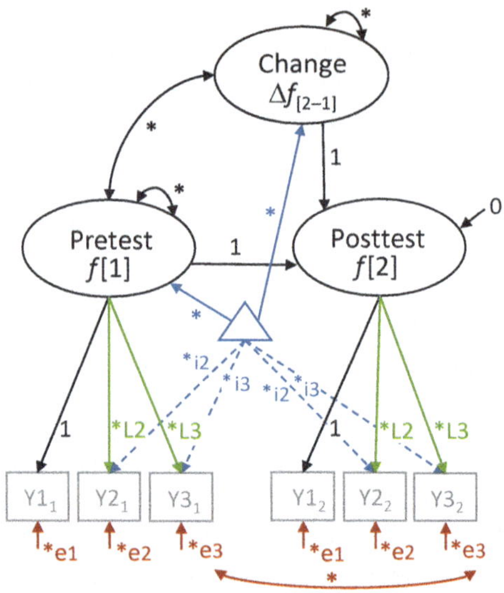

Fig. 2 Latent change model with strict measurement invariance across pretest and posttest. Circles represent latent variables, squares represent observed variables, asterisks represent estimated parameters, and the triangle represents mean- and intercept-information (dashed lines are intercepts). Parameters with the same name are constrained to be equal (are estimated on the same unstandardized value). For model identification, the first factor loading of each latent variable is fixed to one. Correlated error terms of the same indicator across time are allowed (exemplary shown for e3). Factor loadings (L2, L3), intercepts (i2, i3), and error terms (e1, e2, e3) are constrained to be equal across time. Observed indicator variables are named Y1–Y3

+ $\Delta f_{[2-1]}$ (McArdle 2009). The change between two time points ($\Delta f_{[2-1]} = f[2] - f[1]$) is represented as a latent variable with a mean (i.e., average change), a variance (i.e., individual differences in change), a covariance with the initial factor $f[1]$ and, if applicable, covariances with other variables in the model. Such a model allows estimating latent means, latent intraindividual mean changes, and interindividual differences in both. If a latent variable is considered free of measurement error at two time points (e.g., pretest and posttest) then the latent change between both is also considered free of measurement error (cf. McArdle and Prindle 2008). Thus, analyzing latent change scores is preferable to analyzing observed difference scores (Trafimow 2015, for a review of the latter).

As needed, models can include multiple latent change variables, for example, to capture the changes between pretest and posttest (e.g., $\Delta f_{[2-1]}$) and between posttest and follow-up (e.g., $\Delta f_{[3-2]}$). The latent mean score could increase over one period and be stable or even decrease over the next because the direction of change between the measurement occasions is independent. This is especially suitable for cognitive training studies, in which stability, decrease, or increase of transfer effects at follow-up is possible (the latter, for example, due to daily life training benefits). For example, transfer effects of a broad cognitive training were significantly reduced at a 2-year follow-up (in comparison to transfer at posttest) for episodic memory but not for reasoning (Schmiedek et al. 2014), which was analyzed with latent change models. Further, both latent change variables can have differential predictors, which is crucial, because the factors contributing to training-related gains may not be same as the factors contributing to maintenance after training.

As in all structural equation models with latent variables, one must evaluate how well the hypothesized model fits the observed data, usually with a χ^2-test (chi square test) and multiple descriptive fit indices such as the Comparative Fit Index (CFI), the Root Mean Square Error of Approximation (RMSEA), and the Standardized Root Mean Square Residual (SRMR); see West et al. (2012) for details. An introduction to the statistical assumptions of structural equation modeling and common estimation methods (e.g., maximum likelihood estimation) can be found in Kline (2012). Particularly relevant in the context of latent change modeling are the assumptions that indicators are mutually uncorrelated after controlling for their common latent factor (i.e., local independence) and that relations between the indicators and other variables are attributed to relations between the common latent factor and those variables (e.g., Rhemtulla et al. 2019). However, an indicator usually correlates with itself over time over and above common latent factor correlations (e.g., in cognitive tasks due to task-specific effects). Thus, failing to represent these covariances in the model, for example, with correlated error terms (in Fig. 2 exemplary shown for parameter e3) or with method factors for the same indicator across time, can lead to biased estimations of structural relations and decreased model fit (e.g., Pitts et al. 1996). Generally, structural equation models are more flexible in testing and accounting for statistical assumptions than other statistical techniques (e.g., analysis of variance). For example, non-normality in the data distribution can be addressed by using robust estimation methods (Lei and Wu 2012, for details). Measurement invariance (across experimental groups and time points) and statistical power are discussed in later sections of this chapter. More detailed descriptions of latent change models with code examples are available in the literature (e.g., Ghisletta and McArdle 2012; Kievit et al. 2018; Klopack and Wickrama 2019).

Testing the Effectiveness of Training Programs

In randomized controlled trials, group mean differences between experimental and adequate control groups can serve as estimates of average *causal* treatment effects (Holland 1986; Schmiedek, this volume for details). When cognitive training studies are analyzed with multiple-group latent change models (Fig. 3), one can test for any training-related differences by comparing the fit of the model (with a $\Delta\chi^2$-test, i.e., chi square difference test) when a parameter is either *constrained to be equal* or *free to vary* across the training group and an adequate control group (McArdle and Prindle 2008).

One can test for average group effects by constraining the means of the latent change between the pre- and posttest ($\Delta f_{[2-1]}$) of a training or transfer variable to be equal in the training group and control group (for an example see Stine-Morrow et al. 2014). If such a constraint significantly decreases model fit, the groups

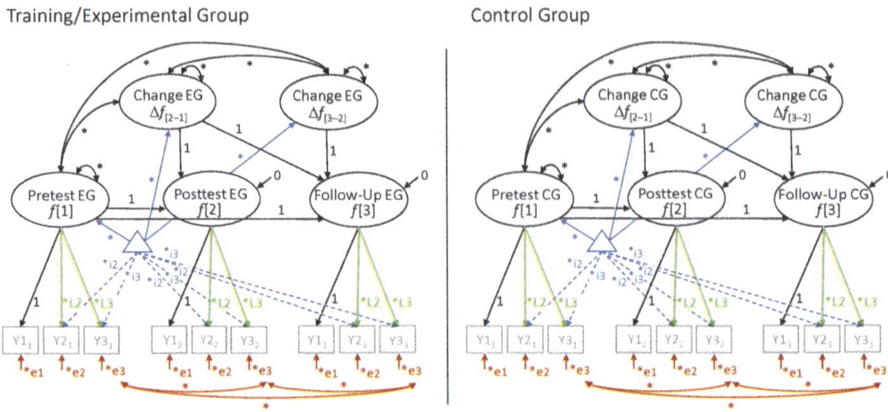

Fig. 3 Multiple-group latent change model with strict measurement invariance across group (training, control) and time (pretest, posttest, follow-up). Circles represent latent variables, squares represent observed variables, asterisks represent estimated parameters, and the triangle represents mean- and intercept-information (dashed lines are intercepts). Parameters with the same name are constrained to be equal (are estimated on the same unstandardized value). For model identification, the first factor loading of each latent variable is fixed to one. Correlated error terms of the same indicator across time are allowed (exemplary shown for e3). Factor loadings (L2, L3), intercepts (i2, i3), and error terms (e1, e2, e3) are constrained to be equal across groups and time. Observed indicator variables are named Y1–Y3

significantly differ in the latent change between the pretest and posttest. Latent effect sizes can be calculated equally as Cohen's d by dividing the latent mean differences by the latent pooled standard deviations at pretest (when analyzing pretest and posttest, for an example see Schmiedek et al. 2014) or posttest (when analyzing posttest and follow-up). Please note that latent standard deviations might not be included in the output of software packages but can be easily calculated based on the provided variances. Alternatively, standardized indicators simplify the interpretation of latent means and latent mean changes (e.g., standardized to a *T* score distribution based on the pretest means and standard deviations as in Stine-Morrow et al. 2014).

It is possible that a training program has significant mean group effects on some indicators of a latent variable, but this effect is not valid on the latent level, which means that the common factor does not capture the effect (e.g., Estrada et al. 2015). Possible explanations can be substantive (e.g., task- or paradigm-specific training effects, such as the development and automatization of specific strategies) or more methodological (e.g., cognitive tasks differ in their reliability and sensitivity to change). At the same time, an effect can be significant on a latent level but not present in all indicators (e.g., Schmiedek et al. 2010). A solution to this issue is to report the average group findings on both a latent and an observed level (e.g., Schmiedek et al. 2010).

After this introduction to testing the effectiveness of training programs with latent change modeling, we also discuss two approaches which are *no* causal tests of cognitive training effects. First, the so-called *responder analyses* allow no causal inferences about cognitive training effects (Tidwell et al. 2014, for details). Applications of responder analyses aim at testing the effectiveness of training regimes for subgroups with specific characteristics. Individuals are classified on posttest (e.g., in high vs. low values) or change scores (e.g., more vs. less improvement, i.e., high and low responders) of an outcome variable and this classification is used as predictor of change in another outcome variable. Although latent change models are generally well-suited for predicting change, caution is necessary with responder analyses. Due to the post-hoc classification, they allow no clear distinction and attribution of cause and effect (see Tidwell et al. 2014 for more information).

Second, it can be informative to test for *correlated gains* on training and transfer scores because it is often reasonable to assume that individuals who benefit the most on the trained tasks are more likely to be the ones who demonstrate transfer to non-trained tasks (e.g., Zelinski et al. 2014). Correlated gains can descriptively support interpretations of training effects established on the mean group level, but they are no test of training effects. Correlated gains can be significant regardless of the group means (i.e., regardless of training-related improvements) because the magnitude of a correlation is invariant to linear transformations of the variables. In line with this, simulation studies demonstrated that transfer can be valid without any correlation in gain scores and correlated gain scores do not necessarily guarantee transfer (Jacoby and Ahissar 2015; Moreau et al. 2016).

Measurement Invariance

To be able to compare scores on a variable such as performance in a cognitive task across experimental groups and time (measurement occasions), the measurement needs to be equivalent (i.e., invariant) across groups and time (e.g., Widaman and Reise 1997). This applies for all types of variables, observed as well as latent variables. In most cases, it can only be assumed when using classical statistical procedures (e.g., analysis of variance) but can be explicitly tested and represented in models with latent variables. In training studies, one would typically test measurement invariance across groups first, separately for each measurement occasion, and then invariance across time (the latter in a multiple-group model were the invariance across groups is held constant). The advantage of this consecutive approach is that findings of non-invariance are directly attributable to either group or time. In a randomized controlled trial, measurement invariance across experimental groups at pretest/baseline is inherently expected due to the random assignment to the groups (Pitts et al. 1996) and any descriptive differences are the result of chance rather than bias (Moher et al. 2010).

The classical procedure of establishing measurement invariance consists of four steps (suggested by Meredith 1993; Widaman and Reise 1997), which are hierarchically ordered and are tested by comparing increasingly constrained models. The procedure is the same regardless of whether invariance across groups or time is investigated (which is why "groups or time" is used in the following). At first, *configural invariance* (the equivalence of model form) is established if the factors across groups or time have the same pattern of fixed and free loadings. Second, *metric invariance* or *weak factorial invariance* (the equivalence of factor loadings) is established if constraining the unstandardized factor loadings (see Fig. 3: parameters L2 and L3) to be equal across groups or time does not result in a substantial drop of model fit compared to a model with only configural invariance. Third, *scalar invariance* or *strong factorial invariance* (the equivalence of intercepts or thresholds) is established if additionally constraining the unstandardized intercepts (Fig. 3: parameters i2 and i3) or thresholds to be equal across groups or time does not result in a substantial drop of model fit compared to a model with only metric invariance (continuous indicators have intercepts, categorical indicators have thresholds). Scalar invariance implies that all substantial mean differences (across groups or time) in the indicators are captured by and attributable to the common latent construct, a necessary condition to compare latent means across groups or time (Widaman and Reise 1997). Fourth, *strict invariance* (the equivalence of residuals) is established if additionally constraining the unstandardized residuals (Fig. 3: parameters e1, e2, and e3) to be equal across groups or time does not result in a substantial drop of model fit compared to a model with only scalar invariance. This implies that all substantial (co)variance differences (across groups or time) in the indicators are captured by and attributable to the common latent construct (Widaman and Reise 1997). Across these four steps, the drop of model fit can be evaluated with a $\Delta\chi^2$-test (chi square difference test) and with descriptive fit indices (e.g., Cheung and Rensvold 2002; Meade et al. 2008).

Taken together, scalar measurement invariance is the necessary condition to compare latent means across groups or time and thus for testing the effectiveness of training programs on a latent level. Strict measurement invariance is even preferable as it implies that all substantial mean and (co)variance differences in the indicators across groups and time are captured by and attributable to the common latent construct, which supports their substantive interpretation. For example, comparing predictors of latent variables across groups or time is strengthened by strict measurement invariance. Finally, the model used for hypotheses testing should include invariance constraints across group and time (e.g., Fig. 3; for an empirical example see Schmiedek et al. 2010).

In case of *violations of invariance* (i.e., non-invariance), one should consider possible reasons for the violations in the given study, which can be practical (e.g., differential recruitment strategies for the training and control group) or theoretical (e.g., the relation of a task with the construct changed because the processes involved in task performance changed during skill acquisition, Ackerman 1988). There is no generally advisable strategy for all training studies, neither dropping the problematic indicator/s or refraining from analyzing the construct nor releasing the invariance constrains on the problematic indicator/s or continuing to impose all invariance constrains. The first two options are a threat to content validity (e.g., Pitts et al. 1996), and the latter two options can result in biased parameter estimates in the model, which are not necessarily indicated by the overall model fit (e.g., Clark et al.

2018). The strategy should depend on the specific research question and the specific measurement instruments used. Most importantly, one should compare and report whether the main findings and conclusions depend on this choice (i.e., are sensitive or not). Finally, the four steps described here are the current standard approach in psychology (Putnick and Bornstein 2016), but several alternatives for testing measurement invariance exist (e.g., Tay et al. 2015 used item-response theory; Van de Schoot et al. 2013 used a Bayesian approach).

Statistical Power Considerations

Simulations with the Monte Carlo method are the state of the art for estimating power in latent change modeling (Muthén and Muthén 2002, for a general introduction; Zhang and Liu 2019, for details on latent change modeling). Easy rule-of-thumbs such as "at least 10 or 20 cases per variable" can be misleading and should not be applied (e.g., Wolf et al. 2013). However, user-friendly online tools have been recently developed for estimating power in latent change modeling (e.g., Brandmaier et al. 2015 [www.brandmaier.de/lifespan]; Zhang and Liu 2019 [https://webpower.psychstat.org]). Still, collecting the basic information needed for power analyses (e.g., information on expected means and co/variances) could be difficult and might require a prestudy. Further, more research on the interplay of factors determining statistical power in latent change models is needed. Most studies investigated latent growth curve models (e.g., Hertzog et al. 2006, 2008; Rast and Hofer 2014), but one cannot generalize findings on statistical power to different classes of developmental models (cf. Hertzog et al. 2006) mostly because of differences in the underlying functions of change. Generally, low power represents not only a reduced chance to find a true effect, but also reduces the likelihood that a statistically significant finding reflects a true effect (cf. Button et al. 2013). Thus, estimating the statistical power of finding the main effects of a study is always worth the effort although this effort is admittedly likely higher for latent change modeling than for traditional approaches such as analysis of variance. Notably, regardless of power, when using frequentist statistics, a nonsignificant finding does not allow to infer the absence of an effect (e.g., Aczel et al. 2018; see De Simoni and Von Bastian 2018, for Bayesian evidence on the absence of effects).

Predictors and Correlates of Change Variables

The effectiveness of training programs is usually the first research question addressed in cognitive training studies. However, as Willis and Schaie (2009) pointed out, "programmatic intervention research should be aimed at the broader goal of answering a series of theoretically important empirical questions" such as "What specific mechanisms, processes, or components of the intervention are responsible for the desired change? What individual difference variables are associated with responsivity to change? How can the change be maintained?" (cf. Willis and Schaie 2009, p. 377). Latent change modeling offers some unique opportunities to address these and related questions because changes between two time points (e.g., pretest and posttest, posttest and follow-up) are represented as latent variables with means (i.e., average changes) and variances (i.e., individual differences in changes). If the variance of a latent change variable is significantly different from zero, it is reasonable to assume that this variance is not only random noise but includes reliable individual differences in change. Analyzing predictors or correlates of latent change variables allows to identify if, for example, some features of an individual or the situation make training or transfer gains more or less likely. Whether predictors or correlates are analyzed should depend on the given research question, but it is important to keep in mind that the mean of a latent change variable, which is predicted by other variables, should be interpreted conditional on the regression paths (i.e., does not represent "raw" mean changes; cf. Kievit et al. 2018).

A typical predictor of change is individual baseline cognitive performance, for example, when testing compensation or magnification effects (see Karbach and Kray, this volume; Katz et al., this volume). A compensation effect predicts that individuals with lower baseline performance tend to profit more from a training (i.e., higher gains over time) whereas a magnification effect predicts that individuals with higher baseline performance tend to profit more (Lövdén et al. 2012, for details). For example, Karbach et al. (2017) found that individuals with lower cognitive performance at baseline showed larger training and transfer benefits of an executive control training. They used multiple-group latent change models and compared models with a $\Delta\chi^2$-test (chi square difference test) in which the relation of baseline performance and change was either *constrained to be equal* or *free to vary* across the training and active control group. The relation of baseline and change was significantly higher in the training group compared to the active control group, which strengthens a substantive interpretation (e.g., because regression to the mean should occur in both groups, see Marsh and Hau 2002, for details on regression to the mean artifacts).

Other possible predictors are, for example, age, years of education, family income, need for cognition, or personality (e.g., Stine-Morrow et al. 2014; Zelinski et al. 2014). One might consider different predictors for different change variables (Fig. 3: $\Delta f_{[2-1]}$ and $\Delta f_{[3-2]}$) because the factors contributing to training-related gains may not be the same as the factors contributing to maintenance after training. Of course, confirmatory and exploratory tests need to be explicitly distinguished, and a suitable correction of the statistical alpha level should be considered if multiple predictors or correlates are tested (e.g., Bonferroni-Holm method).

Further, it can be informative to test for correlated gains on training and transfer scores (e.g., McArdle and Prindle 2008; Zelinski et al. 2014) because it is often reasonable to assume that individuals who benefit the most on the trained tasks could also be the ones who demonstrate transfer to non-trained tasks. For example, Zelinski et al. (2014) analyzed correlations between gains in training and in transfer tasks in older adults with latent change models. Overall, correlations of training and transfer gains were mostly found for tasks with overlapping task demands, which is in line with an overlapping task demand model of transfer (cf. Zelinski et al. 2014). Notably, the effects were valid with and without controlling for covariates (age and education) related to both training and transfer gains. Taken together, predictors and correlates of latent change variables can provide valuable correlative information about possible mechanisms moderating (e.g., compensation or magnification effects) or fostering training outcomes (e.g., overlapping task demands). A more general introduction to analyzing predictors and correlates of intervention-related change is currently under review (Könen & Karbach, 2020).

Conclusion

On the one hand, latent change modeling of cognitive training data is arguably more time consuming than traditional analyses (e.g., analysis of variance), for example, because measurement invariance must be tested and the fit of the hypothesized model to the data must be evaluated. On the other hand, however, latent change modeling offers unique opportunities, which can enhance the practical and theoretical understanding of training and transfer effects. For example, it allows separating task-, paradigm-, and ability-specific effects and testing predictors and correlates of latent change variables. With this, one can not only evaluate whether a training program works on average but also understand which individual and situational characteristics make individual outcomes more likely.

References

Ackerman, P. L. (1988). Determinants of individual differences during skill acquisition: Cognitive abilities and information processing. *Journal of Experimental Psychology: General, 117*, 288–318.

Aczel, B., Palfi, B., Szollosi, A., Kovacs, M., Szaszi, B., Szecsi, P., ... & Wagenmakers, E. J. (2018). Quantifying support for the null hypothesis in psychology: An empirical investigation. *Advances in Methods and Practices in Psychological Science, 1*, 357–366.

Au, J., Sheehan, E., Tsai, N., Duncan, G. J., Buschkuehl, M., & Jaeggi, S. M. (2015). Improving fluid intelligence with training on working memory: A meta-analysis. *Psychonomic Bulletin and Review, 22*, 366–377.

Au, J., Buschkuehl, M., Duncan, G. J., & Jaeggi, S. M. (2016). There is no convincing evidence that working memory training is NOT effective: A reply to Melby-Lervåg and Hulme (2015). *Psychonomic Bulletin and Review, 23*, 331–337.

Baddeley, A. D., & Hitch, G. J. (1994). Developments in the concept of working memory. *Neuropsychology, 8*, 485–493.

Borsboom, D. (2008). Latent variable theory. *Measurement, 6*, 25–53.

Brandmaier, A. M., von Oertzen, T., Ghisletta, P., Hertzog, C., & Lindenberger, U. (2015). LIFESPAN: A tool for the computer-aided design of longitudinal studies. *Frontiers in Psychology, 6*, 272.

Button, K. S., Ioannidis, J. P. A., Mokrysz, C., Nosek, B. A., Flint, J., Robinson, E. S. J., & Munafo, M. R. (2013). Power failure: Why small sample size undermines the reliability of neuroscience. *Nature Reviews Neuroscience, 14*, 365–376.

Cheung, G. W., & Rensvold, R. B. (2002). Evaluating goodness-of fit-indexes for testing measurement invariance. *Structural Equation Modeling, 9*, 233–255.

Clark, D. A., Nuttall, A. K., & Bowles, R. P. (2018). Misspecification in latent change score models: Consequences for parameter estimation, model evaluation, and predicting change. *Multivariate Behavioral Research, 53*, 172–189.

De Simoni, C., & von Bastian, C. C. (2018). Working memory updating and binding training: Bayesian evidence supporting the absence of transfer. *Journal of Experimental Psychology: General, 14*, 829–858.

Estrada, E., Ferrer, E., Abad, F. J., Román, F. J., & Colom, R. (2015). A general factor of intelligence fails to account for changes in tests' scores after cognitive practice: A longitudinal multi-group latent-variable study. *Intelligence, 50*, 93–99.

Ghisletta, P., & McArdle, J. J. (2012). Teacher's corner: Latent curve models and latent change score models estimated in R. *Structural Equation Modeling, 19*, 651–682.

Hertzog, C., & Schaie, K. W. (1986). Stability and change in adult intelligence: Analysis of longitudinal covariance structures. *Psychology and Aging, 1*, 159–171.

Hertzog, C., Lindenberger, U., Ghisletta, P., & von Oertzen, T. (2006). On the power of multivariate latent growth curve models to detect correlated change. *Psychological Methods, 11*, 244–252.

Hertzog, C., von Oertzen, T., Ghisletta, P., & Lindenberger, U. (2008). Evaluating the power of latent growth curve models to detect individual differences in change. *Structural Equation Modeling, 15*, 541–563.

Holland, P. W. (1986). Statistics and causal inference. *Journal of the American Statistical Association, 81*, 945–960.

Jacoby, N., & Ahissar, M. (2015). Assessing the applied benefits of perceptual training: Lessons from studies of training working-memory. *Journal of Vision, 15*, 6.

Karbach, J., & Verhaeghen, P. (2014). Making working memory work: A meta-analysis of executive-control and working memory training in older adults. *Psychological Science, 25*, 2027–2037.

Karbach, J., Könen, T., & Spengler, M. (2017). Who benefits the most? Individual differences in the transfer of executive control training across the lifespan. *Journal of Cognitive Enhancement, 1*, 394–405.

Kievit, R., Brandmaier, A., Ziegler, G., van Harmelen, A.-L., de Mooij, S., Moutoussis, M., the Neuroscience in Psychiatry Network (NSPN) Consortium, Lindenberger, U., & Dolan, R. J. (2018). Developmental cognitive neuroscience using latent change score models: A tutorial and applications. *Developmental Cognitive Neuroscience, 33*, 99–117.

Kline, R. B. (2012). Assumptions in structural equation modeling. In R. H. Hoyle (Ed.), *Handbook of structural equation modeling* (pp. 111–125). New York: The Guilford Press.

Klopack, E. T., & Wickrama, K. A. S. (2019). Modeling latent change score analysis and extensions in Mplus: A practical guide for researchers. *Structural Equation Modeling*. Advance online publication.

Könen, T., & Karbach, J. (2020). *Analyzing individual differences in intervention-related changes*. Manuscript submitted for publication.

Lei, P.-W., & Wu, Q. (2012). Estimation in structural equation modeling. In R. H. Hoyle (Ed.), *Handbook of structural equation modeling* (pp. 164–180). New York: The Guilford Press.

Little, T. D., Lindenberger, U., & Nesselroade, J. R. (1999). On selecting indicators for multivariate measurement and modeling with latent variables: When "good" indicators are bad and "bad" indicators are good. *Psychological Methods, 4*, 192–211.

Lövdén, M., Brehmer, Y., Li, S. C., & Lindenberger, U. (2012). Training induced compensation versus magnification of individual differences in memory performance. *Frontiers in Human Neuroscience, 6*, 141.

Marsh, H. W., & Hau, K. T. (2002). Multilevel modeling of longitudinal growth and change: Substantive effects or regression toward the mean artifacts? *Multivariate Behavioral Research, 37*, 245–282.

McArdle, J. J. (2009). Latent variable modeling of differences and changes with longitudinal data. *Annual Review of Psychology, 60*, 577–605.

McArdle, J. J., & Hamagami, F. (2001). Latent difference score structural models for linear dynamic analyses with incomplete longitudinal data. In L. M. Collins & A. G. Sayer (Eds.), *Decade of behavior. New methods for the analysis of change* (pp. 139–175). Washington: American Psychological Association.

McArdle, J. J., & Prindle, J. J. (2008). A latent change score analysis of a randomized clinical trial in reasoning training. *Psychology and Aging, 23*, 702–719.

Meade, A. W., Johnson, E. C., & Braddy, P. W. (2008). Power and sensitivity of alternative fit indices in tests of measurement invariance. *Journal of Applied Psychology, 93*, 568–592.

Melby-Lervåg, M., & Hulme, C. (2016). There is no convincing evidence that working memory training is effective: A reply to Au et al. (2014) and Karbach and Verhaeghen (2014). *Psychonomic Bulletin and Review, 23*, 324–330.

Meredith, W. (1993). Measurement invariance, factor analysis and factorial invariance. *Psychometrika, 58*, 525–543.

Moher, D., Hopewell, S., Schulz, K., Montori, V., Gøtzsche, P. C., Devereaux, P. J., Elbourne, D., Egger, M., & Altman, D. G. (2010). Consort 2010 explanation and elaboration: Updated guidelines for reporting parallel group randomized trials. *British Medical Journal, 340*, 698–702.

Moreau, D., Kirk, I. J., & Waldie, K. E. (2016). Seven pervasive statistical flaws in cognitive training interventions. *Frontiers in Human Neuroscience, 10*, 153.

Muthén, L. K., & Muthén, B. O. (2002). How to use a Monte Carlo study to decide on sample size and determine power. *Structural Equation Modeling, 9*, 599–620.

Noack, H., Lövdén, M., & Schmiedek, F. (2014). On the validity and generality of transfer effects in cognitive training research. *Psychological Research, 78*, 773–789.

Pitts, S., West, S., & Tein, J. (1996). Longitudinal measurement models in evaluation research: Examining stability and change. *Evaluation and Program Planning, 19*, 333–350.

Putnick, D. L., & Bornstein, M. H. (2016). Measurement invariance conventions and reporting: The state of the art and future directions for psychological research. *Developmental Review, 41*, 71–90.

Rast, P., & Hofer, S. M. (2014). Longitudinal design considerations to optimize power to detect variances and covariances among rates of change: Simulation results based on actual longitudinal studies. *Psychological Methods, 19*, 133–154.

Rhemtulla, M., van Bork, R., & Borsboom, D. (2019). Worse than measurement error: Consequences of inappropriate latent variable measurement models. *Psychological Methods*. Advance online publication.

Salthouse, T. A., Babcock, R. L., & Shaw, R. J. (1991). Effects of adult age on structural and operational capacities in working memory. *Psychology and Aging, 6*, 118–127.

Schmiedek, F., Lövdén, M., & Lindenberger, U. (2010). Hundred days of cognitive training enhance broad cognitive abilities in adulthood: Findings from the COGITO study. *Frontiers in Aging Neuroscience, 2*, 27.

Schmiedek, F., Lövdén, M., & Lindenberger, U. (2014). Younger adults show long-term effects of cognitive training on broad cognitive abilities over 2 years. *Developmental Psychology, 50*, 2304–2310.

Van de Schoot, R., Kluytmans, A., Tummers, L., Lugtig, P., Hox, J., & Muthén, B. O. (2013). Facing off with Scylla and Charybdis: A comparison of scalar, partial, and the novel possibility of approximate measurement invariance. *Frontiers in Psychology, 4*, 770.

Sternberg, R. J. (1978). Intelligence research at the interface between differential and cognitive psychology: Prospects and proposals. *Intelligence, 2*, 195–222.

Stine-Morrow, E. A., Payne, B. R., Roberts, B. W., Kramer, A. F., Morrow, D. G., Payne, L., Hill, P. L., Jackson, J. J., Gao, X., Noh, S. R., Janke, M. C., & Parisi, J. M. (2014). Training versus engagement as paths to cognitive enrichment with aging. *Psychology and Aging, 29*, 891–906.

Tay, L., Meade, A. W., & Cao, M. (2015). An overview and practical guide to IRT measurement equivalence analysis. *Organizational Research Methods, 18*, 3–46.

Tidwell, J. W., Dougherty, M. R., Chrabaszcz, J. R., Thomas, R. P., & Mendoza, J. L. (2014). What counts as evidence for working memory training? Problems with correlated gains and dichotomization. *Psychonomic Bulletin & Review, 21*, 620–628.

Trafimow, D. (2015). A defense against the alleged unreliability of difference scores. *Cogent Mathematics, 2*, 1064626.

West, S. G., Taylor, A. B., & Wu, W. (2012). Model fit and model selection in structural equation modeling. In R. H. Hoyle (Ed.), *Handbook of structural equation modeling* (pp. 209–231). New York: The Guilford Press.

Widaman, K. F., & Reise, S. P. (1997). Exploring the measurement invariance of psychological instruments: Applications in the substance use domain. In K. J. Bryant, M. E. Windle, & S. G. West (Eds.), *The science of prevention: Methodological advances from alcohol and substance abuse research* (pp. 281–324). Washington: American Psychological Association.

Wilhelm, O., Hildebrandt, A., & Oberauer, K. (2013). What is working memory, and how can we measure it? *Frontiers in Psychology, 4*, 433.

Willis, S. L., & Schaie, K. W. (2009). Cognitive training and plasticity: Theoretical perspective and methodological consequences. *Restorative Neurology and Neuroscience, 27*, 375–389.

Wolf, E., Harrington, K. M., Clark, S. L., & Miller, M. W. (2013). Sample size requirements for structural equation models: An evaluation of power, bias, and solution propriety. *Educational and Psychological Measurement, 73*, 913–934.

Zelinski, E. M., Peters, K. D., Hindin, S., Petway, K. T., & Kennison, R. F. (2014). Evaluating the relationship between change in performance on training tasks and on untrained outcomes. *Frontiers in Human Neuroscience, 8*, 617.

Zhang, Z., & Liu, H. (2019). Sample size and measurement occasion planning for latent change score models through Monte Carlo simulation. In E. Ferrer, S. M. Boker, & K. J. Grimm (Eds.), *Longitudinal multivariate psychology* (pp. 189–211). New York: Taylor & Francis.

Neuroplasticity

Elisabeth Wenger and Simone Kühn

Contents

Abstract Our genetic code cannot specify every single connection between individual nerve cells. Hence, every brain has to start as a relatively structureless but extremely flexible network of nerve cells that has the ability to "wire" itself exactly the way in which it is best adjusting to its individual environment with its unique requirements. Neuroplasticity denotes this inherent ability of the brain to adapt with macroscale changes in response to altered environmental demands (Lövdén et al. 2010a). It is therefore an adaptive process triggered by a prolonged mismatch between the functional supply the brain can momentarily provide and the experienced demands the environment currently poses. In this chapter, we first review the accumulated evidence on neuroplasticity, both in animal and human literature. We then turn to biological underpinnings potentially underlying the detectable changes in gray matter structure as visible on magnetic resonance (MR) images. Finally, we review evidence on the sequential progression of structural changes, which has revealed a pattern of expansion followed by renormalization and reiterate the importance of paying close attention to the complex nature of plastic changes.

E. Wenger (✉)
Center for Lifespan Psychology, Max Planck Institute for Human Development, Berlin, Germany
e-mail: wenger@mpib-berlin.mpg.de

S. Kühn
Lise Meitner Group for Environmental Neuroscience, Max Planck Institute for Human Development, Berlin, Germany

Department of Psychiatry and Psychotherapy, University Clinic Hamburg-Eppendorf, Hamburg, Germany

T. Strobach, J. Karbach (eds.), *Cognitive Training*,
https://doi.org/10.1007/978-3-030-39292-5_6

Introduction

The massive amount of connections between neurons cannot be simply inscribed in our genetic code, as the former severely outnumbers the latter. Amongst others, our DNA is assigned to the important task of encoding the variety of nerve cell forms and different neurotransmitters, but it cannot specify the exact connections between individual neurons. Hence, every brain has to start as a relatively structureless, but extremely flexible network of nerve cells that has the inherent ability to "wire" itself exactly the way in which it is best adjusting to its individual environment with its unique requirements. Plasticity is therefore an intrinsic property of the human brain and constitutes evolution's invention to enable the nervous system to escape the restrictions of its own genome and adapt to environmental pressures, physiologic changes, and experiences (Pascual-Leone et al. 2005). Conceivably, there are also quite obvious, definitely indispensable limits to how plastic a brain can be. It has to be stable too, to remain operational and keep and store functions once learned. Plastic changes are also metabolically costly (Kuzawa et al. 2014), a circumstance that becomes increasingly important in systems that have slowly started to accumulate damage (as in aging) and may therefore have literally nothing to spare. So at best, there is a balance between stability and neuroplasticity that results in a brain that is both solid and reliable where it can be, and capable of adaption where it has to be.

The Concept of Plasticity

In the definition we subscribe to, neuroplasticity denotes the inherent ability of the brain to adapt with macroscale changes in response to altered environmental demands (Lövdén et al. 2010a). Within this framework, plasticity is an adaptive process triggered by a prolonged mismatch between the functional supply the brain can momentarily provide and the experienced demands the environment currently poses, as for example when a new cognitive task needs to be accomplished.

In the majority of cases, the brain can meet requirements posed by its environment through neuronal and behavioral variability and flexibility, that is, by optimizing its performance within a given state of resources and using the existing functional repertoire. However, if these processes of flexibility within a given state do not suffice in fulfilling environmental demands – either due to dramatic changes in requirements or due to damaged functionality of the brain following brain injury – then more fundamental change is demanded and can manifest in the form of plasticity (Lövdén et al. 2010a). We have previously raised the notion that flexibility and plasticity may follow different contrary trajectories across the lifespan: While plasticity is the highest in childhood and decreases towards old age, flexibility increases from childhood on, peaks in young adulthood, and decreases thereafter (Kühn and Lindenberger 2016).

Evidence for Neuroplasticity in the Context of Cognitive Training

In the 1960s and 1970s, it was the obligatory animal research providing indications for changes of cortical patterns and the brain structure as a consequence of altered environmental circumstances or experiences. It was firmly established that enriched environments can lead to changes in brain weight (Rosenzweig et al. 1964) and cortical thickness (Rosenzweig et al. 1972). Rats trained in a motor skill task showed increased dendritic branching in the motor cortex (Withers and Greenough 1989) and an increased number of synapses per neuron in the cerebellar cortex (Black et al. 1990).

Altman was the first to report even new neurons and neuroblasts in adult rats (Altman 1962), while Kaplan corroborated these results and also found evidence that complex environments could stimulate neurogenesis in adult visual cortex (Kaplan and Hinds 1977). With the important work by Gould and colleagues who demonstrated adult hippocampal neurogenesis also in mammals that are more related to humans (Gould et al. 1999a), adult mammalian neurogenesis is now firmly established and has been shown to be involved in learning (Gould et al. 1999b). In rodents, environmental enrichment and voluntary exercise have been shown to enhance neurogenesis (Kempermann et al. 1997; van Praag et al. 1999).

For a long time, it was assumed that the brain structure was malleable and amicable to influences from the outside only during critical periods in early life, and to be absent thereafter. And indeed animal models have suggested that age limits the capacity for adaptive changes: The aged rat brain has been found to respond more slowly and to a lesser extent to chemically-induced seizures, suggesting a more stable system (Wagner et al. 2000). Especially changes in thin spine morphology have been amongst possible candidate mechanisms potentially responsible for age-related impairments in learning (Dumitriu et al. 2010). Bloss et al. (2011) reported an experiment using the negative impact of stress on dendritic spines to demonstrate plastic changes in different age groups. In young rats, stress resulted in dendritic spine loss and altered patterns of spine morphology. In contrast, spines from middle-aged and older animals were remarkably stable and did not show evidence of remodeling. The data provide evidence that experience-dependent spine plasticity is altered by aging and, together with other literature on age differences in plastic responses, support a model in which dendritic spines become progressively less plastic and more stable in the aging brain (Grutzendler et al. 2002; Holtmaat and Svoboda 2009; Bloss et al. 2011).

Nevertheless, it has been shown repeatedly that neuroplasticity can take place in the adult brain as well. Animal data show cognitive benefits and neural reorganization in rodents, after long-term voluntary exercise in running wheels (van Praag et al. 2000) and training periods on a treadmill (Aguiar et al. 2011), to name only two examples. In addition to exercise, enrichment of the environment has also been shown to evoke changes in dendrites in middle-aged rats (Green et al. 1983) and in spiny branchlets of cerebellar Purkinje neurons in aged rats

(Greenough et al. 1986). Enriched housing can even return the number of cells of older animals to the level of younger animals living in impoverished cages (Kolb et al. 1998).

These original findings elicited hope for similar effects in humans. Given that it is not possible to measure human brain structure in vitro in healthy, living individuals, advances in noninvasive magnetic resonance imaging (MRI) have opened up new windows into the investigation of changes in the human brain's macrostructure. MRI uses a strong magnetic field to align hydrogen atoms of water molecules in tissue, and radio frequency fields to systematically change this alignment (Huettel et al. 2004). This magnetization results in a rotating magnetic field created by the hydrogen atoms as they return to baseline which can be detected by the MR scanner. The emerging signal can then be used to construct an image of the brain because different tissues have different magnetic properties. On the resulting anatomical images one can then differentiate between gray matter, white matter, and cerebrospinal fluid, which can be quantified in terms of volume by means of manual tracing or automatic segmentation. Even though, conceivably, *in vivo* MRI cannot inform us at the same detailed cellular level as methods in animal research can, it still offers a unique window into brain changes on the macrostructural level.

In the last years, several studies have devoted themselves to investigating instances of neuroplasticity and have identified situations in which plastic changes were observable. Maguire and colleagues (Maguire et al. 2000; Maguire et al. 2006) have found an enlarged region in posterior hippocampus in London taxi drivers in contrast to bus drivers and have observed an enlargement of the same region in the course of becoming a licensed taxi driver, while acquiring London's complex street layout (Woollett and Maguire 2011). Further evidence for changes in gray matter in response to experience comes from musicians (Gaser and Schlaug 2003; see also Swaminathan and Schellenberge, this volume), professional typists (Cannonieri et al. 2007), and medical students preparing for their final exam (Draganski et al. 2006). Also, learning how to juggle for three months has been shown to elicit temporary expansion in temporal lobe and intraparietal sulcus, both in younger (Draganski et al. 2004) as well as in older adults (Boyke et al. 2008), and practicing two weeks of mirror reading has led to reduced activation alongside with an increased volume of gray matter in occipital lobe (Ilg et al. 2008). Eight weeks of memory training using the Method of Loci induced cortical thickness changes in middle-aged and elderly healthy volunteers (Engvig et al. 2010; see also Wenger et al., this volume). 100 days of cognitive training have been shown to evoke plastic changes in white matter in corpus callosum (Lövdén et al. 2010b) and spatial navigation training has led to a deceleration of typical age-related decline in the hippocampal volume (Lövdén et al. 2012), as well as to cortical thickening in precuneus and paracentral lobule in younger adults (Wenger et al. 2012). Also playing a video game that involves navigating in 3D-space has been shown to lead to changes in the right hippocampal formation (Kühn et al. 2014; Bediou et al., this volume; Strobach and Schubert, this volume). Mårtensson et al. (2012) studied changes in the brain structure following three months of intense foreign-language acquisition. Results showed increases in hippocampal volume and in cortical thickness in the left middle frontal gyrus, inferior frontal gyrus, and superior temporal gyrus for military interpreters compared to a control group, whereby some of these regions showed a correlation with behavioral measures of proficiency or struggling. Last but not least, physical exercise has been identified as a powerful agent to influence also adult human brain structure (see also Bherer and Pothier, this volume): it has been shown that the hippocampus size increased after one year of moderate-intensity exercise training (Erickson et al. 2011) and the cerebral blood volume – an indicator of exercise-induced neurogenesis – in dentate gyrus increased after three months of exercising (Pereira et al. 2007). Changes in fitness have been associated with changes in hippocampal perfusion and volume of the hippocampal head in the context of a three-month fitness intervention program (Maass et al. 2015) and with changes in hippocampal microstructure, pointing to a more dense tissue, after six months of ergometer training (Kleemeyer et al. 2016).

Taken together, a considerable number of studies investigating experience-dependent macrostructural changes in human gray matter have accumulated over the last say 20 years (for comprehensive reviews, see May 2011; Zatorre et al. 2012; Lövdén et al. 2013). However, some of these studies do not offer optimal grounds for indisputable conclusions but suffer from various flaws, in both study design as well as statistical analysis regards (Thomas and Baker 2013; see also Cochrane and Green, Schmiedek, this volume). Some of these studies for example lack an appropriate control group against which the results in the experimental group could be compared. This seems especially important in study designs with only two measurement time points where scanner drifts or normal "variability" in brain structure as visible on MR images are hard to distinguish from true effects. Also, so far it seems impossible to state ubiquitously whether experience-dependent brain changes are or are not reduced in aging. With only a few exceptions, there is generally a lack of age-comparative studies that investigate both younger and older adults with the same training paradigm. It remains hard to gauge to which extent the aged brain harbors the potential to exhibit plastic changes relative to the younger brain. More studies using samples with a wide age range or at least two or three age groups plus appropriate control groups are warranted to further explore the premises that need to be fulfilled in different brain regions (e.g., hippocampus vs. cortical regions) within aging brains to set grounds for arising plastic changes.

Microstructural Processes Underlying Changes in Gray Matter

With respect to underlying biological mechanisms, MRI findings do not provide clear evidence on cellular and molecular mechanisms of changes in gray matter. Moreover, phenomena visible with MRI are most likely never the result of a single process happening independently. Instead, they are rather a result of many coordinated structural changes involving various cell types. Candidate mechanisms possibly underlying the visible changes on T_1-weighted images are neurogenesis, synaptogenesis, changes in neuronal morphology, axon sprouting, dendritic branching, glial changes, or angiogenesis (for a summary of possible biological processes see also Zatorre et al. 2012). In the following, we will briefly summarize knowledge of these biological mechanisms.

Neurogenesis denotes the growth of new neurons and has repeatedly been demonstrated in the hippocampus of adult rats, living in an enriched environment (e.g., Kempermann et al. 2002; Kronenberg et al. 2006). As monthly newly produced cells make up only a small part of the total number of hippocampal neurons, neurogenesis is likely a minor factor contributing to changes visible with MRI. Changes observed outside of the hippocampus are probably not due to neurogenesis, as growth of new neurons in adults has only been established in the dentate gyrus and the olfactory bulb (e.g., Ehninger and Kempermann 2008; Huart et al. 2013) and, more recently, in humans in the striatum (Ernst et al. 2014). Whether neurogenesis in the neocortex can occur later in life is still highly controversial (Rakic 2002; Tan and Shi 2013).

Another candidate biological process presumably contributing to MRI volume increases is gliogenesis, referring to an increase in the number of nonneuronal cells (including oligodendrocytes, astrocytes, microglia, and ependymal cells). Glial cells maintain ion homeostasis, regulate blood flow in response to neuronal activity, form myelin, and provide support and protection for neurons (Wang et al. 2009; Brodal 2010). Glial cells are highly plastic and display a number of morphological changes in response to altered experience, including increased cell number, volume fraction, increased cell surface, and proliferation of their processes (Sirevaag and Greenough 1991; Dong and Greenough 2004). Glial processes could in theory increase to support new synapses, or to compensate for neuronal process loss (Anderson 2011). Thus, increases in gliogenesis could to some extent underlie gray matter changes observed with MRI (Zatorre et al. 2012).

Besides neurogenesis and gliogenesis, synaptogenesis and changes in spine morphology have been discussed in the context of learning and gray matter alterations. In animal work, synapse formation has been implicated in supporting learning-dependent changes in cortical function (Kleim et al. 2002; Trachtenberg et al. 2002). Changes in dendritic length and branching or in the actual number of dendritic spines per neuron are likely to contribute to experience-dependent volumetric changes in gray matter (e.g., Kolb et al. 2008; Holtmaat and Svoboda 2009; Fu and Zuo 2011). Additionally, angiogenesis, that is, changes in vasculature, is likely to appear following especially exercise-training (Swain et al. 2003). These changes could support increased energy demands of new or changed neural tissue via a growth of capillaries.

Overall, the bulk of evidence suggests that experience-dependent neuroplasticity may be to a large extent mediated by synaptogenesis (Black et al. 1990; Kleim et al. 2002), changes in dendritic spines/dendritic branching (Trachtenberg et al. 2002; Holtmaat and Svoboda 2009), and changes in nonneural cells like glia

(Dong and Greenough 2004). As glia process growth and retraction in response to manipulations are in general complicated, they might cloud direct relationships between synapse numbers and regional volume (Anderson et al. 1994).

Since the early publications reporting experience-dependent gray matter changes in humans (e.g., Amunts et al. 1997; Maguire et al. 2000; Draganski et al. 2004), speculations about the microstructural biological correlates of these effects have filled paragraphs and paragraphs of countless discussion sections. A future challenge remains the identification of cellular changes underlying the macrostructural changes currently observed with MRI. Meeting this challenge requires employing continuously newly developed MR sequences (Tardif et al. 2016), greater cross-talk between those studying human populations and those working with animals, as well as a greater integration of techniques. This has for example been done in a study by Sagi et al. (2012), where they used diffusion-tensor imaging to investigate changes in the hippocampus in humans following a spatial learning and memory task. They found a significant reduction of mean diffusivity (MD) in hippocampus and parahippocampus after only two hours of training that correlated with behavioral improvement. Additionally, they conducted a supporting rat study with a short-term water maze task to investigate "equivalent" changes on a more detailed microstructural level in the animal brain. Histological analysis of the rat brains indicated that within the regions of MD decrease there was an increase in the number of synaptic vesicles, astrocytic activation, and an increase in BDNF expression (Sagi et al. 2012). Another study focusing on the neural correlates of MRI volume changes found neurogenesis to be the best marker explaining hippocampal gray matter volume after voluntary wheel running in mice (Biedermann et al. 2016). In this study, they compared a group of running mice to sedentary ones, acquired a typically used anatomical MR image and sacrificed the animals immediately after to perform histological analyses. Besides newborn neurons, they also investigated glial cells, microglia, proliferating and pyknotic cells, neuronal activation, blood vessel density and arborization. Interestingly, none of the other above mentioned cell types showed a clear correlation pattern with MR volume changes, even though a marker for astrocytes also showed a significant difference between the two groups (Biedermann et al. 2016). Yet another study by Lerch and colleagues investigated mice trained on different versions of the Morris water maze task (Lerch et al. 2011). Using high resolution MRI, they showed specific volume changes in the hippocampus in mice trained on a spatial variant of the maze, and changes in the striatum after the cued version of the maze. Subsequent immunohistochemistry revealed a correlation between volume increases and a marker for neuronal process remodeling but not with neurogenesis, neuron or astrocyte numbers or sizes (Lerch et al. 2011). Such studies that discuss the biological correlates of structural MR measures, together with advances in MR image acquisition (Hamaide et al. 2016; Lerch et al. 2017) will continue to enable key insights into how neuroplastic changes are implemented.

Time Course of Plastic Changes

As reviewed above, structural brain changes have been observed following many different kinds of skill acquisition and learning. Plastic changes might even emerge much faster than described in the aforementioned studies: Gray matter alterations have been reported after only two weeks of mirror reading training (Ilg et al. 2008), one week of juggling training (Driemeyer et al. 2008), one week of daily pain stimulation (Teutsch et al. 2008), five days of repetitive transcranial magnetic stimulation (May et al. 2007; see also Byrne et al., this volume), and three days of practicing signature writing (Hamzei et al. 2012). Even two sessions of practice in a complex whole-body balancing task (Taubert et al. 2010), two hours (spread out over three days) of learning subcategories of color names (Kwok et al. 2011) or passive viewing of pictures during 263 seconds (Månsson et al. 2020) have led to reports on gray matter alterations.

Many of these studies make use of the classic design, measuring gray matter structure before and after the introduction of a novel experience. It is therefore implicitly assumed that structure is, if at all, developing monotonically during the intervention or training phase. Related work in animals, however, shows initial increases in structure in the beginning of training that are then followed by partial or complete renormalization as experience continues (Dupret et al. 2007; Quallo et al. 2009; Xu et al. 2009; Reed et al. 2011). Quallo et al. (2009) analyzed structural data of three adult macaque monkeys, collected on multiple occasions before, during, and after learning to use a rake to retrieve food. They found learning-related increases in task-relevant brain regions, which also mapped onto the learning curves. Crucially, despite continued training, the observed increased gray matter structure decreased again after the monkey's performance reached asymptote. After training, the volume was still enlarged as compared to before training, but much smaller in magnitude than the peak effect observed before asymptotic performance was reached. Molina-Luna et al. (2008) trained rats to perform a skilled reaching task and found expanded cortical maps after three days of training. After eight days of training, however, these expansions subsided again while behavioral performance remained stable. A very similar pattern was found when investigating postsynaptic dendritic spines (Xu et al. 2009). Mice trained in a reaching task experienced a rapid formation of new dendritic spines within an hour. This rapid increase was then followed by a slower process of elimination of "old" spines that had existed before training, returning the overall number of spines to a comparable pretraining level, despite continuously high performance levels (Xu et al. 2009). Taken together, these results from animal literature and the few reports in humans of structural alterations even after very short periods of time call for a closer investigation of the temporal dynamics of gray matter changes.

Actually already in 1894, Santiago Ramón y Cajal – by many thought to be the father of modern neuroscience – proposed that mental activity might induce "*novel intercellular connections through the new formation of collaterals and protoplasmic expansions*." He then raised the intriguing question: "*How can the volume of brain remain constant if there is a multiplication and even new formation of terminal branches of protoplasmic appendices and nerve collaterals?"* (Azmitia 2007). More than 100 years and numerous studies demonstrating experience-dependent growth of human brain volume later, we are confronted with the same paradox: Is it really feasible to represent the vast amount of knowledge and skills that humans acquire during a whole lifetime as a process of continuous brain volume growth? Importantly, prominent theoretical accounts of plasticity, developmental data, and animal models as described above provide a different account of plasticity, according to which plasticity follows a sequence of expansion, selection, and renormalization.

Informed by this notion, we acquired up to 18 structural MR images over a 7-week period while right-handed adult participants practiced left-hand writing and drawing (Wenger et al. 2017b). We observed that gray matter in primary motor cortices expanded during the first weeks of motor learning and then partially renormalized, in the presence of continued practice and increasing task proficiency. We therefore propose that plastic reorganization processes in the context of skill acquisition consist of an initial but transient phase of brain volume increase followed by partial or even complete return to baseline once optimal rewiring has occurred (Wenger et al. 2017a). Importantly, this pattern of plastic change seems to hold true across different levels of plasticity, so far mostly investigated in animal models: Cortical map plasticity follows a comparable pattern of expansion followed by renormalization during learning (Peters et al. 2014; Albieri et al. 2015; Pruitt et al. 2016). Also work on learning-related changes in dendritic spines is consistent with the hypothesis that the memory trace serving skilled performance is localized in rewired circuitry rather than in any large-scale expansion (Holtmaat and Svoboda 2009; Hofer and Bonhoeffer 2010; Fu and Zuo 2011). Motor sequence learning has been shown to be associated with increasing motor system activity in the early stages of learning, followed by a reduced level of motor system activity during execution of highly practiced motor behavior (Wymbs and Grafton 2015). Metabolic efficiency might be a driving factor behind this pattern (Makino et al. 2016), as learnt information can be represented by a relatively smaller number of spikes or neurons after learning compared to before (Makino and Komiyama 2015; Chu et al. 2016). While during the initial stages of learning, more neurons and synapses are being used, thereby potentially entailing an expansion of tissue in these regions due to metabolic demands, later on, the most efficient wiring is selected, resulting in fewer but specialized and stable neurons and synapses (Makino et al. 2016).

This pattern of experience-dependent initial production of diversity followed by selection and stabilization has the features typically ascribed to Darwinian models of cortical plasticity and neural development (Fernando et al. 2012; Kilgard 2012). Within this framework, plastic changes may be seen like an audition for the cast of a movie. Numerous candidates are first progressively called in, then the best ones are selected and the rest is sent home, that is, "pruned away." Calling in more candidates may possibly improve the outcome; growth can therefore be helpful, but is not the end product.

An expansion-renormalization model of experience-dependent structural brain changes predicts initial learning-dependent volumetric increase of brain structure – reflecting recruitment of additional neural resources and local neural rewiring – followed by a partial or complete return to baseline once optimal rewiring has occurred and the surplus has been eliminated. This way, space restriction within the skull and therefore competition between different brain regions and skills are not an issue. The expansion-renormalization model thus gives motivation for a new look at past findings on experience-dependent plasticity in humans and raises several new research questions and predictions for work on human experience-dependent plasticity probed with MRI (Lindenberger et al. 2017; Wenger et al. 2017a) .

Conclusion

In the last years, evidence has accumulated suggesting that the brain structure can change in response to altered environmental demands. Such structural changes have been observed in rodents after enriched housing, and also in humans for example following intensive studying, musical experience, video game playing, or spatial navigation or the training of a new skill such as juggling, and have also been observed throughout the lifespan.

Much of this research used MRI, which allows for in vivo investigations of the human brain structure with increasingly informative acquisition sequences. A future challenge is and remains to determine the cellular and molecular changes that underlie the macrostructural changes visible on MR images (see also Colzato and Hommel, this volume). Meeting this challenge requires greater exchange between those studying human populations and those working with animal models, and a greater integration of techniques. Animal studies in which both imaging and histological measures can be applied in parallel will be particularly helpful to establish the relative contributions of different cellular processes to the MRI effects. At the same time, one will need to keep in mind that multiple, coordinated cellular processes are most likely associated with changes in a single MR-based variable and that phenomena detected in rodents might not generalize fully to humans or vice versa.

Endless expansion may not be nature's best solution to the phenomenon of lifelong learning when in other parts of evolution processes of trimming and selecting the best among several candidates has proven immensely useful. We have therefore proposed that plastic changes (specifically in the context of skill acquisition) are characterized by a sequence of volume expansion, selection, and renormalization. More complex study designs with at least three or more measurement time points are necessary to make appropriate use of the aforementioned sophisticated hardware and software tools and to eventually gain more knowledge on the phenomenon of brain plasticity, its temporal dynamics, functional relevance, and biological mechanisms.

At a more general level, we hope to have succeeded in affirming the complexity of neuroplasticity. This should not come as a surprise since Pascual-Leone et al. (2005) have already asserted that neuroplasticity constitutes evolution's invention to enable the nervous system to escape the restrictions of its own genome and adapt to environmental demands. Understandably enough, this should indeed constitute a complex and dynamic process that is hard to gauge and that remains highly interesting to study in the future in even more detail.

References

Aguiar, Jr, A. S., Castro, A. A., Moreira, E. L., Glaser, V., Santos, A. R., Tasca, C. I., ... & Prediger, R. D. (2011). Short bouts of mild-intensity physical exercise improve spatial learning and memory in aging rats: Involvement of hippocampal plasticity via AKT, CREB and BDNF signaling. *Mechanisms of Ageing and Development, 132*, 560–567.

Albieri, G., Barnes, S. J., de Celis Alonso, B., Cheetham, C. E., Edwards, C. E., Lowe, A. S., ... & Finnerty, G. T. (2015). Rapid bidirectional reorganization of cortical microcircuits. *Cerebral Cortex, 25*, 3025–3035.

Altman, J. (1962). Are new neurons formed in the brains of adult mammals? *Science, 135*, 1127–1128.

Amunts, K., Schlaug, G., Jäncke, L., Steinmetz, H., Schleicher, A., Dabringhaus, A., & Zilles, K. (1997). Motor cortex and hand motor skills: Structural compliance in the human brain. *Human Brain Mapping, 5*, 206–215.

Anderson, B. J. (2011). Plasticity of gray matter volume: The cellular and synaptic plasticity that underlies volumetric change. *Developmental Psychobiology, 53*, 456–465.

Anderson, B. J., Li, X., Alcantara, A. A., Isaacs, K. R., Black, J. E., & Greenough, W. T. (1994). Glial hypertrophy is associated with synaptogenesis following motor-skill learning, but not with angiogenesis following exercise. *Glia, 11*, 73–80.

Azmitia, E. C. (2007). Cajal and brain plasticity: Insights relevant to emerging concepts of mind. *Brain Research Reviews, 55*, 395–405.

Biedermann, S. V., Fuss, J., Steinle, J., Auer, M. K., Dormann, C., Falfán-Melgoza, C., ... & Weber-Fahr, W. (2016). The hippocampus and exercise: histological correlates of MR-detected volume changes. *Brain Structure and Function, 221*, 1353–1363.

Black, J. E., Isaacs, K. R., Anderson, B. J., Alcantara, A. A., & Greenough, W. T. (1990). Learning causes synaptogenesis, whereas motor activity causes angiogenesis, in cerebellar cortex of adult rats. *Proceedings of the National Academy of Sciences, 87*, 5568–5572.

Bloss, E. B., Janssen, W. G., Ohm, D. T., Yuk, F. J., Wadsworth, S., Saardi, K. M., ... & Morrison, J. H. (2011). Evidence for reduced experience-dependent dendritic spine plasticity in the aging prefrontal cortex. *Journal of Neuroscience, 31*, 7831–7839.

Boyke, J., Driemeyer, J., Gaser, C., Büchel, C., & May, A. (2008). Training-induced brain structure changes in the elderly. *Journal of Neuroscience, 28*, 7031–7035.

Brodal, P. (2010). Glia. In *The central nervous system: Structure and function* (pp. 19–27). Oxford: Oxford University Press.

Cannonieri, G. C., Bonilha, L., Fernandes, P. T., Cendes, F., & Li, L. M. (2007). Practice and perfect: Length of training and structural brain changes in experienced typists. *Neuroreport, 18*, 1063–1066.

Chu, M. W., Li, W. L., & Komiyama, T. (2016). Balancing the robustness and efficiency of odor representations during learning. *Neuron, 92*, 174–186.

Dong, W. K., & Greenough, W. T. (2004). Plasticity of nonneuronal brain tissue: Roles in developmental disorders. *Mental Retardation and Developmental Disabilities Research Reviews, 10*, 85–90.

Draganski, B., Gaser, C., Busch, V., Schuierer, G., Bogdahn, U., & May, A. (2004). Changes in grey matter induced by training. *Nature, 427*, 311–312.
Draganski, B., Gaser, C., Kempermann, G., Kuhn, H. G., Winkler, J., Büchel, C., & May, A. (2006). Temporal and spatial dynamics of brain structure changes during extensive learning. *Journal of Neuroscience, 26*, 6314–6317.
Driemeyer, J., Boyke, J., Gaser, C., Büchel, C., & May, A. (2008). Changes in gray matter induced by learning — Revisited. *PLoS One, 3*, e2669.
Dumitriu, D., Hao, J., Hara, Y., Kaufmann, J., Janssen, W. G., Lou, W., ... & Morrison, J. H. (2010). Selective changes in thin spine density and morphology in monkey prefrontal cortex correlate with aging-related cognitive impairment. *Journal of Neuroscience, 30*, 7507–7515.
Dupret, D., Fabre, A., Döbrössy, M. D., Panatier, A., Rodríguez, J. J., Lamarque, S., ... & Abrous, D. N. (2007). Spatial learning depends on both the addition and removal of new hippocampal neurons. *PLoS Biology, 5*, e214.
Ehninger, D., & Kempermann, G. (2008). Neurogenesis in the adult hippocampus. *Cell and Tissue Research, 331*, 243–250.
Engvig, A., Fjell, A. M., Westlye, L. T., Moberget, T., Sundseth, Ø., Larsen, V. A., & Walhovd, K. B. (2010). Effects of memory training on cortical thickness in the elderly. *NeuroImage, 52*, 1667–1676.
Erickson, K. I., Voss, M. W., Prakash, R. S., Basak, C., Szabo, A., Chaddock, L., ... & Wojcicki, T. R. (2011). Exercise training increases size of hippocampus and improves memory. *Proceedings of the National Academy of Sciences, 108*, 3017–3022.
Ernst, A., Alkass, K., Bernard, S., Salehpour, M., Perl, S., Tisdale, J., ... & Frisén, J. (2014). Neurogenesis in the striatum of the adult human brain. *Cell, 156*, 1072–1083.
Fernando, C. T., Szathmary, E., & Husbands, P. (2012). Selectionist and evolutionary approaches to brain function: a critical appraisal. *Frontiers in Computational Neuroscience, 6*, 24.
Fu, M., & Zuo, Y. (2011). Experience-dependent structural plasticity in the cortex. *Trends in Neurosciences, 34*, 177–187.
Gaser, C., & Schlaug, G. (2003). Gray matter differences between musicians and nonmusicians. *Annals of the New York Academy of Sciences, 999*, 514–517.
Gould, E., Reeves, A. J., Fallah, M., Tanapat, P., Gross, C. G., & Fuchs, E. (1999a). Hippocampal neurogenesis in adult Old World primates. *Proceedings of the National Academy of Sciences, 96*, 5263–5267.
Gould, E., Beylin, A., Tanapat, P., Reeves, A., & Shors, T. J. (1999b). Learning enhances adult neurogenesis in the hippocampal formation. *Nature Neuroscience, 2*, 260–265.
Green, E. J., Greenough, W. T., & Schlumpf, B. E. (1983). Effects of complex or isolated environments on cortical dendrites of middle-aged rats. *Brain Research, 264*, 233–240.
Greenough, W. T., McDonald, J. W., Parnisari, R. M., & Camel, J. E. (1986). Environmental conditions modulate degeneration and new dendrite growth in cerebellum of senescent rats. *Brain Research, 380*, 136–143.
Grutzendler, J., Kasthuri, N., & Gan, W. (2002). Long-term dendritic spine stability in the adult cortex. *Nature, 420*, 812–816.
Hamaide, J., De Groof, G., & Van der Linden, A. (2016). Neuroplasticity and MRI: A perfect match. *NeuroImage, 131*, 13–28.
Hamzei, F., Glauche, V., Schwarzwald, R., & May, A. (2012). Dynamic gray matter changes within cortex and striatum after short motor skill training are associated with their increased functional interaction. *NeuroImage, 59*, 3364–3372.
Hofer, S. B., & Bonhoeffer, T. (2010). Dendritic spines: The stuff that memories are made of? *Current Biology, 20*, 157–159.
Holtmaat, A., & Svoboda, K. (2009). Experience-dependent structural synaptic plasticity in the mammalian brain. *Nature Reviews. Neuroscience, 10*, 647–658.
Huart, C., Rombaux, P., & Hummel, T. (2013). Plasticity of the human olfactory system: The olfactory bulb. *Molecules, 18*, 11586–11600.
Huettel, S. A., Song, A. W., & McCarthy, G. (2004). *Functional Magnetic Resonance Imaging*. Sunderland: Sinauer.

Ilg, R., Wohlschläger, A. M., Gaser, C., Liebau, Y., Dauner, R., Wöller, A., ... & Mühlau, M. (2008). Gray matter increase induced by practice correlates with task-specific activation: A combined functional and morphometric magnetic resonance imaging study. *Journal of Neuroscience, 28*, 4210–4215.

Kaplan, M., & Hinds, J. (1977). Neurogenesis in the adult rat: Electron microscopic analysis of light radioautographs. *Science, 197*, 1092–1094.

Kempermann, G., Kuhn, H. G., & Gage, F. H. (1997). More hippocampal neurons in adult mice living in an enriched environment. *Nature, 386*, 493–495.

Kempermann, G., Gast, D., & Gage, F. H. (2002). Neuroplasticity in old age: Sustained fivefold induction of hippocampal neurogenesis by long-term environmental enrichment. *Annals of Neurology, 52*, 135–143.

Kilgard, M. P. (2012). Harnessing plasticity to understand learning and treat disease. *Trends in Neurosciences, 35*, 715–722.

Kleemeyer, M. M., Kühn, S., Prindle, J., Bodammer, N. C., Brechtel, L., Garthe, A., ... & Lindenberger, U. (2016). Changes in fitness are associated with changes in hippocampal microstructure and hippocampal volume among older adults. *Neuroimage, 131*, 155–161.

Kleim, J. A., Barbay, S., Cooper, N. R., Hogg, T. M., Reidel, C. N., Remple, M. S., & Nudo, R. J. (2002). Motor learning-dependent synaptogenesis is localized to functionally reorganized motor cortex. *Neurobiology of Learning and Memory, 77*, 63–77.

Kolb, B., Forgie, M., Gibb, R., Gorny, G., & Rowntree, S. (1998). Age, experience and the changing brain. *Neuroscience and Biobehavioral Reviews, 22*, 143–159.

Kolb, B., Cioe, J., & Comeau, W. (2008). Contrasting effects of motor and visual spatial learning tasks on dendritic arborization and spine density in rats. *Neurobiology of Learning and Memory, 90*, 295–300.

Kronenberg, G., Bick-Sander, A., Bunk, E., Wolf, C., Ehninger, D., & Kempermann, G. (2006). Physical exercise prevents age-related decline in precursor cell activity in the mouse dentate gyrus. *Neurobiology of Aging, 27*, 1505–1513.

Kühn, S., & Lindenberger, U. (2016). Research on human plasticity in adulthood: A lifespan agenda. In K. W. Schaie & S. L. Willis (Eds.), *Handbook on the psychology of aging* (8th ed., pp. 105–123). Amsterdam: Academic Press.

Kühn, S., Gleich, T., Lorenz, R. C., Lindenberger, U., & Gallinat, J. (2014). Playing Super Mario induces structural brain plasticity: Gray matter changes resulting from training with a commercial video game. *Molecular Psychiatry, 19*, 265–271.

Kuzawa, C. W., Chugani, H. T., Grossman, L. I., Lipovich, L., Muzik, O., Hof, P. R., ... & Lange, N. (2014). Metabolic costs and evolutionary implications of human brain development. *Proceedings of the National Academy of Sciences, 111*, 13010–13015.

Kwok, V., Niu, Z., Kay, P., Zhou, K., Mo, L., Jin, Z., ... & Tan, L. H. (2011). Learning new color names produces rapid increase in gray matter in the intact adult human cortex. *Proceedings of the National Academy of Sciences, 108*, 6686–6688.

Lerch, J. P., Yiu, A. P., Martinez-Canabal, A., Pekar, T., Bohbot, V. D., Frankland, P. W., ... & Sled, J. G. (2011). Maze training in mice induces MRI-detectable brain shape changes specific to the type of learning. *NeuroImage, 54*, 2086–2095.

Lerch, J. P., van der Kouwe, A. J., Raznahan, A., Paus, T., Johansen-Berg, H., Miller, K. L., ... & Sotiropoulos, S. N. (2017). Studying neuroanatomy using MRI. *Nature Neuroscience, 20*, 314–326.

Lindenberger, U., Wenger, E., & Lövdén, M. (2017). Towards a stronger science of human plasticity. *Nature Reviews Neuroscience, 18*, 261–262.

Lövdén, M., Bäckman, L., Lindenberger, U., Schaefer, S., & Schmiedek, F. (2010a). A theoretical framework for the study of adult cognitive plasticity. *Psychological Bulletin, 136*, 659–676.

Lövdén, M., Bodammer, N. C., Kühn, S., Kaufmann, J., Schütze, H., Tempelmann, C., ... & Lindenberger, U. (2010b). Experience-dependent plasticity of white-matter microstructure extends into old age. *Neuropsychologia, 48*, 3878–3883.

Lövdén, M., Schaefer, S., Noack, H., Bodammer, N. C., Kühn, S., Heinze, H. J., ... & Lindenberger, U. (2012). Spatial navigation training protects the hippocampus against age-related changes during early and late adulthood. *Neurobiology of Aging, 33*, 620.e9.

Lövdén, M., Wenger, E., Mårtensson, J., Lindenberger, U., & Bäckman, L. (2013). Structural brain plasticity in adult learning and development. *Neuroscience & Biobehavioral Reviews, 37*, 2296–2310.

Maass, A., Düzel, S., Goerke, M., Becke, A., Sobieray, U., Neumann, K., ... & Ahrens, D. (2015). Vascular hippocampal plasticity after aerobic exercise in older adults. *Molecular Psychiatry, 20*, 585–593.

Maguire, E. A., Gadian, D. G., Johnsrude, I. S., Good, C. D., Ashburner, J., Frackowiak, R. S., & Frith, C. D. (2000). Navigation-related structural change in the hippocampi of taxi drivers. *Proceedings of the National Academy of Sciences, 97*, 4398–4403.

Maguire, E. A., Woollett, K. and Spiers, H. J. (2006). *London taxi drivers and bus drivers: A structural MRI and neuropsychological analysis*, pp. 1091–1101.

Makino, H., & Komiyama, T. (2015). Learning enhances the relative impact of top-down processing in the visual cortex. *Nature Neuroscience, 18*, 1116–1122.

Makino, H., Hwang, E. J., Hedrick, N. G., & Komiyama, T. (2016). Circuit mechanisms of sensorimotor learning. *Neuron, 92*, 705–721.

Månsson, K. N., Cortes, D. S., Manzouri, A., Li, T. Q., Hau, S., & Fischer, H. (2020). Viewing pictures triggers rapid morphological enlargement in the human visual cortex. *Cerebral Cortex, 30*, 851–857.

Mårtensson, J., Eriksson, J., Bodammer, N. C., Lindgren, M., Johansson, M., Nyberg, L., & Lövdén, M. (2012). Growth of language-related brain areas after foreign language learning. *NeuroImage, 63*, 240–244.

May, A. (2011). Experience-dependent structural plasticity in the adult human brain. *Trends in Cognitive Sciences, 15*, 475–482.

May, A., Hajak, G., Gänssbauer, S., Steffens, T., Langguth, B., Kleinjung, T., & Eichhammer, P. (2007). Structural brain alterations following 5 days of intervention: Dynamic aspects of neuroplasticity. *Cerebral Cortex, 17*, 205–210.

Molina-Luna, K., Hertler, B., Buitrago, M. M., & Luft, A. R. (2008). Motor learning transiently changes cortical somatotopy. *NeuroImage, 40*, 1748–1754.

Pascual-Leone, A., Amedi, A., Fregni, F., & Merabet, L. B. (2005). The plastic human brain cortex. *Annual Review of Neuroscience, 28*, 377–401.

Pereira, A. C., Huddleston, D. E., Brickman, A. M., Sosunov, A. A., Hen, R., McKhann, G. M., ... & Small, S. A. (2007). An in vivo correlate of exercise-induced neurogenesis in the adult dentate gyrus. *Proceedings of the National Academy of Sciences, 104*, 5638–5643.

Peters, A. J., Chen, S. X., & Komiyama, T. (2014). Emergence of reproducible spatiotemporal activity during motor learning. *Nature, 510*, 263–267.

Pruitt, D. T., Schmid, A. N., Danaphongse, T. T., Flanagan, K. E., Morrison, R. A., Kilgard, M. P., ... & Hays, S. A. (2016). Forelimb training drives transient map reorganization in ipsilateral motor cortex. *Behavioural Brain Research, 313*, 10–16.

Quallo, M. M., Price, C. J., Ueno, K., Asamizuya, T., Cheng, K., Lemon, R. N., & Iriki, A. (2009). Gray and white matter changes associated with tool-use learning in macaque monkeys. *Proceedings of the National Academy of Sciences, 106*, 18379–18384.

Rakic, P. (2002). Neurogenesis in adult primate neocortex: An evaluation of the evidence. *Nature Reviews. Neuroscience, 3*, 65–71.

Reed, A., Riley, J., Carraway, R., Carrasco, A., Perez, C., Jakkamsetti, V., & Kilgard, M. P. (2011). Cortical map plasticity improves learning but is not necessary for improved performance. *Neuron, 70*, 121–131.

Rosenzweig, M. R., Bennett, E. L., & Krech, D. (1964). Cerebral effects of environmental complexity and training among adult rats. *Journal of Comparative and Physiological Psychology, 57*, 438–439.

Rosenzweig, M. R., Bennett, E. L., & Diamond, M. C. (1972). Brain changes in response to experience. *Scientific American, 226*, 22–29.

Sagi, Y., Tavor, I., Hofstetter, S., Tzur-Moryosef, S., Blumenfeld-Katzir, T., & Assaf, Y. (2012). Learning in the fast lane: New insights into neuroplasticity. *Neuron, 73*, 1195–1203.

Sirevaag, A. M., & Greenough, W. T. (1991). Plasticity of GFAP-immunoreactive astrocyte size and number in visual cortex of rats reared in complex environments. *Brain Research, 540*, 273–278.

Swain, R. A., Harris, A. B., Wiener, E. C., Dutka, M. V., Morris, H. D., Theien, B. E., ... & Greenough, W. T. (2003). Prolonged exercise induces angiogenesis and increases cerebral blood volume in primary motor cortex of the rat. *Neuroscience, 117*, 1037–1046.

Tan, X., & Shi, S.-H. (2013). Neocortical neurogenesis and neuronal migration. *Wiley Interdisciplinary Reviews. Developmental Biology, 2*, 443–459.

Tardif, C. L., Gauthier, C. J., Steele, C. J., Bazin, P. L., Schäfer, A., Schaefer, A., ... & Villringer, A. (2016). Advanced MRI techniques to improve our understanding of experience-induced neuroplasticity. *NeuroImage, 131*, 55–72.

Taubert, M., Draganski, B., Anwander, A., Müller, K., Horstmann, A., Villringer, A., & Ragert, P. (2010). Dynamic properties of human brain structure: Learning-related changes in cortical areas and associated fiber connections. *Journal of Neuroscience, 30*, 11670–11677.

Teutsch, S., Herken, W., Bingel, U., Schoell, E., & May, A. (2008). Changes in brain gray matter due to repetitive painful stimulation. *NeuroImage, 42*, 845–849.

Thomas, C., & Baker, C. I. (2013). Teaching an adult brain new tricks: A critical review of evidence for training-dependent structural plasticity in humans. *NeuroImage, 73*, 225–236.

Trachtenberg, J. T., Chen, B. E., Knott, G. W., Feng, G., Sanes, J. R., Welker, E., & Svoboda, K. (2002). Long-term in vivo imaging of experience-dependent synaptic plasticity in adult cortex. *Nature, 420*, 788–794.

van Praag, H., Kempermann, G., & Gage, F. H. (1999). Running increases cell proliferation and neurogenesis in the adult mouse dentate gyrus. *Nature Neuroscience, 2*, 266–270.

van Praag, H., Kempermann, G., & Gage, F. H. (2000). Neural consequences of environmental enrichment. *Nature Reviews. Neuroscience, 1*, 191–198.

Wagner, A. P., Schmoll, H., Badan, I., Platt, D., & Kessler, C. (2000). Brain plasticity: To what extent do aged animals retain the capacity to coordinate gene activity in response to acute challenges. *Experimental Gerontology, 35*, 1211–1227.

Wang, X., Takano, T., & Needergaard, M. (2009). Astrocytic calcium signaling: Mechanisms and implications for functional brain imaging. In F. Hyder (Ed.), *Methods Molecular Biology* (Vol. 489, pp. 93–109). Totowa: Humana Press (Methods in Molecular Biology).

Wenger, E., Schaefer, S., Noack, H., Kühn, S., Mårtensson, J., Heinze, H. J., ... & Lövdén, M. (2012). Cortical thickness changes following spatial navigation training in adulthood and aging. *NeuroImage, 59*, 3389–3397.

Wenger, E., Brozzoli, C., Lindenberger, U., & Lövdén, M. (2017a). Expansion and renormalization of human brain structure during skill acquisition. *Trends in Cognitive Sciences, 21*, 930–939.

Wenger, E., Kühn, S., Verrel, J., Mårtensson, J., Bodammer, N. C., Lindenberger, U., & Lövdén, M. (2017b). Repeated structural imaging reveals nonlinear progression of experience-dependent volume changes in human motor cortex. *Cerebral Cortex, 27*, 2911–2925.

Withers, G. S., & Greenough, W. T. (1989). Reach training selectively alters dendritic branching in subpopulations of layer II-III pyramids in rat motor-somatosensory forelimb cortex. *Neuropsychologia, 27*, 61–69.

Woollett, K., & Maguire, E. A. (2011). Acquiring "the Knowledge" of London's layout drives structural brain changes. *Current Biology, 21*, 2109–2114.

Wymbs, N. F., & Grafton, S. T. (2015). The human motor system supports sequence-specific representations over multiple training-dependent timescales. *Cerebral Cortex, 25*, 4213–4225.

Xu, T., Yu, X., Perlik, A. J., Tobin, W. F., Zweig, J. A., Tennant, K., ... & Zuo, Y. (2009). Rapid formation and selective stabilization of synapses for enduring motor memories. *Nature, 462*, 915–919.

Zatorre, R. J., Fields, R. D., & Johansen-Berg, H. (2012). Plasticity in gray and white: Neuroimaging changes in brain structure during learning. *Nature Neuroscience, 15*, 528–536.

Cognitive Plasticity and Transcranial Electrical Stimulation

Elizabeth M. Byrne, Camilla L. Nord, and Joni Holmes

Contents

Abstract This chapter provides an introduction to transcranial electrical stimulation (tES), a non-invasive method for modulating activity in the underlying cortex by delivering a weak electrical current through electrodes placed on the scalp. Starting with an introduction to different types of stimulation, we go on to discuss our current understanding of the neurophysiological mechanisms of tES before reviewing its utility as a tool to enhance cognitive function during and after cognitive training. While there is some evidence that tES can be used in conjunction with cognitive training to improve both training gains and transfer to untrained cognitive tasks, the results are mixed and inconclusive with as many studies reporting null effects as those that report positive effects. We discuss possible reasons for these inconsistent results and conclude that to fully understand the potential benefits

E. M. Byrne (✉)
Faculty of Education, University of Cambridge, Cambridge, UK
e-mail: emb72@cam.ac.uk

C. L. Nord · J. Holmes
MRC Cognition and Brain Sciences Unit, University of Cambridge, Cambridge, UK

T. Strobach, J. Karbach (eds.), *Cognitive Training*,
https://doi.org/10.1007/978-3-030-39292-5_7

of tES for enhancing cognitive plasticity we must: (i) develop a better understanding of how the cellular mechanisms of tES contribute to changes at the level of the cortex and (ii) consider optimising tES protocols at the level of the individual.

What Is tES?

Overview

Transcranial electrical stimulation (tES) is a non-invasive brain stimulation technique that delivers a weak electrical current through the scalp to affect processing in the underlying cortex. Over the past decade, a growing body of research has suggested that tES might be an effective tool for enhancing cognitive function in the areas of language learning, working memory (WM), attention and mental arithmetic (for reviews, see Nitsche and Paulus 2011; Kuo and Nitsche 2012; Summers et al. 2015; Xu et al. 2019). Most recently, tES has been combined with other methods of cognitive enhancement, including cognitive training (Elmasry et al. 2015; Mancuso et al. 2016; Nilsson et al. 2017).

The term tES is used to refer to various stimulation protocols. We will discuss three: transcranial direct current stimulation (tDCS), transcranial alternating current stimulation (tACS) and transcranial random noise stimulation (tRNS). tDCS involves delivery of a constant current via an anodal (positive) and cathodal (negative) electrode; certain montages can induce increases or decreases in cortical excitability (Paulus 2011). During tACS, the current is time-dependent, typically oscillating in a sinusoidal shape (although other waveforms are possible), which can interact with ongoing oscillatory rhythms in the brain (Paulus 2011). In tRNS, the current is varied randomly, generating excitability increases in the cortex similar to anodal tDCS (Terney et al. 2008). These protocols are discussed in detail later in the chapter.

Administration

tES is delivered via two or more rubber electrodes positioned on the scalp (with a conductive substance, e.g. gel or saline-soaked sponges). The electrodes are connected to a battery-driven stimulator used to adjust current intensity and stimulation duration (Fig. 1). Stimulation site(s) can be determined using the international 10–20 electroencephalogram (EEG) placement system to locate regions of interest (Fig. 2), or using electric field modelling. tES has relatively poor focal resolution: stimulation typically extends well beyond a targeted region (Woods et al. 2016). Nevertheless, many studies place a so-called 'active' electrode over a target region and a 'return' electrode over the contralateral supraorbital region, or extracephalically (e.g. shoulder).

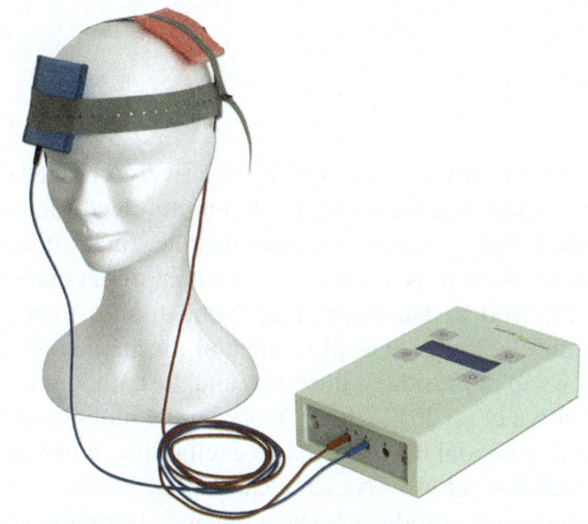

Fig. 1 A neuroConn DC-*STIMULATOR* PLUS (©neuroCare Group GmbH, Munich, Germany; reprinted with permission). This machine is used to deliver transcranial electrical stimulation (tES). For transcranial direct current stimulation (tDCS), where a current flows from the anodal to cathodal electrode, the anode is represented by the red sponge, and the cathode is represented by the blue sponge

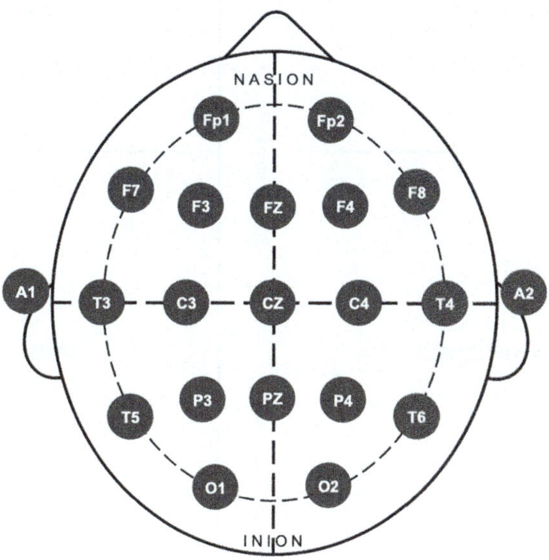

Fig. 2 Electrode locations of the International 10–20 electroencephalography (EEG) system often used to guide electrode placement in transcranial electrical stimulation (tES) studies

Some tES machines have the capacity to run double-blind sham-controlled experiments (i.e. studies in which both participants and experimenters are blind to group allocation), which is difficult with other types of brain stimulation (e.g. transcranial magnetic stimulation; TMS).

tES Protocols

tDCS

All tDCS configurations involve an anodal (positively charged) and a cathodal (negatively charged) electrode (or multiple, in a multi-electrode montage), both of which determine the effects of the stimulation (despite some studies referring to the less interesting electrode as a 'reference'). A constant direct current (see Fig. 3) is sent through the electrodes (via intervening brain tissue) to modulate neuronal excitability under the electrodes. The placement of the anode indicates the location where a current flows to the inside of the body, and the cathode is where the current exits the body (DaSilva et al. 2011). Early work on the motor cortex in non-human animals suggested that anodal tDCS increases excitability, whereas cathodal stimulation decreases neuronal activity (Nitsche and Paulus 2000).

tDCS has the capacity to produce cortical changes lasting beyond the length of stimulation. The duration of physiological after-effects depends on the intensity and duration of the current. When applied to the motor cortex, increasing the current intensity and/or stimulation duration typically results in longer-lasting and stronger

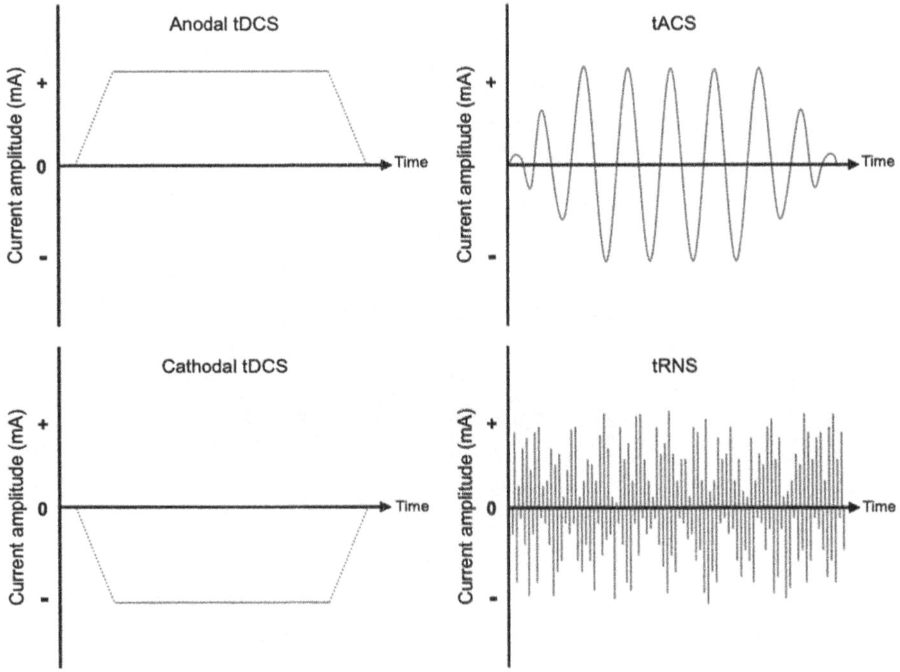

Fig. 3 A simplified illustration of the waveforms of different types of transcranial electrical stimulation (tES): anodal transcranial direct current stimulation (tDCS), cathodal tDCS, transcranial alternating current stimulation (tACS) and transcranial random noise stimulation (tRNS)

after-effects, as measured by motor-evoked potentials (MEPs; Nitsche and Paulus 2000, 2001; Nitsche et al. 2003). A narrow window of current strength exists for inducing tDCS-related after-effects: intensity typically ranges from 0.5 to 2.0 mA (Nitsche and Paulus 2011), with those under 0.5 mA unlikely to produce noticeable effects. Nitsche and Paulus (2000) have shown that a stimulation intensity of at least 0.6 mA is required to produce after-effects in the motor cortex (when applied for 5 min), and that stimulation must be applied for at least 3 min (at 1 mA) to produce noticeable after-effects. Stable after-effects lasting up to an hour have been reported if stimulation is applied for 9–13 min (Nitsche and Paulus 2000, 2001; Nitsche et al. 2003). Typical tDCS montages last 10–20 min (Moreno-Duarte et al. 2014).

tACS

tACS delivers an oscillating current to the brain and has the capacity to interfere with ongoing rhythms in the cortex (see Fig. 3). Like tDCS, it has been shown to influence cortical excitability and activity (Antal et al. 2008; Moliadze et al. 2010; Chaieb et al. 2011; Wach et al. 2013) and facilitate performance on cognitive tasks (e.g. WM; Hoy et al. 2015). It can be used to entrain intrinsic brain oscillations to specific frequency bands (Antal et al. 2008; Paulus et al. 2013; Tavakoli and Yun 2017).

The direction and duration of tACS-induced effects are determined by the frequency, intensity and phase of stimulation (Antal and Paulus 2013). Typically a bidirectional, biphasic current is delivered in sinusoidal waves; stimulation duration ranges from 2–5 min at intensities between 0.25 and 1 mA (Moreno-Duarte et al. 2014). Unlike tDCS, duration-related effects of tACS on MEPs have not been systematically investigated. tACS can be administered in a wide frequency range: usually, it is administered at conventional EEG frequencies (0.1–80 Hz), and in the so-called ripple range of 140 Hz to optimise interactions with ongoing rhythms in the cortex (Moliadze et al. 2010; Antal and Paulus 2013). While some frequencies induce MEP inhibition, others yield excitability increases; some, but not all, result in behavioural improvements (for an overview, see Antal and Paulus 2013). The after-effects of stimulation appear dependent on intensity. In one study, 1 mA of tACS at 140 Hz resulted in significant increases of cortical excitability as measured by MEPs, while reducing the intensity of stimulation to 0.4 mA induced inhibition; the intermediate intensity ranges of 0.6 and 0.8 mA did not produce after-effects (Moliadze et al. 2012).

tRNS

As with the other protocols, tRNS can increase cortical excitability. tRNS is essentially a form of tACS with a white noise characteristic (Terney et al. 2008). Similarly, it is not as well-characterised as tDCS. tRNS is not polarity-specific and can be applied unilaterally. During stimulation, an alternating current is applied along with random amplitudes (see Fig. 3). While tACS uses a fixed frequency, tRNS applies a current within a broad frequency spectrum between 0.1 and 640 Hz with a random noise distribution (Terney et al. 2008; Antal and Paulus 2013).

Numerous stimulation parameters can be altered when using tRNS. A typical tRNS montage involves a randomly alternating level of current between −500 and + 500 μA, with a sampling rate of 1280 samples per second and high-range frequencies between 100 and 640 Hz, providing a current of 1 mA (Terney et al. 2008; Moreno-Duarte et al. 2014). These parameters elicit increased cortical excitability in the motor cortex lasting up to 60 min following 10 min of stimulation (Terney et al. 2008). A minimum of 5 min is necessary to observe effects (Chaieb et al. 2009). tRNS generates behavioural improvements similar to those observed with anodal tDCS (e.g. Romanska et al. 2015) and offers methodological advantages over tDCS. While tDCS is polarity-dependent, tRNS is polarity-independent meaning it can be applied bilaterally. tRNS also has a higher cutaneous perception threshold than tDCS, making it more suitable for blinding groups to stimulation condition (Ambrus et al. 2010).

Neurophysiological Mechanisms of tES

Broadly, anodal polarisation of the cortical surface depolarises membrane potentials of neurons, increasing neuronal excitability and spontaneous firing rates; cathodal polarisation has the opposite effect, hyperpolarising neuronal membranes and decreasing spontaneous firing (Bindman et al. 1964; Bestmann et al. 2015). In the case of an alternating current (tACS; tRNS), ongoing neuronal oscillations are synchronised to the driving frequency via neuronal oscillatory entrainment (Antal and Paulus 2013; Helfrich et al. 2014), which modulates network activity (Fröhlich and McCormick 2010). In all electrical stimulation types, the passage of relatively small amounts of electrical current can elicit long-lasting changes in cortical excitability (Bindman et al. 1964; Nitsche and Paulus 2000; Antal and Paulus 2013).

There is neither a uniform nor localised effect of electrical current on the brain under an electrode (Nitsche et al. 2008). The precise physiological effect of tES depends substantially on the underlying anatomy and physiology of the region being stimulated. The stimulated region also extends beyond the local cortex under the electrode, including deep subcortical regions (Bolzoni et al. 2013). Many factors alter the basic physiological effect of tES: these range from the morphological features of neurons and the orientation of the cell populations, to stimulation parameters including intensity and duration (Bikson et al. 2004; Nitsche et al. 2008; Rahman et al. 2013). For example, 10 but not 20 min of tRNS induces changes in corticospinal excitability (Parkin et al. 2019). The steepness in transient voltage in tACS seems to also affect neural firing as much as the absolute voltage, with the strongest changes in neural firing elicited by steeper transient voltage changes (Fröhlich and McCormick 2010).

The depth of stimulation can drastically alter the resulting physiological effect. In an early in vivo experiment of direct current stimulation, anodal enhanced and cathodal stimulation suppressed cortical neuron activity; this pattern was reversed for neurons more than 3 mm from cortex (Creutzfeldt et al. 1962). This suggests that in human brains, sulci and gyri are differentially affected by electrical current, likely due to the orientation of neurons relative to the electric field (Nitsche et al. 2008). Even distinct cellular compartments (e.g. the soma versus the axon of neurons) respond differently to application of current (de Berker et al. 2013; Rahman et al. 2013; Bestmann et al. 2015). It is probable that tES has both short-term excitatory and inhibitory effects on every neuron stimulated, with the sum determining its net effects on a neuron and its local population (Kabakov et al. 2012).

The acute effects of tES are not thought to underpin the variety of longer-term changes resulting from tES stimulation. Instead, long-lasting effects of tES on cortical excitability are likely driven by synaptic plasticity mechanisms. Sustained increases in firing from direct currents are protein synthesis-dependent (Gartside 1968), and also lead to modification of intracellular cyclic AMP (Hattori et al. 1990) and calcium (Islam et al. 1995). In murine motor cortex slices, tES induces long-term potentiation (LTP), with the behavioural effects of tDCS on motor skill learning dependent on changes in synaptic plasticity (Fritsch et al. 2010).

Human studies are limited in their ability to directly investigate the cellular mechanisms of tES. Nevertheless, non-invasive measures of neurotransmitter concentration such as magnetic resonance spectroscopy (MRS) or brain perfusion (arterial spin labelling; ASL) offer possible avenues for mechanistic research in humans. In one study, reductions in GABA levels elicited by motor cortex tDCS were crucial for behavioural changes, potentially providing a proxy measure of tES-induced effects on plasticity (Stagg et al. 2011). In a second study, prefrontal tDCS was found to evoke local increases in brain perfusion during, but widespread decreases following, stimulation (Stagg et al. 2013). This is in direct contrast to motor cortex tDCS, which induces highly similar excitability changes during and after stimulation (i.e. the expected enhancement of excitability during and after anodal polarisation; Nitsche & Paulus, 2000). The differential effects of prefrontal tDCS on neural mechanisms may underpin the common cognitive finding of differential effects during/after stimulation delivery (Lally et al. 2013; Nord et al. 2013).

Combining tES with Cognitive Training

Many studies have explored whether the application of tES during a learning or cognitive task enhances brain plasticity. In the cognitive training field, where participants practice on increasingly demanding cognitive tasks, tES has been evaluated as a modulatory tool to boost the efficacy and generalisability of training and transfer effects. The effects of different tES protocols have been tested in different cognitive training domains; however, the largest literature focusses on coupling tDCS with WM training. Key findings are summarised below, with details of the training tasks and stimulation montages used reported in Table 1.

Table 1 tES and training studies

Study	Training	Stimulation	Region	Control	Results (positive effects of stimulation vs. sham)
tDCS and WM training					
Au et al. (2016)	7 sessions, *n*-back	Anodal tDCS (2 mA, 25 min)	Left or right DLPFC	Sham stimulation	Enhanced learning on training task, sustained for 7 mths. Enhanced performance on an untrained *n*-back task, but not on other WM or inhibition tests
Byrne et al. (in press)	3 sessions, backward digit span	Anodal tDCS (1 mA, 10 min)	Left DLPFC	Sham stimulation	Null effects
Jones et al. (2015)	10 sessions, operation span and visuo-spatial object recognition	Anodal tDCS (1.5 mA, 10 min)	Right prefrontal, parietal, or alternating prefrontal-parietal regions	Sham stimulation	No immediate effects. Gains on training and transfer tasks 1 mth after training
Lally et al. (2013)	2 sessions, *n*-back	Anodal tDCS (1 mA, 10 min)	Left DLPFC	Sham stimulation	Enhancements to training in the first session, but not the second
Martin et al. (2014)	4 sessions, dual *n*-back	Anodal tDCS (2 mA, 30 min)	Left DLPFC	Sham stimulation	Null effects
Nilsson et al. (2017)	20 sessions, *n*-back, running span, and switching (task/rule-switching)	Anodal tDCS (2 mA, 25 min)	Left DLPFC	Sham stimulation	Null effects

Park et al. (2014)	10 sessions, digit span	Anodal tDCS (2 mA, 30 min)	Bilateral DLPFC (using two stimulators)	Sham stimulation	Enhanced learning on training task. Enhanced performance on untrained *n*-back task, sustained for 1 mth
Richmond et al. (2014)	10 sessions, operation or symmetry span	Anodal tDCS (1.5 mA, 15 min)	Left DLPFC	Sham stimulation	Enhanced performance on operation span training task, but not symmetry span
Ruf et al. (2017)	3 sessions, *n*-back	Anodal tDCS (1 mA, 20 min)	Left or right DLPFC	Sham stimulation	Steeper learning curve during training, effects sustained for 9 mths. Enhanced performance on untrained *n*-back task
Talsma et al. (2017)	3 sessions, *n*-back	Anodal tDCS (1 mA, 20 min)	Left DLPFC	Sham stimulation	Enhancements to training in the first session, but not the final two sessions
tDCS and training					
Ditye et al. (2012)	4 sessions, inhibition	Anodal tDCS (1.5 mA, 15 min)	Right inferior frontal gyrus	None	Null effects
Dockery et al. (2009)	3 sessions, tower of London	Anodal and cathodal tDCS (1 mA, 15 min)	Left DLPFC	Sham stimulation	Enhanced training, sustained for 1 yr
Meinzer et al. (2014)	5 sessions, language learning	Anodal tDCS (1 mA, 20 min)	Left posterior temporo-parietal junction	Sham stimulation	Steeper learning curve on trained tasks. Enhanced learning at the end of training, sustained for 1 wk
tDCS and training in clinical populations					
Cotelli et al. (2014a)	10 sessions, language training	Anodal tDCS (2 mA, 25 min)	Left DLPFC	Sham stimulation	Enhanced learning on training task, sustained for 6 mths

(continued)

Table 1 (continued)

Study	Training	Stimulation	Region	Control	Results (positive effects of stimulation vs. sham)
Cotelli et al. (2014b)	10 sessions, memory training of face-name associations	Anodal tDCS (2 mA, 25 min)	Left DLPFC	Sham stimulation	Null effects
Meinzer et al. (2016)	8 sessions, language/naming	Anodal tDCS (1 mA, 20 min)	Left primary motor cortex	Sham stimulation	Transfer effects to untrained items maintained only for the anodal stimulation group at 6 mth follow-up
tRNS and training					
Brem et al. (2018)	9 sessions, multiple executive function tasks	tRNS (1 mA, 30 min)	Bilateral DLPFC	None	Null effects
Cappelletti et al. (2013)	5 sessions, numerosity discrimination	tRNS (1 mA, 20 min)	Bilateral parietal cortex	Sham stimulation, active tRNS over a control cortical location (motor area), active parietal tRNS alone	Steeper learning curves on trained task, gains sustained for 4 mths. Enhanced performance on untrained magnitude judgment tasks, but not attention, executive function, or visual pattern recognition tasks
Cappelletti et al. (2015)	5 sessions, numerosity discrimination	tRNS (1 mA, 20 min)	Bilateral parietal cortex	Sham stimulation, active tRNS over a control cortical location (motor area)	Boost to learning rates during training, sustained for 16 wks
Holmes, et al. (2016)	10 sessions, multiple WM tasks	tRNS (1 mA, 10 min)	Bilateral DLPFC	Sham stimulation	Null effects

Looi et al. (2017)	9 sessions, arithmetic	tRNS (0.75 mA, 20 min)	Bilateral DLPFC	Sham stimulation	Enhanced accuracy and steeper learning curves during training. Enhanced gains on an untrained mathematics test, but not a WM test
Popescu et al., 2016)	5 sessions, arithmetic	tRNS (1 mA, 20 min)	Bilateral DLPFC	Sham stimulation	Enhanced training rates. Enhanced performance on untrained arithmetic task, but not WM or attention tasks
Snowball et al. (2013)	5 sessions, arithmetic	tRNS (1 mA, 20 min)	Bilateral DLPFC	Sham stimulation	Enhanced performance on untrained mathematical problems, sustained for 6 mths
tACS and training					
Brem et al. (2018)	9 sessions, multiple executive function tasks	Multi-focal tACS in the gamma band (40 Hz, 30 min)	Bilateral prefrontal and parietal-occipital regions	None	Null effects

N.B. This is not a comprehensive or systematic review of all studies in the field. It serves to illustrate the variety of training paradigms and tES protocols that have been employed. *tDCS* transcranial direct current stimulation, *tRNS* transcranial random noise stimulation, *tACS* transcranial alternating current stimulation, *DLPFC* dorsolateral prefrontal cortex, *WM* working memory

tDCS and Training

Working Memory Training and tDCS

There is evidence from sham-controlled studies that coupling different types of WM training (see Könen et al., this volume) with tDCS improves performance on trained and untrained tasks in healthy adult populations. Au et al. (2016) found enhanced rates of learning on a visuo-spatial *n*-back training task (i.e. steeper rates of improvement) and better performance an untrained *n*-back task for participants receiving anodal tDCS relative to those receiving sham tDCS. Similarly, Ruf et al. (2017) found that anodal tDCS enhanced learning rates during verbal and spatial *n*-back training, and also led to greater improvements on an untrained version of *n*-back relative to sham stimulation. Gains on this task were sustained up to 9 mths post-training. Positive effects have also been reported for healthy older adults. Anodal versus sham tDCS applied during digit span training enhanced training gains and modulated transfer to an untrained *n*-back WM measure (Park et al. 2014). Gains on the transfer task were sustained 1 mth later (Park et al. 2014). In another sham-controlled study, there were no immediate effects of anodal tDCS during training (on operation span and a visuo-spatial object recognition task; Jones et al. 2015). However, participants who received training with active stimulation demonstrated superior performance on trained and untrained tasks 1 mth after training.

However, results are mixed: some studies report enhanced training gains for active versus sham stimulation, but no group differences on transfer tasks. Richmond et al. (2014) found that anodal tDCS enhanced on-task training gains on a verbal, but not spatial, complex span task relative to sham stimulation. Stimulation enhanced transfer to untrained WM tasks, but this was only true for the active stimulation with training group compared to a no-intervention group. Critically, no significant differences were found between the training groups with active and sham tDCS in terms of transfer. Consequently, this effect can be attributed to training alone. Other studies report initial benefits on the training task(s), but no after-effects or cumulative effect (Lally et al. 2013; Talsma, Lotte et al. 2017), while still others find no benefits of active versus sham altogether (Martin et al. 2014; Nilsson et al. 2017).

Overall, there is no consistent evidence that tDCS enhances the transfer of training gains. Consistent with this, a recent study employing a sham-controlled design found no evidence that tDCS promotes the transfer of WM training gains, even when the training and transfer tasks were distinguished by only a single task feature (e.g. paradigm or stimuli). Byrne et al. (in press) systematically manipulated the degree of overlap between trained and untrained tasks in terms of isolated task features (i.e. type of stimuli, stimuli modality, WM paradigm), to examine the magnitude and distance over which tDCS influences transfer following training. There was no evidence that anodal tDCS enhanced learning on a backward digit recall WM training task relative to sham stimulation. Training was associated with paradigm-specific transfer effects (i.e. gains on backward recall with novel stimuli). However, tDCS did not enhance these effects, nor did it promote transfer to a novel WM paradigm (*n*-back).

Other Cognitive Training Paradigms and tDCS

Evidence for the impact of tDCS on cognitive training in other domains is limited. In language learning, active stimulation results in steeper learning curves on trained tasks, and enhanced learning at the end of training, with gains sustained up to 1 wk post-training (Meinzer et al. 2014). Results for executive function training (see also Karbach and Kray, this volume) are mixed. tDCS combined with a planning task resulted in training task gains that were sustained up to 1 year later (Dockery et al. 2009). In contrast, tDCS combined with inhibition training enhanced performance during training, but this was not maintained on a fifth day without stimulation and there was no sham control group (Ditye et al. 2012).

Clinical Applications of Cognitive Training Coupled with tDCS

When combined with cognitive training, tDCS has shown promise as a tool for cognitive rehabilitation in clinical and atypical populations (e.g. for the amelioration of depressive symptoms; Brunoni & Vanderhasselt 2014; Manenti et al. 2018; Segrave et al. 2014). tDCS coupled with cognitive training has been shown to improve cognitive function in stroke patients with aphasia, but the evidence is inconsistent. Language training resulted in enhanced naming ability in post-stroke aphasic patients, but no differences were found between groups who received sham or anodal tDCS (Meinzer et al. 2016). However, patients who received active stimulation outperformed the sham group 6 mths later. In another study, post-stroke patients showed improvements during language training, and those who received anodal tDCS made additional gains relative to those who received sham stimulation, which were sustained 12 wks later (Cotelli et al. 2014a). For patients with Alzheimer's dementia, greater improvements in memory performance were found for those receiving anodal tDCS combined with memory training relative to those receiving anodal stimulation with motor training. However, there were no group differences when the anodal tDCS group receiving memory training were compared to patients who received memory training with sham stimulation (Cotelli et al. 2014b). This indicates that the memory training alone may have been sufficient to enhance memory performance.

tRNS and Training

The potential for tRNS to enhance the effects of cognitive training has been explored most widely in the numerical domain (see Johann and Karbach, Schaeffner et al., this volume). Overall, the effects are positive, suggesting that tRNS enhances on-task numerical or arithmetic training (steeper learning curves) in younger and older adults, and children, with gains sustained up to 4 mths later (Cappelletti et al.

2013, 2015; Looi et al. 2017). Additionally, there is some evidence that these gains benefit untrained mathematical/numerical tasks (Cappelletti et al. 2013; Snowball et al. 2013; Looi et al. 2017), which are sustained up to 6 mths later. These effects do not, however, appear to extend to other transfer tasks such as WM or attention (Popescu et al. 2016). This suggests that the benefits of tRNS on mathematical training are relatively task-specific. The effects of tRNS on cognitive training in other domains have been less promising. To date, only one study has investigated the effects of combining tRNS with WM training (see Könen et al., this volume): active stimulation failed to enhance WM training or transfer relative to sham controlled training (Holmes et al. 2016). Brem et al. (2018) reported benefits of combining tRNS with a variety of executive function training activities for transfer tests of fluid intelligence. However, these benefits were only present when those receiving stimulation and training were compared to a no-contact control group who did not receive training or sham stimulation. These effects could therefore reflect the benefits of training alone.

tACS and Training

Some studies show improved performance on cognitive tasks from tACS (for a review, see Antonenko et al. 2016). For example, theta-range tACS improved WM, fluid reasoning and language learning when applied over bilateral dorsolateral prefrontal cortex (DLPFC), left parietal and left temporo-parietal regions, respectively (Meiron and Lavidor 2014; Pahor and Jaušovec 2014; Antonenko et al. 2016). However, the impact of tACS on cognitive performance over multiple sessions of practice has rarely been examined. The only study in this field to date, conducted by Brem et al. (2018), reported that tACS did not improve on-task training gains on a range of executive function measures, nor did it enhance post-training performance relative to a no-contact control group on untrained tasks.

Why Are the Results Inconsistent?

The cognitive effects of tES are highly inconsistent between studies (Tremblay et al. 2014), leading to doubts regarding its overall efficacy. A quantitative review even suggested there was no effect of tDCS on cognition in healthy individuals (Horvath et al. 2015; though for a critique of this study, see Price and Hamilton 2015). Possible reasons for these inconsistencies are discussed below.

Methodological Inconsistencies

One complication in interpreting the tES literature is that methodologies vary widely, including using different control groups, as well as variable blinding of participants and experimenters (see Cochrane and Green, Schmiedek et al., this volume). The tES field desperately needs standard methodological practices. First, appropriate control groups are required to ensure that participants are matched on motivation and expectancy effects, and to ensure any effects can be attributed both to the stimulation montage and site, and to the specific training task (Morrison and Chein 2011; Parkin et al. 2015). Ideally, the following control groups should be included: (i) a sham stimulation control group that completes the same training activity as the active stimulation group, (ii) a control cortical site receiving active stimulation; and (iii) a control training task combined with active stimulation. Second, participant and investigator blinding is recommended where possible. Many tES machines can be programmed in advance to deliver double-blind active or sham stimulation. For research involving cognitive training, this is trickier. Using appropriate control groups, participants should be naïve to their condition. However, unless separate investigators are used to deliver the training and transfer sessions, they will be aware of group allocation. In any case, researchers must also randomly assign participants to groups to reduce bias (Simons et al. 2016).

Individual Differences in Response to tES

A second cause of inconsistency in findings is individual differences (see Karbach and Kray, Katz et al., Könen and Auerswald, this volume) in response to tES. The majority of experimental studies collapse across all participants and report group effects of tES (or lack thereof). This pragmatic approach tends to ignore known contributions of age, gender, current brain state, head/tissue morphology, hormonal levels and circadian factors on individual differences in pre-existing regional excitability (Bikson et al. 2012; Krause and Cohen Kadosh 2014), in spite of the role of pre-existing excitability in modulating the outcome of tES (Antal et al. 2007; Krause and Cohen Kadosh 2014). Unaccounted-for individual differences could mask true effects of stimulation on cognition and behaviour by treating participants as a uniform group. Cluster analyses reveal a statistically-bimodal distribution of response to tDCS, even when applied to the motor cortex; only 45% of subjects show typical effects of tDCS on corticospinal excitability, while the rest show reduced, or even reversed, responses (López-Alonso et al. 2014).

One approach to address the plethora of factors affecting tES response is measuring baseline variables that might index regional excitability and attempt to predict, post-hoc, a relationship with cognitive or behavioural effects of stimulation. For example, GABA levels in the motor cortex (measured with MRS) predicted behavioural response to anodal tDCS over primary motor cortex (O'Shea et al.

2014). A trial combining tDCS with cognitive behavioural therapy for depression found baseline DLPFC activation (measured with functional magnetic resonance imaging; fMRI) predicted response to anodal tDCS over DLPFC (Nord et al. 2019). DLPFC tDCS response may also be affected by genetic factors (see Colzato and Hommel, this volume): only individuals homozygous for the Val-allele of the COMT Val(108/158)Met polymorphism showed impairments in response inhibition following cathodal stimulation (Nieratschker et al. 2015). As yet, however, no reliable index of tES effects on cognition has been established.

The utility of individual difference measures in tES may extend beyond the binary responder/non-responder classification (a paradigm borrowed from pharmacology, where mechanism and dose-response curve is often clearer). Few tES studies have methodically investigated response to a series of stimulation montages in the same participants. A rare exception reported findings that were contrary to most of the field's assumptions: at the group level, only 1 mA of unilateral anodal stimulation increased corticospinal excitability (and 1 mA of cathodal produced inhibition), with no effect of 2 mA or bilateral stimulation (despite both being commonly used in cognitive experiments) (Parkin et al. 2019). Similarly, unilateral, but not bilateral, tRNS altered corticospinal excitation.

Considered at the level of the individual, stimulation montage may also play a crucial role in determining who responds (or fails to respond) to a given intervention. tDCS is unusual in the wider field of brain stimulation in that every participant receives an identical amount of stimulation. TMS studies, in contrast, calibrate each individual's stimulation according to excitability of their motor cortex (Pascual-Leone et al. 1994). Likewise, tACS studies use participants' individual oscillatory frequencies (measured with EEG) to modulate endogenous oscillatory frequency. The lack of individual calibration in the most common form of tES, tDCS, implies some participants are receiving non-optimal amplitudes of stimulation, which might occlude its true efficacy.

Conclusions and Future Directions

The effects of tES on cognitive training are currently unclear (Elmasry et al. 2015; Mancuso et al. 2016; Nilsson et al. 2017). Inconsistent results are likely a result of methodological differences across studies and limited consideration of individual differences. The most important gap in our understanding of tES on behaviour is how cellular and molecular mechanisms of tES map on to changes at the cortical level and to alterations in cognition. It is not an accurate reflection of the physiology of tES to claim a direct link between anodal polarisation, regional excitation and cognitive improvement (though many do, such is the appeal of its simplicity). Instead, a physiologically plausible model of tES must incorporate findings from each level of observation. One proposal is that computational modelling approaches could leverage data from the microscopic, mesoscopic and cognitive levels to bridge this gap in understanding (Bestmann et al. 2015). In support of this, a biophysically-informed model of decision-making accurately predicted differential cognitive effects of medial and lateral prefrontal cortex tDCS (Hämmerer et al. 2016). Improving the design and interpretation of cognitive tES experiments will necessitate an ameliorated understanding of how the cellular mechanisms of tES contribute to changes at the level of the cortex, and ultimately at the level of cognition.

The tES field must also progress toward individual-level optimisation of tES protocols. Software to simulate current distributions is now employed in numerous experimental (Hämmerer et al. 2016) and clinical (Brunoni et al. 2017) tES studies, giving researchers the ability to model the physiological consequences of stimulation in an individual brain, and optimise targeting of specific regions, certainly an improvement over the 10–20 EEG system of localisation. Novel trial designs could test whether proxies for regional excitability (e.g. fMRI and MRS) could help establish the optimal tES dose for a given participant (calibrating voltage, number of sessions and/or stimulation duration). Finally, experimental designs should endeavour to test multiple montages, as well as incorporate measures of potential sources of response variability to allow individualised refinement of tES delivery. Advancing the field in these different ways will enable to us to understand more about the mechanisms of tES, and its potential benefits for enhancing cognitive function.

References

Ambrus, G. G., Paulus, W., & Antal, A. (2010). Cutaneous perception thresholds of electrical stimulation methods: Comparison of tDCS and tRNS. *Clinical Neurophysiology, 121*, 1908–1914.

Antal, A., & Paulus, W. (2013). Transcranial alternating current stimulation (tACS). *Frontiers in Human Neuroscience, 7*, 1–4.

Antal, A., Terney, D., Poreisz, C., & Paulus, W. (2007). Towards unravelling task-related modulations of neuroplastic changes induced in the human motor cortex. *The European Journal of Neuroscience, 26*, 2687–2691.

Antal, A., Boros, K., Poreisz, C., Chaieb, L., Terney, D., & Paulus, W. (2008). Comparatively weak after-effects of transcranial alternating current stimulation (tACS) on cortical excitability in humans. *Brain Stimulation, 1*, 97–105.

Antonenko, D., Faxel, M., Grittner, U., Lavidor, M., & Flöel, A. (2016). Effects of transcranial alternating current stimulation on cognitive functions in healthy young and older adults. *Neural Plasticity, 2016*, 1–13.

Au, J., Katz, B., Buschkuehl, M., Bunarjo, K., Senger, T., Zabel, C., ... & Jonides, J. (2016). Enhancing working memory training with transcranial direct current stimulation. *Journal of Cognitive Neuroscience, 28*, 1419–1432.

Bestmann, S., de Berker, A. O., & Bonaiuto, J. (2015). Understanding the behavioural consequences of noninvasive brain stimulation. *Trends in Cognitive Sciences, 19*, 13–20.

Bikson, M., Inoue, M., Akiyama, H., Deans, J. K., Fox, J. E., Miyakawa, H., & Jefferys, J. G. (2004). Effects of uniform extracellular DC electric fields on excitability in rat hippocampal slices *in vitro*. *Journal of Physiology, 557*, 175–190.

Bikson, M., Rahman, A., & Datta, A. (2012). Computational models of transcranial direct current stimulation. *Clinical EEG and Neuroscience, 43*, 176–183.

Bindman, L. J., Lippold, O. C. J., & Redfearn, J. W. T. (1964). The action of brief polarizing currents on the cerebral cortex of the rat (1) during current flow and (2) in the production of long-lasting after-effects. *Journal of Physiology, 172*, 369–382.

Bolzoni, F., Bączyk, M., & Jankowska, E. (2013). Subcortical effects of transcranial direct current stimulation in the rat. *Journal of Physiology, 591*, 4027–4042.

Brem, A. K., Almquist, J. N. F., Mansfield, K., Plessow, F., Sella, F., Santarnecchi, E., ... & Yeung, N. (2018). Modulating fluid intelligence performance through combined cognitive training and brain stimulation. *Neuropsychologia, 118*, 107–114.

Brunoni, A. R., & Vanderhasselt, M.-A. (2014). Working memory improvement with non-invasive brain stimulation of the dorsolateral prefrontal cortex: A systematic review and meta-analysis. *Brain and Cognition, 86*, 1–9.

Brunoni, A. R., Moffa, A. H., Sampaio-Junior, B., Borrione, L., Moreno, M. L., Fernandes, R. A., ... & Chamorro, R. (2017). Trial of electrical direct-current therapy versus escitalopram for depression. *New England Journal of Medicine, 376*, 2523–2533.

Byrne, E. M., Ewbank, M. P., Gathercole, S. E., Holmes, J. (in press). The effects of transcranial direct current stimulation on within- and cross-paradigm transfer following multi-session backward recall training. Brain and Cognition. https://doi.org/10.1016/j.bandc.2020.105552.

Cappelletti, M., Gessaroli, E., Hithersay, R., Mitolo, M., Didino, D., Kanai, R., ... & Walsh, V. (2013). Transfer of cognitive training across magnitude dimensions achieved with concurrent brain stimulation of the parietal lobe. *Journal of Neuroscience, 33*, 14899–14907.

Cappelletti, M., Pikkat, H., Upstill, E., Speekenbrink, M., & Walsh, V. (2015). Learning to integrate versus inhibiting information is modulated by age. *Journal of Neuroscience, 35*, 2213–2225.

Chaieb, L., Kovacs, G., Cziraki, C., Greenlee, M., Paulus, W., & Antal, A. (2009). Short-duration transcranial random noise stimulation induces blood oxygenation level dependent response attenuation in the human motor cortex. *Experimental Brain Research, 198*, 439–444.

Chaieb, L., Antal, A., & Paulus, W. (2011). Transcranial alternating current stimulation in the low kHz range increases motor cortex excitability. *Restorative Neurology and Neuroscience, 29*, 167–175.

Cotelli, M., Manenti, R., Petesi, M., Brambilla, M., Cosseddu, M., Zanetti, O., ... & Borroni, B. (2014a). Treatment of primary progressive aphasias by transcranial direct current stimulation combined with language training. *Journal of Alzheimer's Disease, 39*, 799–808.

Cotelli, M., Manenti, R., Brambilla, M., Petesi, M., Rosini, S., Ferrari, C., ... & Miniussi, C. (2014b). Anodal tDCS during face-name associations memory training in Alzheimer's patients. *Frontiers in Aging Neuroscience, 6*, 38.

Creutzfeldt, O. D., Fromm, G. H., & Kapp, H. (1962). Influence of transcortical d-c currents on cortical neuronal activity. *Experimental Neurology, 5*, 436–452.

DaSilva, A. F., Volz, M. S., Bikson, M., & Fregni, F. (2011). Electrode positioning and montage in transcranial direct current stimulation. *Journal of Visualized Experiments*. https://doi.org/10.3791/2744.

de Berker, A. O., Bikson, M., & Bestmann, S. (2013). Predicting the behavioral impact of transcranial direct current stimulation: Issues and limitations. *Frontiers in Human Neuroscience, 7*, 613.

Ditye, T., Jacobson, L., Walsh, V., & Lavidor, M. (2012). Modulating behavioral inhibition by tDCS combined with cognitive training. *Experimental Brain Research, 219*, 363–368.

Dockery, C. A., Hueckel-Weng, R., Birbaumer, N., & Plewnia, C. (2009). Enhancement of planning ability by transcranial direct current stimulation. *Journal of Neuroscience, 29*, 7271–7277.

Elmasry, J., Loo, C., & Martin, D. M. (2015). A systematic review of transcranial electrical stimulation combined with cognitive training. *Restorative Neurology and Neuroscience, 33*, 263–278.

Fritsch, B., Reis, J., Martinowich, K., Schambra, H. M., Ji, Y., Cohen, L. G., & Lu, B. (2010). Direct current stimulation promotes BDNF-dependent synaptic plasticity: Potential implications for motor learning. *Neuron, 66*, 198–204.

Fröhlich, F., & McCormick, D. A. (2010). Endogenous electric fields may guide neocortical network activity. *Neuron, 67*, 129–143.

Gartside, I. B. (1968). Mechanisms of sustained increases of firing rate of neurones in the rat cerebral cortex after polarization: Role of protein synthesis (24). *Nature, 220*, 383–384.

Hämmerer, D., Bonaiuto, J., Klein-Flügge, M., Bikson, M., & Bestmann, S. (2016). Selective alteration of human value decisions with medial frontal tDCS is predicted by changes in attractor dynamics. *Scientific Reports, 6*, 25160.

Hattori, Y., Moriwaki, A., & Hori, Y. (1990). Biphasic effects of polarizing current on adenosine-sensitive generation of cyclic AMP in rat cerebral cortex. *Neuroscience Letters, 116*, 320–324.

Helfrich, R. F., Schneider, T. R., Rach, S., Trautmann-Lengsfeld, S. A., Engel, A. K., & Herrmann, C. S. (2014). Entrainment of brain oscillations by transcranial alternating current stimulation. *Current Biology, 24*, 333–339.

Holmes, J., Byrne, E. M., Gathercole, S. E., & Ewbank, M. P. (2016). Transcranial random noise stimulation does not enhance the effects of working memory training. *Journal of Cognitive Neuroscience, 28*, 1–13.

Horvath, J. C., Forte, J. D., & Carter, O. (2015). Quantitative review finds no evidence of cognitive effects in healthy populations from single-session transcranial direct current stimulation (tDCS). *Brain Stimulation, 8*, 535–550.

Hoy, K. E., Bailey, N., Arnold, S., Windsor, K., John, J., Daskalakis, Z. J., & Fitzgerald, P. B. (2015). The effect of γ-tACS on working memory performance in healthy controls. *Brain and Cognition, 101*, 51–56.

Islam, N., Aftabuddin, M., Moriwaki, A., Hattori, Y., & Hori, Y. (1995). Increase in the calcium level following anodal polarization in the rat brain. *Brain Research, 684*, 206–208.

Jones, K. T., Stephens, J. A., Alam, M., Bikson, M., & Berryhill, M. E. (2015). Longitudinal neurostimulation in older adults improves working memory. *PLoS One, 10*, e0121904.

Kabakov, A. Y., Muller, P. A., Pascual-Leone, A., Jensen, F. E., & Rotenberg, A. (2012). Contribution of axonal orientation to pathway-dependent modulation of excitatory transmission by direct current stimulation in isolated rat hippocampus. *Journal of Neurophysiology, 107*, 1881–1889.

Krause, B., & Cohen Kadosh, R. (2014). Not all brains are created equal: The relevance of individual differences in responsiveness to transcranial electrical stimulation. *Frontiers in Systems Neuroscience, 8*, 25.

Kuo, M.-F., & Nitsche, M. A. (2012). Effects of transcranial electrical stimulation on cognition. *Clinical EEG and Neuroscience, 43*, 192–199.

Lally, N., Nord, C.L., Walsh, V., Roiser, J.P. (2013). Does excitatory fronto-extracephalic tDCS lead to improved working memory performance? *F1000Research*. https://doi.org/10.12688/f1000research.2-219.v1.

Looi, C. Y., Lim, J., Sella, F., Lolliot, S., Duta, M., Avramenko, A. A., & Kadosh, R. C. (2017). Transcranial random noise stimulation and cognitive training to improve learning and cognition of the atypically developing brain: A pilot study. *Scientific Reports, 7*, 1–10.

López-Alonso, V., Cheeran, B., Río-Rodríguez, D., & Fernández-del-Olmo, M. (2014). Inter-individual variability in response to Non-invasive brain stimulation paradigms. *Brain Stimulation, 7*, 372–380.

Mancuso, L. E., Ilieva, I. P., Hamilton, R. H., & Farah, M. J. (2016). Does transcranial direct current stimulation improve healthy working memory?: A meta-analytic review. *Journal of Cognitive Neuroscience, 28*, 1063–1089.

Manenti, R., Cotelli, M. S., Cobelli, C., Gobbi, E., Brambilla, M., Rusich, D., ... & Cotelli, M. (2018). Transcranial direct current stimulation combined with cognitive training for the treatment of Parkinson disease: A randomized, placebo-controlled study. *Brain Stimulation, 11*, 1251–1262.

Martin, D. M., Liu, R., Alonzo, A., Green, M., & Loo, C. K. (2014). Use of transcranial direct current stimulation (tDCS) to enhance cognitive training: Effect of timing of stimulation. *Experimental Brain Research, 232*, 3345–3351.

Meinzer, M., Jähnigen, S., Copland, D. A., Darkow, R., Grittner, U., Avirame, K., ... & Flöel, A. (2014). Transcranial direct current stimulation over multiple days improves learning and maintenance of a novel vocabulary. *Cortex, 50*, 137–147.

Meinzer, M., Darkow, R., Lindenberg, R., & Flöel, A. (2016). Electrical stimulation of the motor cortex enhances treatment outcome in post-stroke aphasia. *Brain, 139*, 1152–1163.

Meiron, O., & Lavidor, M. (2014). Prefrontal oscillatory stimulation modulates access to cognitive control references in retrospective metacognitive commentary. *Clinical Neurophysiology, 125*, 77–82.

Moliadze, V., Antal, A., & Paulus, W. (2010). Boosting brain excitability by transcranial high frequency stimulation in the ripple range. *Journal of Physiology, 588*, 4891–4904.

Moliadze, V., Atalay, D., Antal, A., & Paulus, W. (2012). Close to threshold transcranial electrical stimulation preferentially activates inhibitory networks before switching to excitation with higher intensities. *Brain Stimulation, 5*, 505–511.

Moreno-Duarte, I., Gebodh, N., Schestatsky, P., Guleyupoglu, B., Reato, D., Bikson, M., & Fregni, F. (2014). Transcranial electrical stimulation: Transcranial direct current stimulation (tDCS), transcranial alternating current stimulation (tACS), transcranial pulsed current stimulation (tPCS), and transcranial random noise stimulation (tRNS). In R. C. Kadosh (Ed.), *The stimulated brain: Cognitive enhancement using non-invasive brain stimulation* (pp. 35–59). London, UK: Academic Press.

Morrison, A. B., & Chein, J. M. (2011). Does working memory training work? The promise and challenges of enhancing cognition by training working memory. *Psychonomic Bulletin & Review, 18*, 46–60.

Nieratschker, V., Kiefer, C., Giel, K., Krüger, R., & Plewnia, C. (2015). The COMT Val/met polymorphism modulates effects of tDCS on response inhibition. *Brain Stimulation, 8*, 283–288.

Nilsson, J., Lebedev, A. V., Rydström, A., & Lövdén, M. (2017). Direct-current stimulation does little to improve the outcome of working memory training in older adults. *Psychological Science, 28*, 907–920.

Nitsche, M. A., & Paulus, W. (2000). Excitability changes induced in the human motor cortex by weak transcranial direct current stimulation. *Journal of Physiology, 527*, 633–639.

Nitsche, M. A., & Paulus, W. (2001). Sustained excitability elevations induced by transcranial DC motor cortex stimulation in humans. *Neurology, 57*, 1899–1901.

Nitsche, M. A., & Paulus, W. (2011). Transcranial direct current stimulation - update 2011. *Restorative Neurology and Neuroscience, 29*, 463–492.

Nitsche, M. A., Nitsche, M. S., Klein, C. C., Tergau, F., Rothwell, J. C., & Paulus, W. (2003). Level of action of cathodal DC polarisation induced inhibition of the human motor cortex. *Clinical Neurophysiology, 114*, 600–604.

Nitsche, M. A., Cohen, L. G., Wassermann, E. M., Priori, A., Lang, N., Antal, A., ... & Pascual-Leone, A. (2008). Transcranial direct current stimulation: State of the art 2008. *Brain Stimulation, 1*, 206–223.

Nord, C. L., Lally, N., & Charpentier, C. J. (2013). Harnessing electric potential: DLPFC tDCS induces widespread brain perfusion changes. *Frontiers in Systems Neuroscience, 7*, 99.

Nord, C. L., Halahakoon, D. C., Limbachya, T., Charpentier, C., Lally, N., Walsh, V., ... & Roiser, J. P. (2019). Neural predictors of treatment response to brain stimulation and psychological therapy in depression: a double-blind randomized controlled trial. *Neuropsychopharmacology, 44*, 1613–1622.

O'Shea, J., Boudrias, M. H., Stagg, C. J., Bachtiar, V., Kischka, U., Blicher, J. U., & Johansen-Berg, H. (2014). Predicting behavioural response to TDCS in chronic motor stroke. *NeuroImage, 85*, 924–933.

Pahor, A., & Jaušovec, N. (2014). The effects of theta transcranial alternating current stimulation (tACS) on fluid intelligence. *International Journal of Psychophysiology, 93*, 322–331.

Park, S. H., Seo, J. H., Kim, Y. H., & Ko, M. H. (2014). Long-term effects of transcranial direct current stimulation combined with computer-assisted cognitive training in healthy older adults. *Neuroreport, 25*, 122–126.

Parkin, B. L., Ekhtiari, H., & Walsh, V. F. (2015). Non-invasive human brain stimulation in cognitive neuroscience: A primer. *Neuron, 87*, 932–945.

Parkin, B. L., Bhandari, M., Glen, J. C., & Walsh, V. (2019). The physiological effects of transcranial electrical stimulation do not apply to parameters commonly used in studies of cognitive neuromodulation. *Neuropsychologia, 128*, 332–339.

Pascual-Leone, A., Valls-Solé, J., Wassermann, E. M., & Hallett, M. (1994). Responses to rapid-rate transcranial magnetic stimulation of the human motor cortex. *Brain, 117*, 847–858.

Paulus, W. (2011). Transcranial electrical stimulation (tES - tDCS; tRNS, tACS) methods. *Neuropsychological Rehabilitation, 21*, 602–617.

Paulus, W., Antal, A., & Nitsche, M. A. (2013). Physiological basis and methodological aspects of transcranial electric stimulation (tDCS, tACS, and tRNS). In C. Miniussi, W. Paulus, & P. M. Rossini (Eds.), *Transcranial brain stimulation* (pp. 93–111). Boca Raton, FL: CRC Press.

Popescu, T., Krause, B., Terhune, D. B., Twose, O., Page, T., Humphreys, G., & Kadosh, R. C. (2016). Transcranial random noise stimulation mitigates increased difficulty in an arithmetic learning task. *Neuropsychologia, 81*, 255–264.

Price, A. R., & Hamilton, R. H. (2015). A re-evaluation of the cognitive effects from single-session transcranial direct current stimulation. *Brain Stimulation, 8*, 2014–2016.

Rahman, A., Reato, D., Arlotti, M., Gasca, F., Datta, A., Parra, L. C., & Bikson, M. (2013). Cellular effects of acute direct current stimulation: Somatic and synaptic terminal effects. *Journal of Physiology, 591*, 2563–2578.

Richmond, L. L., Wolk, D., Chein, J. M., & Olson, I. R. (2014). Transcranial direct current stimulation enhances verbal working memory training performance over time and near-transfer outcomes. *Journal of Cognitive Neuroscience, 26*, 2443–2454.

Romanska, A., Rezlescu, C., Susilo, T., Duchaine, B., & Banissy, M. J. (2015). High-frequency transcranial random noise stimulation enhances perception of facial identity. *Cerebral Cortex, 25*, 4334–4340.

Ruf, S. P., Fallgatter, A. J., & Plewnia, C. (2017). Augmentation of working memory training by transcranial direct current stimulation (tDCS). *Scientific Reports, 7*, 876.

Segrave, R. A., Arnold, S., Hoy, K., & Fitzgerald, P. B. (2014). Concurrent cognitive control training augments the antidepressant efficacy of tDCS: A pilot study. *Brain Stimulation, 7*, 325–331.

Simons, D. J., Boot, W. R., Charness, N., Gathercole, S. E., Chabris, C. F., Hambrick, D. Z., & Stine-Morrow, E. A. (2016). Do "brain-training" programs work?. *Psychological Science in the Public Interest, 17*, 103–186.

Snowball, A., Tachtsidis, I., Popescu, T., Thompson, J., Delazer, M., Zamarian, L., ... & Kadosh, R. C. (2013). Long-term enhancement of brain function and cognition using cognitive training and brain stimulation. *Current Biology, 23*, 987–992.

Stagg, C. J., Bachtiar, V., & Johansen-Berg, H. (2011). The role of GABA in human motor learning. *Current Biology, 21*, 480–484.

Stagg, C. J., Lin, R. L., Mezue, M., Segerdahl, A., Kong, Y., Xie, J., & Tracey, I. (2013). Widespread modulation of cerebral perfusion induced during and after transcranial direct current stimulation applied to the left dorsolateral prefrontal cortex. *Journal of Neuroscience, 33*, 11425–11431.

Summers, J. J., Kang, N., & Cauraugh, J. H. (2015). Does transcranial direct current stimulation enhance cognitive and motor functions in the ageing brain? A systematic review and meta-analysis. *Ageing Research Reviews, 25*, 42–54.

Talsma, L. J., Kroese, H. A., & Slagter, H. A. (2017). Boosting cognition: Effects of multiple-session transcranial direct current stimulation on working memory. *Journal of Cognitive Neuroscience, 29*, 755–769.

Tavakoli, A. V., & Yun, K. (2017). Transcranial alternating current stimulation (tACS) mechanisms and protocols. *Frontiers in Cellular Neuroscience, 11*, 1–10.

Terney, D., Chaieb, L., Moliadze, V., Antal, A., & Paulus, W. (2008). Increasing human brain excitability by transcranial high-frequency random noise stimulation. *Journal of Neuroscience, 28*, 14147–14155.

Tremblay, S., Lepage, J. F., Latulipe-Loiselle, A., Fregni, F., Pascual-Leone, A., & Théoret, H. (2014). The uncertain outcome of prefrontal tDCS. *Brain Stimulation, 7*, 773–783.

Wach, C., Krause, V., Moliadze, V., Paulus, W., Schnitzler, A., & Pollok, B. (2013). Effects of 10Hz and 20Hz transcranial alternating current stimulation (tACS) on motor functions and motor cortical excitability. *Behavioural Brain Research, 241*, 1–6.

Woods, A. J., Antal, A., Bikson, M., Boggio, P. S., Brunoni, A. R., Celnik, P., ... & Knotkova, H. (2016). A technical guide to tDCS, and related non-invasive brain stimulation tools. *Clinical Neurophysiology, 127*, 1031–1048.

Xu, Y., Qiu, Z., Zhu, J., Liu, J., Wu, J., Tao, J., & Chen, L. (2019). The modulation effect of non-invasive brain stimulation on cognitive function in patients with mild cognitive impairment: A systematic review and meta-analysis of randomized controlled trials 11 medical and health sciences 1103 clinical sciences 11 Medica. *BMC Neuroscience, 20*, 2.

Individual Differences in Cognitive Training Research

Benjamin Katz, Masha R. Jones, Priti Shah, Martin Buschkuehl, and Susanne M. Jaeggi

Contents

Abstract Asking whether a cognitive training program "works" does less to further research in the field than asking *why* a participant does or does not improve on training or transfer measures following the intervention. To better understand the divergent results between many cognitive training studies, it is likely necessary to investigate the individual difference factors that might influence the outcome of training. The present chapter covers a range of factors that have been examined in

COI Statement: MB is an employee at the MIND Research Institute, Irvine, CA, whose interests are related to this work. SMJ has an indirect financial interest in the MIND Research Institute.

B. Katz (✉)
Virginia Tech, Blacksburg, VA, USA
e-mail: katzben@vt.edu

M. R. Jones
WestEd, Los Alamitos, CA, USA

P. Shah
University of Michigan, Ann Arbor, MI, USA
e-mail: priti@umich.edu

M. Buschkuehl
MIND Research Institute, Irvine, CA, USA
e-mail: mbuschkuehl@mindresearch.org

S. M. Jaeggi
University of California-Irvine, Irvine, CA, USA
e-mail: smjaeggi@uci.edu

T. Strobach, J. Karbach (eds.), *Cognitive Training*,
https://doi.org/10.1007/978-3-030-39292-5_8

cognitive training research, including baseline performance, age, motivation, personality, and socioeconomic status. While baseline performance and age have received the most attention, a growing body of cognitive training studies have incorporated other individual difference factors in their analyses as well. Given this development, we include a discussion of how researchers might more effectively incorporate individual difference variables into their studies. Continued, high quality research into the role of these factors in the outcome of training remains an important step in developing training interventions that are broadly effective for individuals from a variety of backgrounds.

Introduction

The study of individual differences in cognitive training research dates back nearly as far as the field itself. Sometime after the publication of his early attention training study, *The Effect of Practice Upon Visual Apprehension in School Children* in 1914, Karl Dallenbach, then a young professor of psychology at Cornell University, faced something of a dilemma. In his initial study – one of the first to examine something that might be recognizable to modern researchers as "cognitive training" – he found, in addition to a fairly robust transfer effect to an early Binet test of attention, an interesting pattern of performance across his participants (Dallenbach 1914; see Katz et al. 2018b, for more details on this early cognitive training work). When he stratified the schoolchildren in his sample by initial performance on the training task, Dallenbach found that those students with the lowest baseline performance made a steady, gradual improvement throughout the course of the training, while those with the highest initial performance improved rapidly at first – but then only a little later on. The students in Dallenbach's sample would be considered "typically developing" in modern work, but he knew the students with lower baseline performance on his tasks were also generally lower in academic achievement. Dallenbach realized that, in order to determine whether the differential training performance was a function of baseline performance, or of academic standing, he would need to collect additional data that was not affected by those confounding variables. He did exactly that and embarked on one of the earliest recorded investigations specifically focused on individual difference factors in cognitive training research (Dallenbach 1919).

We include this example in our chapter, written on the 100th anniversary of the publication of Dallenbach's study, to point out that individual differences have been considered an important moderator of cognitive training since the beginning of research in this domain. We are not the only group who has noted this: Karbach et al. (2017) recently discussed the long history of aptitude by treatment interaction (ATI) research that extends back to the middle part of the twentieth century. For example, in a well-cited book from 1977, Cronbach and Snow discussed how the relationship between intervention task demands and learner aptitudes may be highly complex (1977). Yet the number of studies that systematically examine how variables such as baseline performance, age, motivation, personality, and other factors that might influence the outcome of training, in studies well-powered enough to do so, remains fairly limited. Cronbach and Snow's summary of ATI research in the 1970s remains applicable to individual differences in cognitive training work as well: "The studies are relatively expensive, how to conduct them is unclear, and substantive ideas that could guide such research are little better than speculative" (1977, p. xxx). Numerous individual difference variables may impact the outcome of an intervention by themselves, but may also interact with each other. In some cases, these factors might be responsible for the divergent results across similar training studies.

If some consensus has formed around whether cognitive training "works," it can likely be summed up as follows: most participants are able to improve on training tasks, and small transfer effects are often found on measures similar in nature to those used in the training (see also Guye et al., Rueda et al., Karbach and Kray, Könen et al., Umanath et al., this volume). There is less consistent evidence for far transfer on measures distal to the training, and less still for real-world outcomes such as academic achievement (see also Johann and Karbach, Falkenstein and Gajewski, Schaeffner et al., this volume). However, even meta-analytic studies differ in their conclusions regarding transfer, depending on the types of training tasks used and the parameters of each individual study. In our chapter in the first edition of this book (Strobach and Karbach 2016), we asked whether examining individual difference factors might be one way of understanding the divergent conclusions reached by the existing meta-analytic work (Soveri et al. 2017; Melby-Lervåg et al. 2016; Schwaighofer et al. 2015; Au et al. 2014; Karbach and Verhaeghen 2014; Melby-Lervåg and Hulme 2013). Three years following the publication of the first edition, we again offer an overview of individual differences and motivational factors, now updated with many studies published since then. While there continues to be a debate that aims to answer the seemingly simple yes-or-no question regarding whether cognitive training works, an answer in either direction from any individual study, review, or meta-analysis would not do much more than dismiss a substantial body of research that is not in line with the verdict, whatever it might be. We suggest that the consideration of individual differences provides an opportunity to investigate the effects of cognitive training in a more nuanced way. We also believe there is growing agreement that individual difference factors are worth considering in the context of this research (Green et al. 2019; Redick 2019; see also Cochrane and Green, Könen et al., this volume).

In this chapter, we focus on transfer outcomes from executive function and working memory (WM) interventions in typically developing populations (see also Karbach and Kray, Könen et al., this volume). We also discuss performance on the training itself, which is often relevant given that the level of training improvements is sometimes related to transfer gains (Jaeggi et al. 2011; Jones et al. 2020). We will discuss the roles of baseline performance, age, personality, motivational factors, and socio-economic status as individual difference factors, but without a doubt there are other factors that may be considered, including culture, sex, race, and many others. Our selected topics reflect a compromise between those factors that seem most likely to exert an influence on the outcome of cognitive training and those that have been reasonably well covered in published work. We conclude with recommendations that might improve the examination of individual differences within cognitive training research, with the hope of promoting more research in this domain. Work that systematically investigates individual difference factors accomplishes two goals, after all: not only it might help elucidate why cognitive training works only for some individuals in some studies but it also may serve to improve the development and implementation of future training programs.

Baseline Performance

The relation between one's initial performance on a cognitive measure and training gains or transfer performance remains one of the most studied individual difference factors in cognitive training research (see Verhaeghen et al. 1992, Willis 1989, Ackerman 1986, Snow 1991, and also Dallenbach 1919, as discussed earlier; Karbach and Kray, Könen et al., this volume). This is unsurprising, given that investigating the role of initial performance (on both the training and transfer tasks) serves to improve our understanding of both how cognitive training works for whom it might be most useful for whom. For example, consider the possible explanations for why better initial performance might be associated with better outcomes in an individual study. One possibility is that individuals with better initial performance experience greater gains (sometimes referred to as magnification effects); that is, people who are better at some cognitive task are also better at learning that task (or learning in general) and thus more likely to demonstrate improvements (see Lövdén et al. (2012) and also Borella et al. (2017) for a recent discussion in the context of a WM training study). Another possibility is that lower performing individuals may simply need more practice at easier levels of the task or scaffolding to enable improvement. The same reasoning may be used if *lower* baseline performance is associated with greater training gains (sometimes called compensation effects). This scenario might be due to ceiling effects or simply the fact that the training or transfer task doesn't offer a high enough level of difficulty. However, the cause of this could also be "regression to the mean." This does not mean, in this scenario, that there isn't a real possibility that lower-performers actually benefit more from the training than higher-performers – just that these alternative explanations need to be fully investigated. To improve the methodology of training studies, researchers should focus on understanding *why* a particular association is observed in any individual study. For example, if those with better initial performance actually experience less transfer, whether it is "regression to the mean" or a function of difficulty level on the task becomes quite important. The former may indicate that there has been no real improvement following training, while the latter may mean that, although improvements in some untrained skill did occur, the measures might need to be adjusted in order to observe them.

An additional challenge in examining the effects of baseline performance on the outcome of training (as with all individual difference factors) is that these effects may themselves be associated with other factors, such as age, education, or training compliance, that may or may not be examined in any specific study (Jaeggi et al. 2014). It is worth remembering that there is likely not a *consistent* effect of baseline performance across the wide field of cognitive training work. The role of initial skill in the outcome of any individual study is likely dependent on the design of the training and the measures used.

Given this, it should be unsurprising that extant research remains conflicted on whether baseline performance is positively or negatively associated with training gains. While two studies by Zinke et al. (2012, 2014) suggest that those with worse baseline performance improved more on WM and executive control training paradigms, they did not examine how baseline performance on untrained measures influenced transfer improvements. However, a more recent study by Karbach et al. (2017) with a sample of children, young adults, and older adults completing an executive control training paradigm also found that individuals in the training group with lower cognitive abilities at baseline improved more on the training, as well as on transfer tasks. Additionally, a study by Carretti and colleagues with children found that individuals with the lowest baseline performance on a WM training task showed greater improvements at posttest and at follow-up (Carretti et al. 2017). Furthermore, in some studies the extent of improvements on the training task itself has been found to be associated with transfer to untrained measures of fluid intelligence (Jaeggi et al. 2011).

Lower initial performance has not always been associated with higher training performance over the course of an intervention, however. One study using the dual n-back task –where the ceiling on task difficulty is likely very high – found that higher baseline task performance was associated with *higher* training task performance following the training (Rhodes and Katz 2017). Others using similar WM tasks have also found that higher baseline WM performance was associated with greater training gains (Foster et al. 2017; Wiemers et al. 2019). We also note that another recent study, using latent growth curve modeling (see also Könen and Auerswald, this volume), found that lower baseline cognitive performance was actually associated with lower gains on a WM training intervention, particularly within younger adults (Guye et al. 2017).

There are relatively few experiments that have specifically examined how baseline performance on untrained transfer measures might be related to transfer gain following training. Research from three larger-scale training datasets has found that those with lower pretest performance may also experience greater gains on the transfer tasks following training (in addition to Karbach et al. 2017; mentioned above, Hardy et al. 2015; Willis and Caskie 2013). Two other recent studies – in one case using executive function training and language assessments as transfer measures and in the other kindergarten-aged children completing WM training and using math achievement transfer measures – also suggest that lower achieving individuals stand to gain the most on untrained measures following an intervention (Ramani et al. 2017; Wang et al. 2019). Again, it is sometimes difficult to tell if these findings are related to ceiling effects or factors related to the design of the intervention (such as adaptivity).

Whether compensation or magnification effects are found in any individual study is potentially related to the precise design of the individual training paradigm– for example, magnification effects may be more likely if the training design enables the use of strategies by higher performing individuals – as well as other factors, such as age (discussed below). One clear takeaway from this work, however, is that researchers should carefully consider (a) whether ceiling effects exist in both training performance and untrained measures and (b) whether transfer measures are sensitive enough to detect improvements, especially at higher performance levels.

Age

Given the current focus on mitigating age-related cognitive decline, there is a substantial reason to develop interventions for older adults who may benefit from cognitive training (Richmond et al. 2011). We note that age effects continue to be examined largely in older adult versus younger adult (generally college-age) populations. With the exception of Borella et al. (2014), which compared old–old adults to young–old adults, most training studies do not examine differences within age bands, or in comparison to middle-aged adults. Middle-aged populations (such as those individuals between 25 and 60 years of age) remain understudied across cognitive psychology research and within the cognitive training subfield.

Researchers have long known that age is closely linked to baseline cognitive performance on a sizeable set of executive function-related tasks (Salthouse 1996), but there is also a growing body of evidence that supports age-related differences in cognitive plasticity (e.g., Guye et al., Karbach and Kray, Rueda et al., Umanath et al., Wenger et al., this volume). As with baseline performance, however, the research on how age impacts one's likelihood of experiencing training improvements is somewhat mixed. One study covered earlier, using dual n-back training in a lifespan sample (Rhodes and Katz 2017), found that age exerted an effect on training performance such that older adults experienced more gradual improvements, and peaked at lower levels of performance, even after accounting for initial WM performance. However, another recent WM training study from Borella et al. (2014) with a sample of older adults found that the oldest individuals actually experienced *greater* gains in training performance than the young–old adults, although all participants were aged 60 or older. Given that this study also found that higher baseline performance on crystallized intelligence measures was associated with *lower* training progression, this highlights the importance of considering baseline performance together with any examination of age, as the two are likely closely interrelated.

In terms of transfer, studies from Brehmer et al. (2012), Schmiedek et al. (2010), and Zinke et al. (2014) have found that older adults experienced smaller pre/post gains than younger ones. However, Karbach et al. (2017) found that children and older adults actually experienced more transfer gain relative to young adults following an executive control intervention, as did earlier work from Bherer et al. (2008), Cepeda et al. (2001), Karbach and Kray (2009), Kramer et al. (1995), Kray et al. (2008), and Kray and Lindenberger (2000). Meta-analytic work in this space is similarly mixed: Karbach and Verhaeghen (2014) did not identify any differences in transfer improvements between older and younger individuals, while a meta-analysis by Wass et al. (2012) found that younger adults improved more on untrained measures than older ones. We note that these meta-analyses differ in several ways from each other, with the inclusion criteria that define the range of sample ages being one key disparity.

Motivational Factors

Motivational factors refer to an umbrella of related constructs that might influence the outcome of a cognitive training program. These can include components such as expectancy (or a participant's belief that a given intervention might impact a particular outcome) or the use of extrinsic reward for participation in training. These factors may also be linked to other motivation-adjacent factors that could influence performance on training or transfer batteries, such as cognitive fatigue (see Green et al. 2019). Thus we note that this review of motivational factors is not meant to be comprehensive. Here we discuss three constructs that have received some focus in training research: expectancy effects, extrinsic reward, and game-like elements incorporated in training.

Many training studies tell study participants during recruitment that the training might improve cognition (e.g., Jaeggi et al. 2008; Klingberg et al. 2005). It is also questionable what can be gained from concealment or misleading of participants as to the goal of the study (or even if concealment is possible over a long-term study), particularly if an active control condition is included in the study design (such as in Katz et al. 2018a). Nevertheless, to explore potential expectancy effects in training, a small set of studies have limited the provided instruction to language around "practicing computerized tasks" rather than focus on an improvement outcome (e.g., Redick et al. 2013).

Evidence from one such study suggests that beliefs about the malleability of intelligence may influence the level of transfer demonstrated following training (Jaeggi et al. 2014). Participants who thought that cognitive ability could be improved demonstrated greater transfer gains after the training paradigm than those who thought cognition was more fixed. While the interaction effect of intervention (WM training versus knowledge trainer control) was not significant, this does provide preliminary evidence that these beliefs may influence transfer. Another study of expectancy effects by Foroughi et al. (2016) found that, after just one session of n-back "training," transfer was greater among a group of participants recruited with brain-improvement messaging versus those recruited without an improvement focus. However, a multisession study of expectancy effects found that participants improved on a near-transfer untrained task regardless of an expectancy manipulation (Tsai et al. 2018). In this study, participants viewed a recorded presentation designed to induce expectancy for transfer either between different domains, or within only the domain trained, followed by seven sessions of either adaptive n-back training or a knowledge trainer program (for the active control condition). Regardless of the type of presentation viewed, the individuals who completed the n-back training improved on an untrained n-back near transfer measure, while the individuals within the active control group, regardless of the type of presentation viewed, did not. In aggregate, this work suggests that expectancy effects, while potentially influential in the short-term, may not be as influential in longer-term interventions.

Payment and reward – as a proxy for extrinsic motivation – have also been theorized to influence the outcome of training. One recent study found that while levels of compensation –approximately $350 versus a nominal reward – were linked to baseline differences on a fluid intelligence composite and some thinking disposition measures, they were not linked to improvements on untrained tasks (Katz et al. 2018a). Also, a meta-analysis has examined this in the context of WM training and found that greater amounts of remuneration were related to lower transfer improvements following training, but this finding did not survive the removal of outliers (Au et al. 2014).

A participant's effort on a particular training paradigm might also be related to the outcome of an intervention. If improvement on a training program is necessary to demonstrate gains on untrained tasks, a certain level of effort or engagement may be required on the training task itself. Thus, it is unsurprising that many studies include features thought to drive engagement, such as a points system, game-like feedback, and game-like themes and animations (for example, Jaeggi et al. 2011; Klingberg et al. 2005). One study looking at a brief WM training paradigm for children suggests that the addition of game-like elements – particularly those that might distract from moment-to-moment training, like a persistently updating score –may actually reduce training and transfer performance (Katz et al. 2014). However, some of us have argued that if implemented thoughtfully (i.e., by avoiding factors that distract from the core elements of the task), the implementation of game-like elements can have beneficial effects supporting engagement and enjoyment, and potentially training performance, if not necessarily transfer (Mohammed et al. 2017; Deveau et al. 2015). Careful, measured consideration should be given to including these elements in training: "gamification" is likely not a panacea for improving cognitive training outcomes. Finally, we note that while we are focused on executive function or WM training in typically developing populations within the current chapter, there is a growing body of research with specific populations, such as individuals with a diagnosis of ADHD, that has also examined the role of these motivational elements within cognitive training (for example, see Prins et al. 2011; see also Johann and Karbach, deVries and Geurts, this volume).

Personality

Another salient question is whether personality factors may be related to the outcome of cognitive training. Of the five-factor personality inventory (Hendriks et al. 1999; openness, narcissism, conscientiousness, agreeableness, extraversion), there is some evidence that conscientiousness in particular may be related to transfer improvements. Studer-Luethi et al. (2012) found that individuals with higher levels of conscientiousness had greater improvements on near-transfer measures but also, somewhat surprisingly, lower levels of transfer on far-transfer measures following a cognitive training paradigm. Studer-Luethi and colleagues posit that this may be because individuals with higher levels of conscientiousness may develop nontransferable, task-specific skills that facilitated success on both the training task itself as well as closely related near-transfer measures. However, when faced with far transfer measures, it may be that highly conscientious individuals are negatively impacted by higher evaluation apprehension that could actually impair performance on these more distal tests. As with many cognitive training studies, this experiment utilized a fairly limited sample, and thus any conclusions are preliminary.

Another study by Studer-Luethi et al. (2016) suggests that another construct related to conscientiousness, effortful control, may also be related to training outcomes. Effortful control refers to a participant's ability to self-regulate behavior and emotion in the context of current and future goals. Their training (consisting of a single n-back task and a WM span task) study found that higher ratings of effortful control, and lower ratings of neuroticism, were predictive of transfer effects (Studer-Luethi et al. 2016). This provides some evidence that self-regulation may be important, especially in children, in facilitating transfer gains. Additionally, Urbánek and Marček (2016) found that participants in one training group who scored higher on the rhapsodic scale of the Personality Styles and Disorders Inventory (Urbánek and Marček 2016), were less likely to experience gains on transfer following training. Together, these findings suggest that emotional regulation and processing style might be important factors in the outcome of cognitive training, particularly in certain populations where they might be linked to compliance in the intervention.

Other recent studies that examined the association of conscientiousness with training, however, have failed to find a consistent association between this personality factor and the outcome of executive function or WM training and transfer. For example, Thompson et al. (2013) did not find an association between conscientiousness and training performance or transfer within an n-back training group. Another WM training study failed to establish links between both conscientiousness and neuroticism and change in training performance over the course of an intervention (Guye et al. 2017). Additional research is needed to investigate whether these or other personality factors might be linked to training outcomes. For example, openness to experience has itself been improved as an *outcome* following cognitive training (Jackson et al. 2012), but the underlying mechanisms supporting this improvement remain largely unexplored.

Socioeconomic Status

Socioeconomic status (SES) refers to a construct that incorporates a variety of measures, including parental education, income, and workplace attainment, that have also shown to be associated with executive function and WM (Hackman and Farah 2009). In general, cross-sectional studies suggest that lower socioeconomic status is also associated with lower performance on measures of executive function. Researchers in this area (such as Hackman and Farah 2009) strongly promote the inclusion of socioeconomic indicators in studies so that they might be, at the least, controlled as confounding variables. Researchers have also suggested that training paradigms might be used to address SES-related disparities in cognitive functioning (Raizada and Kishiyama 2010). Given that higher-SES individuals may have more access to cognitively enriching experiences and technology, it is possible that they might be more likely to receive benefit from such interventions (and, indeed, may even be more likely to have access to interventions at all). However, it is also possible that individuals from lower-SES backgrounds might have more room to improve from such interventions.

While some work has established that executive function training may be able to improve untrained measures of cognitive function and academic achievement for low-SES individuals (Goldin et al. 2014), there are only a small number of studies examining *variable* levels of SES in the context of cognitive training. One recent study examined the influence of SES-related factors in two studies with adolescents and an executive function training program, one with school-level free/reduced-price lunch and another with school-level and individual free/reduced-price lunch status (Katz and Shah 2017). In this study, while the SES variable was associated with the amount of improvement on untrained executive function tasks following training, such that greater improvement on the outcome measures was associated with higher-SES, there was no interaction between SES and condition (Katz and Shah 2017). This suggests that there were no strong SES-related differences in how participants benefited from the intervention. However, another study, albeit with a younger population using the *Tools of the Mind* program, found that many transfer effects were limited to individuals from lower-SES schools (Blair and Raver 2014). In yet another study with Argentinean children, not only were SES-related factors such as having a dual-parent household or having a parent with a better occupational background associated with higher executive function measures at baseline, but higher ratings on these factors as well as better housing conditions, were associated with higher improvement trajectories on certain cognitive performance measures, such as WM (Segretin et al. 2014).

Given the significant differences between these training programs (such as age of population and in-person versus computer-based training), as well as the possible interactions between SES and baseline performance, it is difficult to draw strong conclusions about the role of SES on the outcome of cognitive training. While there is little consistency in how SES acts on the outcome of training across these three studies, Segretin et al. (2014) does highlight how SES itself is not a single unified construct, but rather refers to a set of experiences, environmental factors, and financial constraints that may exert influence on how someone responds to a given cognitive training intervention. Like Hackman and Farah (2009), we agree that it is important to include SES variables in a cognitive training study but also that it is necessary to carefully consider which SES-related indicators might be most relevant to a given intervention.

Improving the Study of Individual Difference Factors in Cognitive Training Research

It is perhaps an understatement to say that the association between individual difference factors and the outcome of cognitive training is complex. Each factor may exert independent effects on training and transfer measures, but may also interact with one or more other individual difference factors in influencing results. Thus researchers may find themselves in a bit of a bind while approaching this work: substantial time and effort is required to conduct a cognitive training study and small sample sizes are common – even today, many studies are conducted with as few as 20 individuals within each condition. While these studies may be powered well enough to include basic mean-difference analyses, they are very likely underpowered to examine the influence of even a few additional individual difference variables, especially if they act as moderating effects.

However, we argue that this issue does not mean one should exclude measuring such variables in training studies. Many of these studies are exploratory in nature and include a variety of secondary baseline measures. Even within clinical work, in early phases, there is a need to account for and closely examine such factors through exploratory analyses. Furthermore preregistered research, using platforms such as that provided through the Open Science Foundation (Munafò et al. 2017) may include planned analyses with exploratory, secondary measures. Researchers should also be up-front about whether they intended to examine them when disseminating their work. Recent consensus statements (such as Green et al. 2019) do not suggest eliminating exploratory analyses, but rather appropriately defining which analyses are established a priori versus those chosen post hoc.

We believe that trying to answer the question of whether cognitive training "works," without also exploring these factors, is probably not the most useful, or most interesting, approach to cognitive training research. Even the "gold-standard" meta-analytic work has not definitively addressed the question of whether these interventions are effective (Pergher et al. 2019), but if more studies include measures of these individual difference variables, it will also be possible to include them in future reviews and meta-analyses. These studies, while not a replacement for adequately powered experiments that systematically investigate one or more of these factors, may nonetheless help to establish which individual difference variables play a meaningful role in the outcome of executive function training work.

Thus researchers should carefully consider how to measure these factors in each study. If possible, validated, reliable measures without problematic ceiling or floor effects are preferred.

Finally, one other point is worth discussing as one makes decisions about how to incorporate individual difference measures within cognitive training research. Some variables, such as baseline performance and SES, are recorded at a single time point and are thus included as a single covariate or moderator within an individual model for analysis. Others, like daily engagement, may be collected repeatedly throughout the course of training. Enjoyment of the task and desire to engage may change in a meaningful way that may also be associated with actual differences in training task performance from session to session. Training research focused on exploring the relation between motivation and the outcome of training should be followed-up with analytical models, using structural equation modeling or multilevel modeling, that allow one to more fully account for intraindividual differences in these factors, and how they might be related to condition or cognitive variables (see Schmiedek, Könen and Auerwald, this volume). Rather than collapsing variables such as training performance or engagement ratings over time, these methods allow researchers to comprehensively explore the development of these variables throughout the course of an intervention. One recent paper provides more detail on how one might approach these sorts of analyses to accomplish this (Könen and Karbach 2015).

Conclusion

There is growing consensus that individual difference factors play an important role in the outcome of cognitive training. Without measuring and examining these variables in a study, it may be that one finds a positive effect (or a negative one) that is the result of some factor other than the training condition. In these scenarios it is difficult to draw appropriate conclusions about the efficacy of an intervention. However, even if these factors are measured carefully and in adequately powered samples, it remains possible that the true extent that certain factors influence the outcome of a study might remain unknown. Furthermore, multicollinearity of individual difference variables is likely, given that many of these factors are often interrelated to some degree. We argue that, despite these concerns, it remains necessary to include these measures in training studies, and to think carefully, before conducting training research, about how they might be examined following the experiment.

By examining individual difference factors, a host of new possibilities may be available for future research. For example, if researchers have confidence that previous experience with technology is related to the outcome of a tablet-based cognitive training paradigm, they might offer a different version of the training program to individuals with less technology experience to train them with the tablet before starting the training. And, while many cognitive training programs adjust difficulty to some degree to match performance, new adaptivity methodologies (see, in particular, SMART design trials), may offer new means of improving their efficacy

(Lei et al. 2012). A multimodal intervention could offer, for example, a new type of task or activity whenever a decline in participant engagement was observed. Other techniques – such as machine learning – might offer new ways to examine multiple individual difference factors together in creating profiles of participants across a variety of dimensions (Rennie et al. 2019).

In our chapter for the first edition of this book (Strobach and Karbach 2016), we referenced how therapists and clinical psychologists are often careful to offer different treatments, or to personalize treatment, based on individual difference factors (Snow 1991). The same is often true of teachers and in medicine. If cognitive training is to "work" in improving a particular untrained outcome measure, we must follow this sort of example in incorporating individual difference factors into our training studies. As we suggested in our initial chapter, there is no "one-size-fits-all" in cognitive training research. Our participants come from a variety of backgrounds and come equipped with different experiences that may influence how they respond to training. By more closely examining these measures, we are only continuing in a long tradition – at least as old as Karl Dallenbach's 1919 study – that has investigated the extent to which they might facilitate improvements in training and transfer.

References

Ackerman, P. L. (1986, September). Skill acquisition, individual differences, and human abilities. *In Proceedings of the Human Factors Society Annual Meeting, 30*, 270–274.

Au, J., Sheehan, E., Tsai, N., Duncan, G. J., Buschkuehl, M., & Jaeggi, S. M. (2014). Improving fluid intelligence with training on working memory: A meta-analysis. *Psychonomic Bulletin & Review, 22*, 366–377.

Bherer, L., Kramer, A. F., Peterson, M. S., Colcombe, S., Erickson, K., & Becic, E. (2008). Transfer effects in task-set cost and dual-task cost after dual-task training in older and younger adults: Further evidence for cognitive plasticity in attentional control in late adulthood. *Experimental Aging Research, 34*, 188–219.

Blair, C., & Raver, C. C. (2014). Closing the achievement gap through modification of neurocognitive and neuroendocrine function: Results from a cluster randomized controlled trial of an innovative approach to the education of children in kindergarten. *PLoS ONE, 9*, e112393.

Borella, E., Carretti, B., Cantarella, A., Riboldi, F., Zavagnin, M., & De Beni, R. (2014). Benefits of training visuospatial working memory in young–old and old–old. *Developmental Psychology, 50*, 714–727.

Borella, E., Carbone, E., Pastore, M., De Beni, R., & Carretti, B. (2017). Working memory training for healthy older adults: The role of individual characteristics in explaining short-and long-term gains. *Frontiers in Human Neuroscience, 11*, 99.

Brehmer, Y., Westerberg, H., & Bäckman, L. (2012). Working-memory training in younger and older adults: training gains, transfer, and maintenance. *Frontiers in Human Neuroscience, 6*, 63.

Carretti, B., Borella, E., Elosúa, M. R., Gómez-Veiga, I., & García-Madruga, J. A. (2017). Improvements in reading comprehension performance after a training program focusing on executive processes of working memory. *Journal of Cognitive Enhancement, 1*, 268–279.

Cepeda, N. J., Kramer, A. F., & Gonzalez de Sather, J. (2001). Changes in executive control across the life span: Examination of task-switching performance. *Developmental Psychology, 37*, 715.

Cronbach, L. J., & Snow, R. E. (1977). *Aptitudes and instructional methods: A handbook for research on interactions*. Oxford, England: Irvington.

Dallenbach, K. M. (1914). The effect of practice upon visual apprehension in school children. *Part I. Journal of Educational Psychology, 5*, 321–334.
Dallenbach, K. M. (1919). The effect of practice upon visual apprehension in the feeble-minded. *Journal of Educational Psychology, 10*, 61–82.
Deveau, J., Jaeggi, S. M., Zordan, V., Phung, C., & Seitz, A. R. (2015). How to build better memory training games. *Frontiers in Systems Neuroscience, 8*, 1662–5137.
Foroughi, C. K., Monfort, S. S., Paczynski, M., McKnight, P. E., & Greenwood, P. M. (2016). Placebo effects in cognitive training. *Proceedings of the National Academy of Sciences, 113*, 7470–7474.
Foster, J. L., Harrison, T. L., Hicks, K. L., Draheim, C., Redick, T. S., & Engle, R. W. (2017). Do the effects of working memory training depend on baseline ability level?. *Journal of Experimental Psychology: Learning, Memory, and Cognition, 43*, 1677–1689.
Goldin, A. P., Hermida, M. J., Shalom, D. E., Costa, M. E., Lopez-Rosenfeld, M., Segretin, M. S., ... & Sigman, M. (2014). Far transfer to language and math of a short software-based gaming intervention. *Proceedings of the National Academy of Sciences, 111*, 6443–6448.
Green, C. S., Bavelier, D., Kramer, A. F., Vinogradov, S., Ansorge, U., Ball, K. K., & Facoetti, A. (2019). Improving methodological standards in behavioral interventions for cognitive enhancement. *Journal of Cognitive Enhancement, 3*, 2–29.
Guye, S., De Simoni, C., & Von Bastian, C. C. (2017). Do individual differences predict change in cognitive training performance? A latent growth curve modeling approach. *Journal of Cognitive Enhancement, 1*, 374–393.
Hackman, D. A., & Farah, M. J. (2009). Socioeconomic status and the developing brain. *Trends in Cognitive Sciences, 13*, 65–73.
Hardy, J. L., Nelson, R. A., Thomason, M. E., Sternberg, D. A., Katovich, K., Farzin, F., & Scanlon, M. (2015). Enhancing cognitive abilities with comprehensive training: a large, online, randomized, active-controlled trial. *PloS one, 10*, e0134467.
Hendriks, A. J., Hofstee, W. K., & De Raad, B. (1999). The five-factor personality inventory (FFPI). *Personality and individual differences, 27*, 307–325.
Jackson, J. J., Hill, P. L., Payne, B. R., Roberts, B. W., & Stine-Morrow, E. A. (2012). Can an old dog learn (and want to experience) new tricks? Cognitive training increases openness to experience in older adults. *Psychology and aging, 27*, 286–292.
Jaeggi, S. M., Buschkuehl, M., Jonides, J., & Perrig, W. J. (2008). Improving fluid intelligence with training on working memory. *Proceedings of the National Academy of Sciences, 105*, 6829–6833.
Jaeggi, S. M., Buschkuehl, M., Jonides, J., & Shah, P. (2011). Short-and long-term benefits of cognitive training. *Proceedings of the National Academy of Sciences, 108*, 10081–10086.
Jaeggi, S. M., Buschkuehl, M., Shah, P., & Jonides, J. (2014). The role of individual differences in cognitive training and transfer. *Memory & Cognition, 42*, 464–480.
Jones, M. R., Katz, B., Buschkuehl, M., Jaeggi, S. M., & Shah, P. (2020). Exploring N-Back cognitive training for children with ADHD. *Journal of Attention Disorders, 24*, 704–719.
Karbach, J., & Kray, J. (2009). How useful is executive control training? Age differences in near and far transfer of task-switching training. *Developmental Science, 12*, 978–990.
Karbach, J., & Verhaeghen, P. (2014). Making working memory work a meta-analysis of executive-control and working memory training in older adults. *Psychological Science, 25*, 2027–2037.
Karbach, J., Könen, T., & Spengler, M. (2017). Who benefits the most? Individual differences in the transfer of executive control training across the lifespan. *Journal of Cognitive Enhancement, 1*, 394–405.
Katz, B., & Shah, P. (2017). The role of child socioeconomic status in cognitive training outcomes. *Journal of Applied Developmental Psychology, 53*, 139–150.
Katz, B., Jaeggi, S., Buschkuehl, M., Stegman, A., & Shah, P. (2014). Differential effect of motivational features on training improvements in school-based cognitive training. *Frontiers in Human Neuroscience, 8*, 242.

Katz, B., Jaeggi, S. M., Buschkuehl, M., Shah, P., & Jonides, J. (2018a). The effect of monetary compensation on cognitive training outcomes. *Learning and Motivation, 63*, 77–90.

Katz, B., Shah, P., & Meyer, D. E. (2018b). How to play 20 questions with nature and lose: Reflections on 100 years of brain-training research. *Proceedings of the National Academy of Sciences, 115*, 9897–9904.

Klingberg, T., Fernell, E., Olesen, P. J., Johnson, M., Gustafsson, P., Dahlström, K., Gillberg, C., Forssberg, H., & Westerberg, H. (2005). Computerized training of working memory in children with ADHD-a randomized, controlled trial. *Journal of the American Academy of Child & Adolescent Psychiatry, 44*, 177–186.

Könen, T., & Karbach, J. (2015). The benefits of looking at intraindividual dynamics in cognitive training data. *Frontiers in Psychology, 6*, 615.

Kramer, A. F., Larish, J. F., & Strayer, D. L. (1995). Training for attentional control in dual task settings: A comparison of young and old adults. *Journal of Experimental Psychology: Applied, 1*, 50–76.

Kray, J., & Lindenberger, U. (2000). Adult age differences in task switching. *Psychology and Aging, 15*, 126–147.

Kray, J., Eber, J., & Karbach, J. (2008). Verbal self-instructions in task switching: A compensatory tool for action-control deficits in childhood and old age? *Developmental Science, 11*, 223–236.

Lei, H., Nahum-Shani, I., Lynch, K., Oslin, D., & Murphy, S. A. (2012). A "SMART" design for building individualized treatment sequences. *Annual Review of Clinical Psychology, 8*, 21–48.

Lövdén, M., Brehmer, Y., Li, S. C., & Lindenberger, U. (2012). Training-induced compensation versus magnification of individual differences in memory performance. *Frontiers in Human Neuroscience, 6*, 141.

Melby-Lervåg, M., & Hulme, C. (2013). Is working memory training effective? A meta-analytic review. *Developmental Psychology, 49*, 270–291.

Melby-Lervåg, M., Redick, T. S., & Hulme, C. (2016). Working memory training does not improve performance on measures of intelligence or other measures of "far transfer" evidence from a meta-analytic review. *Perspectives on Psychological Science, 11*, 512–534.

Mohammed, S., Flores, L., Deveau, J., Hoffing, R. C., Phung, C., Parlett, C. M., ... & Zordan, V. (2017). The benefits and challenges of implementing motivational features to boost cognitive training outcome. *Journal of Cognitive Enhancement, 1*, 491–507.

Munafò, M. R., Nosek, B. A., Bishop, D. V., Button, K. S., Chambers, C. D., Du Sert, N. P., Simonsohn, U., Wagenmakers, E.-J., Ware, J. J., & Ioannidis, J. P. (2017). A manifesto for reproducible science. *Nature Human Behaviour, 1*, 0021.

Pergher, V., Shalchy, M. A., Pahor, A., Van Hulle, M. M., Jaeggi, S. M., & Seitz, A. R. (2019). Divergent research methods limit understanding of working memory training. *Journal of Cognitive Enhancement*. https://doi.org/10.1007/s41465-019-00134-7.

Prins, P. J., Dovis, S., Ponsioen, A., Ten Brink, E., & Van Der Oord, S. (2011). Does computerized working memory training with game elements enhance motivation and training efficacy in children with ADHD? *Cyberpsychology, Behavior and Social Networking, 14*, 115–122.

Raizada, R. D., & Kishiyama, M. M. (2010). Effects of socioeconomic status on brain development, and how cognitive neuroscience may contribute to leveling the playing field. *Frontiers in Human Neuroscience, 4*, 3.

Ramani, G. B., Jaeggi, S. M., Daubert, E. N., & Buschkuehl, M. (2017). Domain-specific and domain-general training to improve Kindergarten children's mathematics. *Journal of Numerical Cognition, 3*, 468–495.

Redick, T. S. (2019). The hype cycle of working memory training. *Current Directions in Psychological Science*. https://doi.org/10.1177/0963721419848668.

Redick, T. S., Shipstead, Z., Harrison, T. L., Hicks, K. L., Fried, D. E., Hambrick, D. Z., ... & Engle, R. W. (2013). No evidence of intelligence improvement after working memory training: a randomized, placebo-controlled study. *Journal of Experimental Psychology: General, 142*, 359–379.

Rennie, J. P., Zhang, M., Hawkins, E., Bathelt, J., & Astle, D. E. (2019). Mapping differential responses to cognitive training using machine learning. *Developmental Science*, e12868.

Rhodes, R. E., & Katz, B. (2017). Working memory plasticity and aging. *Psychology and Aging, 32*, 51–79.

Richmond, L. L., Morrison, A. B., Chein, J. M., & Olson, I. R. (2011). Working memory training and transfer in older adults. *Psychology and Aging, 26*, 813–822.

Salthouse, T. A. (1996). The processing-speed theory of adult age differences in cognition. *Psychological Review, 103*, 403–428.

Schmiedek, F., Lövdén, M., & Lindenberger, U. (2010). Hundred days of cognitive training enhance broad cognitive abilities in adulthood: Findings from the COGITO study. *Frontiers in Aging Neuroscience, 2*, 27.

Schwaighofer, M., Fischer, F., & Bühner, M. (2015). Does working memory training transfer? A meta-analysis including training conditions as moderators. *Educational Psychologist, 50*, 138–166.

Segretin, M. S., Lipina, S. J., Hermida, M. J., Sheffield, T. D., Nelson, J. M., Espy, K. A., & Colombo, J. A. (2014). Predictors of cognitive enhancement after training in preschoolers from diverse socioeconomic backgrounds. *Frontiers in Psychology, 5*, 205.

Snow, R. E. (1991). Aptitude-treatment interaction as a framework for research on individual differences in psychotherapy. *Journal of Consulting and Clinical Psychology, 59*, 205–216.

Soveri, A., Antfolk, J., Karlsson, L., Salo, B., & Laine, M. (2017). Working memory training revisited: A multi-level meta-analysis of n-back training studies. *Psychonomic Bulletin & Review, 24*, 1077–1096.

Strobach, T., & Karbach, J. (2016). *Cognitive training: An overview of features and applications*. New York: Springer. https://doi.org/10.1177/0963721419848668.

Studer-Luethi, B., Jaeggi, S. M., Buschkuehl, M., & Perrig, W. J. (2012). Influence of neuroticism and conscientiousness on working memory training outcome. *Personality and Individual Differences, 53*, 44–49.

Studer-Luethi, B., Bauer, C., & Perrig, W. J. (2016). Working memory training in children: Effectiveness depends on temperament. *Memory & Cognition, 44*, 171–186.

Thompson, T. W., Waskom, M. L., Garel, K. L., Cardenas-Iniguez, C., Reynolds, G. O., Winter, R., Chang, P., Pollard, K., Lala, N., Alvarez, G. A., & Gabrieli, J. D. (2013). Failure of working memory training to enhance cognition or intelligence. *PLoS ONE, 8*, e63614.

Tsai, N., Buschkuehl, M., Kamarsu, S., Shah, P., Jonides, J., & Jaeggi, S. M. (2018). (Un) Great expectations: The role of placebo effects in cognitive training. *Journal of Applied Research in Memory and Cognition, 7*, 564–573.

Urbánek, T., & Marček, V. (2016). Investigating the effectiveness of working memory training in the context of personality systems interaction theory. *Psychological Research Psychologische Forschung, 80*, 877–888.

Verhaeghen, P., Marcoen, A., & Goossens, L. (1992). Improving memory performance in the aged through mnemonic training: a meta-analytic study. *Psychology and Aging, 7*, 242–251.

Wang, C., Jaeggi, S. M., Yang, L., Zhang, T., He, X., Buschkuehl, M., & Zhang, Q. (2019). Narrowing the achievement gap in low-achieving children by targeted executive function training. *Journal of Applied Developmental Psychology, 63*, 87–95.

Wass, S. V., Scerif, G., & Johnson, M. H. (2012). Training attentional control and working memory–Is younger, better? *Developmental Review, 32*, 360–387.

Wiemers, E. A., Redick, T. S., & Morrison, A. B. (2019). The influence of individual differences in cognitive ability on working memory training gains. *Journal of Cognitive Enhancement, 3*, 174–185.

Willis, S. L. (1989). Improvement with cognitive training: Which old dogs learn what tricks? In L. W. Poon, D. C. Rubin, & B. A. Wilson (Eds.), *Everyday cognition in adulthood and late life* (Vol. xii,. 708, pp. 545–569). New York: Cambridge University Press. https://doi.org/10.1017/CBO9780511759390.

Willis, S. L., & Caskie, G. I. (2013). Reasoning training in the active study how much is needed and who benefits? *Journal of Aging and Health, 25*, 43S–64S.

Zinke, K., Zeintl, M., Eschen, A., Herzog, C., & Kliegel, M. (2012). Potentials and limits of plasticity induced by working memory training in old-old age. *Gerontology, 58*, 79–87.
Zinke, K., Zeintl, M., Rose, N. S., Putzmann, J., Pydde, A., & Kliegel, M. (2014). Working memory training and transfer in older adults: Effects of age, baseline performance, and training gains. *Developmental Psychology, 50*, 304–315.

Part II
Plasticity in Different Age Groups

Cognitive Training in Childhood and Adolescence

M. Rosario Rueda, Lina M. Cómbita, and Joan P. Pozuelos

Contents

Abstract In this chapter we discuss the importance of promoting cognitive skills during development. Mental skills are key to learning and socialization and predict success in a wide range of life outcomes. In the past years, a great bulk of research has examined the extent to which cognitive skills can be enhanced through training interventions during development. We present different approaches to training and the results of a sample of studies showing great promise to the goal of promoting children mental capacities. Many studies demonstrate that training cognitive skills, such as working memory, executive attention, and cognitive flexibility, leads to gains in performance of tasks that entail these very same skills (near transfer) and often extend to untrained domains (far transfer). Benefits of training seem to be larger for children with lower initial levels of cognitive skills. Although many questions remain to be answered about individual differences in training susceptibility, neural underpinnings, and generalization of training to life outcomes, we argue about the significance of this research for both school and clinic.

M. R. Rueda (✉) · L. M. Cómbita · J. P. Pozuelos
Department of Experimental Psychology & Center for Research on Mind, Brain, & Behavior (CIMCYC), Universidad de Granada, Granada, Spain
e-mail: rorueda@ugr.es

T. Strobach, J. Karbach (eds.), *Cognitive Training*,
https://doi.org/10.1007/978-3-030-39292-5_9

Introduction: Promoting Cognitive Skills During Development

Fostering the mental and emotional potential of people is an important endeavor of developmental psychologists and other professionals in the field of human development and education. There is great consensus on the idea that a society is in the tracks toward economic and social flourishing when it can provide the means to make the most of cognitive abilities and emotional well-being of its members (Beddington et al. 2008).

Throughout generations of researchers, the question of whether experience and/or environmental factors influence children's cognitive capacities has been fundamental in the study of human development. For many years, this question brought about an intense debate on the extent to which the development of cognitive capacities is determined by genes or else depends on experience, the so-called nature vs. nurture debate. However, scientific advancements in the field of genetics in the past decades hurt this debate to death. Genes are not expressed in a vacuum milieu. Instead, epigenetic research has shown that gene expression gets turned on and off or up and down by environmental factors and that particular nurturing or lack of nurturing conditions during development can impact gene expression permanently in the life of the individual (Zhang and Meaney 2010). This research shows that heritability of cognitive skills is a flawed concept because it does not take into account this complex genes × environment interactions. Thus, the question of whether cognitive skills can be improved by training or not turns out to be a matter of finding the conditions and/or experiences that optimize cognitive development.

In the past decades, there have been an increasing number of studies aiming at examining the impact of training programs in children's cognitive capacities. In the light of the substantial evidence provided by this research, the query of whether it is possible to enhance children's cognitive and emotional capacities becomes a question of what are the most beneficial methods as well as a question of what are the periods of development in which intervention may be more effective.

In this chapter, we first discuss about the importance of early interventions in relation to possible developmental differences in brain plasticity. Next, we present an overview of the multiple studies that have been conducted in the past years to examine training-related gains in diverse cognitive domains through randomized controlled trials. In the last section of the chapter, we discuss the relevance of this line of research for education and clinical practice.

Plasticity During Development

Cognitive training thrives on the lure of the plastic nature of the brain. It is well known that the brain changes in response to experience or environmental stimulation. Many studies have shown the impact of family/school environment on a variety of cognitive skills, including executive attention, working memory, and intelligence, as well as the function and structure of the brain networks supporting them (Hackman et al. 2010). All these skills are crucial to school learning, and their vulnerability to poverty very likely explains the largely documented robust association between low family socioeconomic status and children's poor academic outcomes (see Schaeffner et al., this volume). Yet, the same plastic nature of the cognitive system that enables negative experience to undermine cognitive skills also opens a window to beneficial effects of positive environment and developmental intervention. A large bulk of evidence shows that a good number of factors, from lifestyle (e.g. exercise, sleep, exposure to nature) to intervention and education, cause physiological, structural, and functional changes in the brain, which promote the development and enhancement of cognitive processes (Beddington et al. 2008).

In humans, the development of brain structures underlying superior cognitive abilities shows a developmental trajectory that extends during the first and a large proportion of the second decades of life. Developmental trajectories are not equal for brain regions supporting different cognitive skills. While structures that support more basic perceptual and sensory processes develop earlier in life, structures that support more complex processes (e.g. language, executive functions, social cognition) continue developing during late childhood and adolescence (Shaw et al. 2008). Due to this principle of brain development, the potential for brain plasticity varies over development and across brain regions. Sensitive periods of development (i.e., times during which a neural system is maximally sensitive to environmental influences) have been long believed to run in the first and second years of life. Although this might be the case for sensory systems such as vision, hearing, and aspects of language, sensitive periods for higher cognitive functions that rely on prefrontal regions are thought to stretch late into childhood and adolescence (Rice and Barone 2000). For these reasons, a child's brain is believed to be more plastic than an adult's; however, it is not the case that inputs to the system after the end of sensitive periods can no longer influence cognition. There are examples in different domains of a high degree of plasticity outside the sensitive period. In fact, some authors consider that the protracted development of the neural system constitutes a sustained sensitive period where environmental influences support the fine tuning and shaping of cortical circuits that underlies higher-order cognitive processes (Johnson 2011).

Training of Diverse Cognitive Skills in Childhood and Adolescence

A large bulk of studies have been carried out in the past decade in order to examine the potential benefits of cognitive training programs on the development of cognitive skills and the brain mechanisms that support them. Although studies often differ in methods, length, and intensity of intervention, most of them have targeted cognitive processes that fall under the umbrella of executive functions (EFs), namely working memory (WM), inhibitory control, executive attention, and cognitive flexibility (see Karbach and Kray, Könen et al., Strobach and Schubert, this volume).

The development of EFs enables the top-down coordination and regulation of thoughts, behaviors, and emotions necessary to flexibly adapt to the demands of a changing environment. EFs have been particularly stressed in training research because of the role they play on several aspects of children's and adolescents' development such as social adjustment, academic competence, and mental health (Checa et al. 2008; Rothbart and Posner 2006; Moffitt et al. 2011; see Johann and Karbach, this volume).

Generally, cognitive training refers to programs designed to improve the efficiency of cognitive and brain mechanisms through practice and/or intentional instruction. Most training studies have taken a process-based approach, which consists on training specific cognitive processes by means of practicing with tasks that entail such processes. A different training strategy consists on providing instructions to develop metacognitive knowledge about task relevant procedures (see Schaeffner et al., this volume), an approach that reminds of the Vygotskian concept of scaffolding, or providing information about particular strategies that may enhance task performance (e.g., using visuo-spatial cues to improve memory; Karbach and Unger 2014).

Mostly the effects of cognitive training are studied on the performance of tasks that tap the same process or processes targeted with the intervention (near-transfer) although often effects are also measured in the performance of tasks that engage processes different from, albeit related to, those being trained (far-transfer). For instance, given that EFs are central to the development of higher-order executive functions such as reasoning, problem solving, and planning, several studies have addressed the generalization of EF training to other functions such as fluid intelligence, schooling skills, or the improvement of symptoms in the case of children with ADHD (see also De Vries and Geurts, Johann and Karbach, Katz et al., Könen et al., Schaeffner et al., this volume).

In the following sections, we provide an overview of the empirical evidence derived from studies that have examined the impact of cognitive training in infants, children, and adolescents in the past decade. We mostly cover studies using a process-based approach, a large amount of which used computer-based training programs. Also, we describe other studies that have used noncomputerized programs, some of which used scaffolding or a different type of coaching (e.g., mindfulness; see Verhaeghen, this volume). The evidence reviewed covers behavioral and neuroimaging data of the impact of training programs on three main domains of the EFs in typically developing children and clinical populations.

Working Memory

WM is perhaps the EF domain with the largest amount of training studies (see Katz et al., Könen et al., this volume). Most studies involve the use of computer-based programs aiming at practicing the ability to monitor, update, and manipulate information in memory for short periods of time by performing n-back or memory span tasks with increased levels of difficulty. In a typical n-back exercise, children are presented with sequences of stimuli and their task is to report whenever the current stimulus is similar to the one presented *n* items back in some particular dimension (e.g., location, color, sound, etc.; Jaeggi et al. 2011). Memory span exercises require children to retain a series of visuospatial or verbal stimuli in memory and repeat them after a brief delay either in the same or the reversed order of presentation. Using these types of exercises for training, several studies have demonstrated enhancements of WM capacities in typically developing children (Alloway et al. 2013) as well as in children and adolescents diagnosed with attention-deficit hyperactivity disorder (ADHD; Holmes et al. 2010; Stevens et al. 2016).

Besides near-transfer effects, there is evidence that WM training also translates into significant benefits in different domains of children's lives. With children and adolescents (7–15 year-olds) diagnosed with ADHD, Klingberg and colleagues have shown significant improvements on measures of nonverbal reasoning ability and inhibitory control in trained children compared to an active control group (i.e., children who only performed the initial levels of the training program; Klingberg et al. 2002, 2005). Using the same or similar training protocol, parents of children who received the treatment reported amelioration on the severity of inattentive and impulsivity/hyperactive symptoms exhibited by their children (Klingberg et al. 2005; Stevens et al. 2016; see also de Vries and Geurts, this volume).

Additionally, a few studies have tested the generalization of WM training effects into measures of verbal competence and reading performance in typically developing children (see Johann and Karbach, this volume). In one of such studies, Alloway et al. (2013) reported higher scores on fluid intelligence as well as a significant improvement on measures of verbal competence and spelling following 32 sessions of WM training. Importantly, WM gains and transfer to verbal competence and spelling were still maintained in a follow-up assessment carried out 8 months after the intervention. Transfer of WM training to reading performance has also been reported with shorter interventions (Karbach et al. 2015; Loosli et al. 2012), suggesting that the length of the program may not determine the generalization of WM training to children's reading competence.

Some studies have also explored the neural mechanisms that underlie training-related improvements of WM in children and adolescents. Jolles et al. (2012) found that, after 6 weeks of WM training, children showed significant pre- to posttraining increases of activation in fronto-parietal structures. Likewise, in a different study conducted with adolescents diagnosed with ADHD, it was found that the magnitude of the pre- to posttraining increase of fronto-parietal activation predicted participants' gains in WM following training. More importantly, the observed changes in neural activation were distinctly correlated with the reduction of inattention and hyperactive/impulsive symptoms after training (Stevens et al. 2016). Additionally, there is evidence that intrinsic functional connectivity between these fronto-parietal circuits and other brain regions increase after WM training over several weeks (Astle et al. 2015). Taken together, these data suggest that training may produce a broad impact in the efficacy of activation and communication between distant brain areas involved in maintaining and updating relevant information in memory.

Executive Attention and Inhibitory Control

Because of its involvement in perceptual processing and behavioral regulation, attention is central to most of our activities in daily life. Out of the broad concept of attention, executive attention refers to goal-directed behavior and action regulation and involves processes such as inhibitory control, conflict resolution, and attentional flexibility. Given that executive attention strongly develops during the first years of life (Rueda 2014), many training studies have focused on the behavioral and neural effects of cognitive interventions during the preschool years.

Although the number of studies targeting executive attention processes is still small (see Karbach and Kray, this volume), the evidence that has been gathered in the past decade suggest that these interventions translate into near (Thorell et al. 2008) and far transfer effects, particularly to measures of fluid intelligence (Liu et al. 2015; Rueda et al. 2012). In order to assess the influence of training in the plasticity of brain dynamics, some of these studies have also recorded brain activity measures using electroencephalography (EEG). Results show that training induces enhanced amplitude of attention-related ERP components (Liu et al. 2015) as well as a reduction in latency of brain responses while performing executive attention tasks (Rueda et al. 2005, 2012). Importantly, these effects are still observed 2 months after intervention without further training (Rueda et al. 2012). In a more recent study, it has been shown that training executive attention accompanied by metacognitive scaffolding provided by an adult boosts transfer of training to fluid intelligence in 5-year-old children and that the fluid IQ gain following training is predicted by changes in conflict-related brain activation in the frontal midline (Pozuelos et al. 2019; see also Schaeffner et al., this volume). This indicates that the extent of posttraining changes in the patterns of brain function is related to the generalization of training effects to other cognitive domains. However, additional research covering different age groups is needed in order to characterize the possible differences in training effects along development.

Cognitive Flexibility

Cognitive flexibility is the ability to change the course of action to adapt effectively to the changing demands of a given task or situation. This skill greatly relies on the capacity to update information in WM and implement attentional and behavioral control mechanisms. In fact, developmental studies have shown that among the executive domains, cognitive flexibility emerges later and exhibits a more protracted development, extending to late adolescence (Cepeda et al. 2001).

Most of the training studies on cognitive flexibility have been carried out with groups of school-aged children, usually starting at the age of 7 years, using a variety of exercises based on the classical task-switching paradigm. Switching tasks often involve responding to stimuli according to particular rules, which can change from one trial to the next. For example, a series of numbers are presented, and the participant is asked to indicate whether the number is odd or even (task A) if printed in red ink (cue for task A) or indicate whether the number is larger or smaller than 5 (task B) if printed in blue ink (cue for task B). The task requires flexibly switching between set of rules and adjusting response-mapping representations accordingly.

Although the number of studies is still small, evidence indicates that after switching training, children and adolescents show improvements in cognitive flexibility measures as well as far transfer effects to other cognitive domains. For instance, Karbach and Kray (2009) reported that task-switching training resulted in significant transfer to measures of response inhibition, verbal and spatial WM, and fluid intelligence. Similar results have also been reported in studies that trained children diagnosed with ADHD. Together with improvements on cognitive flexibility, children trained in task switching showed better performance on measures of inhibitory control and verbal WM (Kray et al. 2012) as well as faster choice reaction times and a tendency toward faster responses when performing an updating task (Zinke et al. 2012), compared to children who received different training protocols.

Despite of the small number of studies that have been conducted, the evidence presented here suggests that cognitive flexibility can be enhanced during development and that such beneficial effects translate into the improvement of other cognitive functions. However, given the lack of studies that investigate changes of brain function following switching-based interventions, information necessary to characterize the neural mechanisms that underlie the observed behavioral effects is lacking.

Multidomain Training

In view of the overlapping neuroanatomy of executive functions in the prefrontal cortex, a number of studies have approached cognitive training implementing a multidomain strategy. For example, Wass et al. (2011) studied the influence of a multidomain training protocol based on a number of gazed-contingent exercises that aimed to train executive attention (focused/selective attention, interference resolution, visual search) as well as WM and cognitive flexibility in infants. They found that infants significantly improved their performance on measures of cognitive control, sustained attention, and attentional control following training although no gains were found in WM.

Also, given that children diagnosed with ADHD exhibit cognitive and behavioral symptoms related to the different EF domains (see de Vries and Geurts, this volume), several studies have implemented training protocols that target two or more executive-related processes. In one of the studies, near transfer effects were limited to measures of visuospatial WM and inhibitory control while no significant differences were observed for measures of verbal WM and cognitive flexibility (Dovis et al. 2015). Furthermore, training also led to the amelioration of the frequency and severity of the ADHD symptoms (Johnstone et al. 2012). Transfer between EF tasks in multidomain training is expected given the overlapping neuroanatomy yet further research is needed for a detailed understanding of the neural dynamics underlying training benefits.

Noncomputerized Training Programs

Until now, we have described studies using process-based training interventions mostly based on computerized exercises designed to target specific cognitive functions. However, other studies have examined the effects of interventions implemented in the classroom either as incorporated to the school curricula or as extra-curricular activities performed in the school context.

An example of school curricula that incorporates exercises aimed at increasing EFs is the so-called *Tools of the Mind* program (Bodrova and Leong 2007). This is a program based on Vygotsky's insights into development of high cognitive functions, emphasizing training of EFs through guided social interactions in the classroom. Some studies were able to evaluate the impact of the *Tools* program in children's EF skills during the second year of preschool in comparison to a different curriculum implemented by the school district, which had the same academic content but did not emphasize EFs. Data revealed better performance of children in the *Tools* program in executive control tasks (Barnett et al. 2008), an effect that was bigger in task conditions with higher executive demands (Diamond et al. 2007).

Hermida and colleagues (2015) took a somewhat different approach in a recent study. They trained teachers to include activities to promote executive functions (WM, attention, inhibitory control, and planning) in the classroom and tested both near and far transfer effects of intervention to behavioral tests of EF and academic achievement, respectively. Results failed to show significant differences in EF performance between children in the intervention and control groups; however, they found significant differences in four of the six academic achievement areas evaluated: language and mathematics, autonomy, and socialization with peers. On a similar approach, Neville et al. (2013) implemented a family-based training program for preschoolers from low socioeconomic backgrounds. The program consisted on training sessions for parents combined with attention training for children. They found significant benefits in reducing the stress of parents as well as behavioral, social, and cognitive (language and fluid IQ) improvements in children. Also, children showed better auditory selective attention skills in a task involving brain measurements.

A different approach to promoting self-regulatory skills at school that is generating promising results is mindfulness practice (see Verhaeghen, this volume). Mindfulness is a contemplative exercise that aims at improving the ability to have a nonjudgmental awareness that arises by paying attention to the present moment (Malinowski 2013). In a randomized control study with 7- to 9-year-old children, Flook and colleagues (2010) examined the effects of mindful awareness practice on parent- and teacher-report measures of EF. They reported gains in behavioral regulation, metacognition, and executive control scores after mindfulness training for children who were less regulated before intervention. In this study, improvements were found with both teachers and parent-reported measures, suggesting that benefits of practice in children's regulation generalized across different settings. However, other studies have reported either weak or no effects of meditation on attention and self-regulation; thus further research is needed before reaching grounded conclusions (Goyal et al. 2014).

Overall, these interventions show promising results and point to the importance of incorporating interventions to promote cognitive and self-regulation skills into school curricula (see Alloway, Robinson, and Frankenstein, this volume) and with families. Importantly, stronger effects of interventions are consistently found in children with greater difficulties. This suggests that there exist individual differences in windows of improvement. Knowing whether upper boundaries of these improvement windows depend on the developmental stage of the individual remains a future research question.

Implications of Training for the School and Clinic

Education

Children's academic learning and school adjustment are supported by cognitive abilities such as attention, memory, and intelligence (see also Johann and Karbach, this volume). We know that attention and self-regulation skills are key to school readiness because of their power to predict later achievement in school (Duncan et al. 2007) and many other life outcomes (Moffitt et al. 2011). Age (developmental stage) and constitution (temperament and genes) are two important sources of interindividual variability that are to be taken into account to optimize learning and adjustment in schools. Abundant evidence presents attention as an integral component in the academic success of children. Variability in attentiveness and self-regulation accounts for differences in learning and socio-emotional competencies displayed in the classroom (Checa et al. 2008) as well as learning of curricular contents such as maths (Checa and Rueda 2011) and language (Franceschini et al. 2012). This evidence speaks up for the importance of promoting children's cognitive capacities as part of the educational curricula.

The usefulness of training tools for education will increase to the extent that their development is evidence-based and guided by scientific principles. Hence, the design of training programs must align with known processes of children's learning and cognitive development. Literature in psychological science suggests that children learn best when they are cognitively active and engaged when learning experiences are meaningful and socially interactive, and when learning is guided by a specific goal (Hirsh-Pasek et al. 2015). With the foundation of these learning pillars, psychologists and educators can take a proactive approach to the development and evaluation of intervention tools aimed to enhance children's odds to successful learning and socio-emotional outcomes.

Prevention and Intervention

The understanding of pathophysiological mechanisms of developmental diseases offers a way through which a particular pathology may be changed. The development of efficient treatments is greatly facilitated by knowing the pathological mechanisms of diseases because once pathological mechanisms are identified, they become putative targets of intervention (see Boller et al., this volume).

Comorbidities are common in developmental disorders. For instance, deficits of executive attention appear to underlie both autism spectrum disorder (ASD) and attention deficit hyperactivity disorder (ADHD) (Van Der Meer et al. 2012; see de Vries and Geurts, this volume). In this context, process-based training may constitute a suitable method for disease prevention and treatment. Early intervention to train executive attention in children at risk for developing these disorders may act as a general positive or protective factor such that children with strong executive attention skills have better developmental outcomes. This approach has already proven to be beneficial for children with ADHD. As discussed earlier, several studies have shown that working memory and executive control training in children with ADHD improve performance and increase neural efficiency in relevant brain circuits, although with limited transfer to behavioral symptoms and academic outcomes.

In addition, studying the impact of training on targeted function at the brain and behavior levels, and the subsequent relationship of the training effect to the clinical outcome, facilitates an understanding of mechanisms of action of particular interventions. Importantly, effectiveness of treatment has to be tested with randomized trials including treatment and placebo groups (see Cochrane and Green, Schmiedek, this volume). In such studies, interventions can be considered efficient to the extent that they revert or palliate pathological mechanisms. In turn, information on individual differences in effectiveness has the potential to help building more potent, personalized interventions.

Conclusions and Future Research

One of the greater challenges of modern societies is to find methods to foster children's cognitive and emotional skills. In an increasingly technological world, nurturing mental wealth is a major way to prosper both economically and socially. To accomplish this objective, psychologists and educators must work together to provide individuals with tools that can optimize cognitive skills and prevent or palliate the development of psychopathologies.

Future research will be key for identifying risk factors and behavioral, cognitive, and neural markers of learning difficulties and neurodevelopmental disorders as well as to studying the factors (e.g. genetic, temperamental, etc.; see Colzato and Hommel, this volume) that determine the effectiveness of interventions designed to fight these harmful conditions. Crucially, multidisciplinary longitudinal studies are needed in order to deepen our understanding of the complex processes supporting typical and atypical development and use this knowledge to improve developmental interventions.

Acknowledgments This work was supported by a grant from the Spanish National Research Agency awarded to MMR (rf. PSI2017-82670-P).

References

Alloway, T. P., Bibile, V., & Lau, G. (2013). Computerized working memory training: Can it lead to gains in cognitive skills in students? *Computers in Human Behavior, 29*, 632–638.

Astle, D. E., Barnes, J. J., Baker, K., Colclough, G. L., & Woolrich, M. W. (2015). Cognitive training enhances intrinsic brain connectivity in childhood. *Journal of Neuroscience, 35*, 6277–6283.

Barnett, W. S., Jung, K., Yarosz, D. J., Thomas, J., Hornbeck, A., Stechuk, R., & Burns, S. (2008). Educational effects of the tools of the mind curriculum: A randomized trial. *Early Childhood Research Quarterly, 23*, 299–313.

Beddington, J., Cooper, C. L., Field, J., Goswami, U., Huppert, F. A., Jenkins, R., ... & Thomas, S. M. (2008). The mental wealth of nations. *Nature, 455*, 1057–1060.

Bodrova, E., & Leong, D. J. (2007). Tools of the Mind: The Vygotskian Approach to Early Childhood Education (2nd ed.). New York: Merrill/Prentice-Hall.

Cepeda, N. J., Kramer, A. F., & Gonzalez de Sather, J. C. M. (2001). Changes in executive control across the life span: Examination of task-switching performance. *Developmental Psychology, 37*, 715–730.

Checa, P., & Rueda, M. R. (2011). Behavioral and brain measures of executive attention and school competence in late childhood. *Developmental Neuropsychology, 36*, 1018–1032.

Checa, P., Rodríguez-Bail6n, R., & Rueda, M. R. (2008). Neurocognitive and temperamental systems of self-regulation and early adolescents' social and academic outcomes. *Mind, Brain, and Education, 2*, 177–187.

Diamond, A., Barnett, W. S., Thomas, J., & Munro, S. (2007). Preschool program improves cognitive control. *Science, 318*, 1387–1388.

Dovis, S., Van der Oord, S., Wiers, R. W., & Prins, P. J. M. (2015). Improving executive functioning in children with ADHD: Training multiple executive functions within the context of a computer game. A randomized double-blind placebo controlled trial. *PLoS ONE, 10*, e0121651.

Duncan, G. J., Dowsett, C. J., Claessens, A., Magnuson, K., Huston, A. C., Klebanov, P., ... & Sexton, H. (2007). School readiness and later achievement. *Developmental Psychology, 43*, 1428–1446.

Flook, L., Smalley, S. L., Kitil, M. J., Galla, B. M., Kaiser-Greenland, S., Locke, J., Ishijima, E., Kasari, C. (2010). Effects of mindfull awareness practices on executive functions in elementary school children. *Journal of Applied School Psychology, 26*, 70–95.

Franceschini, S., Gori, S., Ruffino, M., Pedrolli, K., & Facoetti, A. (2012). A causal link between visual spatial attention and reading acquisition. *Current Biology, 22*, 814–819.

Goyal, M., Singh, S., Sibinga, E. M., Gould, N. F., Rowland-Seymour, A., Sharma, R., Berger, Z., Sleicher, D., Maron, D. D., Shihab, H. M., Ranasinghe, P. D., Linn, S., Saha, S., Bass, E. B., & Haythornthwaite, J. A. (2014). Meditation programs for psychological stress and Well-being: A systematic review and meta-analysis. *JAMA Internal Medicine, 174*, 357–368.

Hackman, D. A., Farah, M. J., & Meaney, M. J. (2010). Socioeconomic status and the brain: Mechanistic insights from human and animal research. *Nature Review Neuroscience, 11*, 651–659.

Hermida, M. J., Segretin, M. S., Prats, L. M., Fracchia, C. S., Colombo, J. A., & Lipina, S. J. (2015). Cognitive neuroscience, developmental psychology, and education: Interdisciplinary development of an intervention for low socioeconomic status kindergarten children. *Trends in Neuroscience and Education, 4*, 15–25.

Hirsh-Pasek, K., Zosh, J. M., Golinkoff, R. M., Gray, J. H., Robb, M. B., & Kaufman, J. (2015). Putting education in "educational" apps: Lessons from the science of learning. *Psychological Science in the Public Interest, 16*, 3–34.

Holmes, J., Gathercole, S. E., Place, M., Dunning, D. L., Hilton, K. A., & Elliott, J. G. (2010). Working memory deficits can be overcome: Impacts of training and medication on working memory in children with ADHD. *Applied Cognitive Psychology, 24*, 827–836.

Jaeggi, S. M., Buschkuehl, M., Jonides, J., & Shah, P. (2011). Short- and long-term benefits of cognitive training. *Proceedings of the National Academy of Sciences, 108*, 10081–10086.

Johnson, M. H. (2011). Interactive specialization: A domain-general framework for human functional brain development? *Developmental Cognitive Neuroscience, 1*, 7–21.

Johnstone, S. J., Roodenrys, S., Blackman, R., Johnston, E., Loveday, K., Mantz, S., & Barratt, M. F. (2012). Neurocognitive training for children with and without AD/HD. *Attention Deficit and Hyperactivity Disorders, 4*, 11–23.

Jolles, D. D., van Buchem, M. A., Rombouts, S. A. R. B., & Crone, E. A. (2012). Practice effects in the developing brain: A pilot study. *Developmental Cognitive Neuroscience, 2*, S180–S191.

Karbach, J., & Kray, J. (2009). How useful is executive control training? Age differences in near and far transfer of task-switching training. *Developmental Science, 12*, 978–990.

Karbach, J., & Unger, K. (2014). Executive control training from middle childhood to adolescence. *Frontiers in Psychology, 5*, 390.

Karbach, J., Strobach, T., & Schubert, T. (2015). Adaptive working-memory training benefits reading, but not mathematics in middle childhood. *Child Neuropsychology, 21*, 285–301.

Klingberg, T., Forssberg, H., & Westerberg, H. (2002). Training of working memory in children with ADHD. *Journal of Clinical and Experimental Neuropsychology, 24*, 781–791.

Klingberg, T., Fernell, E., Olesen, P. J., Johnson, M., Gustafsson, P., Dahlström, K., ... & Westerberg, H. (2005). Computerized training of working memory in children with ADHD-a randomized, controlled trial. *Journal of the American Academy of Child and Adolescent Psychiatry, 44*, 177–186.

Kray, J., Karbach, J., Haenig, S., & Freitag, C. (2012). Can task-switching training enhance executive control functioning in children with attention deficit/−hyperactivity disorder? *Frontiers in Human Neuroscience, 5*, 1–9.

Liu, Q., Zhu, X., Ziegler, A., & Shi, J. (2015). The effects of inhibitory control training for preschoolers on reasoning ability and neural activity. *Scientific Reports, 5*, 14200.

Loosli, S. V., Buschkuehl, M., Perrig, W. J., & Jaeggi, S. M. (2012). Working memory training improves reading processes in typically developing children. *Child Neuropsychology, 18*, 62–78.

Malinowski, P. (2013). Neural mechanisms of attentional control in mindfulness meditation. *Frontiers in Neuroscience, 7*: 8.

Moffitt, T. E., Arseneault, L., Belsky, D., Dickson, N., Hancox, R. J., Harrington, H., ... & Sears, M. R. (2011). A gradient of childhood self-control predicts health, wealth, and public safety. *Proceedings of the National Academy of Sciences, 108*, 2693–2698.

Neville, H. J., Stevens, C., Pakulak, E., Bell, T. A., Fanning, J., Klein, S., & Isbell, E. (2013). Family-based training program improves brain function, cognition, and behavior in lower socioeconomic status preschoolers. *Proceedings of the National Academy of Sciences, 110*, 12138–12143.

Pozuelos, J. P., Combita, L. M., Abundis, A., Paz-Alonso, P. M., Conejero, Á., Guerra, S., & Rueda, M. R. (2019). Metacognitive scaffolding boosts cognitive and neural benefits following executive attention training in children. *Developmental Science, 22*, e12756.

Rice, D., & Barone, S. (2000). Critical periods of vulnerability for the developing nervous system: Evidence from humans and animal models. *Environmental Health Perspectives, 108*, 511–533.

http://www.pubmedcentral.nih.gov/articlerender.fcgi?artid=1637807&tool=pmcentrez&rendertype=abstract. Accessed 27 Aug 2015.
Rothbart, M. K., & Posner, M. I. (2006). Temperament, attention, and developmental psychopathology. In D. Cicchetti & D. Cohen (Eds.), *Developmental psychopathology, Vol 2: Developmental neuroscience* (2nd ed., pp. 465–501). Hoboken: John Wiley & Sons.
Rueda, M. R. (2014). Development of attention. *The Oxford Handbook of Cognitive Neuroscience, 1*, 296–318.
Rueda, M. R., Rothbart, M. K., McCandliss, B. D., Saccomanno, L., & Posner, M. I. (2005). Training, maturation, and genetic influences on the development of executive attention. *Proceedings of the National Academy of Sciences, 102*, 14931–14936.
Rueda, M. R., Checa, P., & Cómbita, L. M. (2012). Enhanced efficiency of the executive attention network after training in preschool children: Immediate changes and effects after two months. *Developmental Cognitive Neuroscience, 2*, S192–S204.
Shaw, P., Kabani, N. J., Lerch, J. P., Eckstrand, K., Lenroot, R., Gogtay, N., ... & Giedd, J. N. (2008). Neurodevelopmental trajectories of the human cerebral cortex. *Journal of Neuroscience, 28*, 3586–3594.
Stevens, M. C., Gaynor, A., Bessette, K. L., & Pearlson, G. D. (2016). A preliminary study of the effects of working memory training on brain function. *Brain Imaging and Behavior, 10*, 387–407.
Thorell, L. B., Lindqvist, S., Bergman Nutley, S., Bohlin, G., & Klingberg, T. (2008). Training and transfer effects of executive functions in preschool children. *Developmental Science, 12*, 106–113.
Van Der Meer, J. M., Oerlemans, A. M., Van Steijn, D. J., Lappenschaar, M. G., De Sonneville, L. M., Buitelaar, J. K., & Rommelse, N. N. (2012). Are autism spectrum disorder and attention-deficit/hyperactivity disorder different manifestations of one overarching disorder? Cognitive and symptom evidence from a clinical and population-based sample. *Journal of the American Academy of Child & Adolescent Psychiatry, 51*, 1160–1172.
Wass, S., Porayska-Pomsta, K., & Johnson, M. H. (2011). Training attentional control in infancy. *Current Biology, 21*, 1543–1547.
Zhang, T., & Meaney, M. J. (2010). Epigenetics and environmental regulation of the genome and its function. *Annual Review of Psychology, 61*, 439–466.
Zinke, K., Einert, M., Pfennig, L., & Kliegel, M. (2012). Plasticity of executive control through task switching training in adolescents. *Frontiers in Human Neuroscience, 6*, 41.

Cognitive Training Across the Adult Lifespan

Sabrina Guye, Christina Röcke, Susan Mérillat, Claudia C. von Bastian, and Mike Martin

Contents

Abstract Normal aging is generally associated with deterioration in a number of cognitive abilities, although large individual differences in size and progression of age-related cognitive change exist. Cognitive training interventions have become an increasingly important object of research, aiming at the stabilization and improvement of cognitive abilities in old age. However, training gains tend to be of small to moderate magnitude compared to no training both on the behavioral and the brain level, but are small or disappear when compared to active control conditions. Across the different types of training interventions, mainly near transfer effects of small to moderate size have been documented. To gain further insights into the mechanism and boundaries of cognitive plasticity, we argue that future research should focus on

S. Guye (✉) · C. Röcke · S. Mérillat
University Research Priority Program (URPP) "Dynamics of Healthy Aging", University of Zurich, Zurich, Switzerland
e-mail: sabrina.guye@uzh.ch

C. C. von Bastian
Department of Psychology, University Sheffield, Sheffield, UK

M. Martin
University Research Priority Program (URPP) "Dynamics of Healthy Aging", University of Zurich, Zurich, Switzerland

Department of Psychology, Division of Gerontopsychology and Gerontology and Gerontology Center, University of Zurich, Zurich, Switzerland

T. Strobach, J. Karbach (eds.), *Cognitive Training*,
https://doi.org/10.1007/978-3-030-39292-5_10

investigating more thoroughly the cognitive processes involved in cognitive training, on identifying training contexts that may maximize training and transfer effects, and on quantifying potential impacts on meaningful real-life outcomes.

Introduction

There is robust longitudinal evidence for age-related decline in cognitive abilities. Fluid abilities are affected earlier than crystallized abilities, but with varying onset and slope between individuals (e.g., Salthouse 2010). These negative age-related changes have sparked early interest in the possibility of preventing or counteracting this decline and thus maintaining cognitive health into later life with cognitive training interventions. In this chapter, we review the literature regarding training-induced plasticity in healthy older adults.

Many of the early training interventions focused on improving (episodic) memory ability, given that subjective changes in one's memory functioning are frequently voiced concerns from older adults (see also Wenger et al., this volume). Using a testing-the-limits paradigm, these training interventions typically instructed participants in a specific memory strategy, such as the Method of Loci, trying to uncover the strategy-independent latent performance potential and the boundary conditions for such latent reserve capacity of the aging cognitive system. The second generation of cognitive training interventions consisted of process-based approaches that focused on broader, more basic cognitive processes including working memory (see also Könen et al., this volume) or executive functions (see also Karbach and Kray, this volume). As a special form of process-based training, newer approaches target multiple cognitive domains simultaneously to achieve broader and larger transfer and greater ecological validity.

In the first part of this chapter, we review empirical evidence regarding the benefits of cognitive training interventions in healthy older adults separately for training gains, transfer, and their maintenance, as well as findings regarding brain structure and function. In the second part, we will outline the key points to consider in future research to design more effective training interventions for healthy older adults to help maintain cognitive functioning.

Benefits of Cognitive Training Interventions

Cognitive training studies differ on a multitude of design choices (e.g., type of training and its administration, cognitive domain, setting, intensity and duration, type of control group, and type of outcome measure to assess training effectiveness; see Cochrane and Green, this volume). In addition, the systematic reviews and meta-analyses available also differ substantially in their scope and inclusion criteria, and whether they distinguish between training gain and transfer effects and between different types of control groups. Thus, conclusions from these overview analyses are not straightforward to compare.

Evidence for Training Effects

Training Gains: Passive vs. Active Controls Training effects are typically operationalized as pre to posttraining performance increases on the trained tasks compared to pre to posttraining performance changes in *passive* (i.e., with no instructed activity) or *active* control groups (i.e., with an instructed activity, but clear differentiation in the involved cognitive processes; Shipstead et al. 2012). Findings across different kinds of interventions indicate cognitive plasticity in terms of training gains (e.g., Baltes and Kliegl 1992). For example, in their meta-analysis on process-based working memory and executive functioning training, Karbach and Verhaeghen (2014) reported raw training gains of 0.9 *SD*, which remained almost equal in size when compared to passive controls (0.8 *SD*; see also Kelly et al. 2014 for similar effect sizes in working memory and speed training interventions). Interestingly, however, training gains were found as reduced to 0.5 *SD* (Karbach and Verhaeghen 2014) or even zero (Martin et al. 2011, see also Kelly et al. 2014 for replication) after comparing to active controls. Promising training gains emerge for multidomain training interventions (Park et al. 2014, see also Bediou et al., this volume).

Age-Related Differences in Training Gains In contrast to findings from strategy-based training interventions indicating a magnification of age differences in cognitive performance and limits to training-induced plasticity in the very old (e.g., Verhaeghen and Marcoen 1996), no such age differences in training gains were observed for process-based working memory and executive functioning training interventions (Karbach and Verhaeghen 2014). The implementation of complex cognitive strategies may require a higher level of cognitive functioning than is true for the elementary cognitive processes targeted in process-based training interventions (Verhaeghen 2014). While research concerning multidomain training is still in its infancy, there is initial evidence for greater video game training benefits for older-old compared to younger-old adults, but the underlying reasons are yet poorly understood (see also Bediou et al., Strobach and Schubert, this volume).

Moderators of Training Effectiveness Group-based lab settings show greater effects than home-based training interventions (Kelly et al. 2014; Lampit et al. 2014), but it remains unclear whether these differences are due to formal vs. informal instruction or to the social setting vs. being alone. The same is true for training frequency and duration, where there is conflicting evidence about whether shorter or longer duration is the most beneficial (Karbach and Verhaeghen 2014; Kelly et al. 2014).

Evidence for Transfer Effects

As discussed in the paragraphs above, training interventions improve performance on the trained task, with greater gains compared to passive than active controls, and more robust effects for process- than strategy-based training interventions. Some of the training gains reported were of equivalent size as normal age-related declines across various cognitive domains, suggesting that training interventions likely help to reverse age-related declines and thus to stabilize cognitive functioning (Ball et al. 2002). The question is, however, if these improvements transfer to untrained tasks measuring either the same ability (i.e., near transfer) or to tasks measuring different abilities sharing underlying cognitive processes (i.e., far transfer; see e.g., Noack et al. 2009; Shipstead et al. 2012; see also Schmiedek, Taatgen, Chochrane and Green, this volume).

Transfer to Other Cognitive Tasks Assessed in the Laboratory For strategy-based trainings, little to no transfer effects have been found (e.g., Martin et al. 2011). It has been argued, though, that in contrast to the acquisition of specific memory strategies, practice effects from process-based training would be more prone to induce transfer to other cognitive tasks sharing the same core processes as the ones targeted in the intervention (e.g., Shipstead et al. 2012). Indeed, some process-based cognitive training interventions have been shown to induce small to moderate near transfer effects when the intervention is adaptive and of longer duration (Kelly et al. 2014, but see Guye and von Bastian 2017; von Bastian et al. 2019). For training interventions targeting working memory and executive functioning, for example, Karbach and Verhaeghen's (2014) meta-analysis indicated a net gain in near transfer tasks compared to active controls of 0.5 *SD*. However, far transfer effects were very small (net far transfer effects 0.2 *SD* in Karbach and Verhaeghen 2014; see also Könen et al., Karbach and Kray, this volume). The few available multidomain training interventions including cognitively complex group activities (e.g., Park et al. 2014), problem-solving (Stine-Morrow et al. 2008), or video games (see Bediou et al., Strobach and Schubert, this volume) have also shown small to moderate transfer effects to some cognitive functions, including executive functioning, episodic memory or processing speed. However, in order to design effective training interventions in the future, the understanding of the underlying processes, the cognitive functions targeted, and a high degree of ecological validity are necessary (see also Binder et al. 2015).

Transfer to Everyday Life Transfer to everyday life has been examined in only few studies, and some recent reviews have even excluded studies with everyday transfer from their analysis (e.g., Lampit et al. 2014). Studies in which everyday life was examined typically focused on self-reported basic or instrumental activities of daily living (BADL/IADL). However, these measures are designed to assess rather severe impairments in everyday life competence and, thus, are not necessarily optimal indicators of everyday functioning in healthy older adults due to ceiling effects. In the ACTIVE trial, the speed of information processing in everyday life was assessed by tasks such as looking up a telephone number, finding a respective food item on the supermarket shelf, identifying the ingredients on food labels, as well as self-reported driving ability. Not surprisingly, little to no evidence of transfer of the memory, reasoning, and processing speed training interventions to impairments in everyday functioning has been found immediately after training (Ball et al. 2002).

Evidence for Maintenance Effects

Most studies assess pre and immediate posttest performance and transfer, but do not follow-up on these effects over extended periods of time. Many studies examine maintenance only across a few months, even though it has been proposed that a three-year interval provides a more sensitive test of maintenance, differential stability, and change (Salthouse 2006).

Maintenance of Training Gains Kelly et al. (2014) report maintenance effects examined after up to 6 months, indicating maintenance of training gains following executive functioning and memory training interventions. Longer follow-up intervals have been tested in selected studies, such as the ACTIVE trial (Rebok et al. 2014; Willis et al. 2006) that assessed maintenance effects 5 years and 10 years posttraining. In the ACTIVE study, training gains observed in each training group were maintained over 5 years, with positive additive effects through intermediate booster training (Willis et al. 2006). After 10 years, training effects were maintained in the reasoning and processing speed domains, but not in the episodic memory domain (Rebok et al. 2014).

Maintenance of Transfer Effects Even though there were no immediate or shorter-term effects after 2 years in the ACTIVE trial (Ball et al. 2002), promising transfer effects to everyday functioning for particular training conditions and everyday outcomes were found after longer periods: (process-based) speed training was related to better driving performance and self-reported driving experience after up to 6-year intervals (Ball et al. 2010). In addition, there is evidence for effects of training on the slope of change trajectories in everyday functioning: Across a 5-year interval, participants in the (strategy-based) reasoning training group showed less steep declines in BADL/IADL competence and a 50% reduced risk of experiencing a car accident compared to the passive control participants (Willis et al. 2006). After an extended time period of 10 years, ACTIVE data showed transfer to everyday functioning in terms of BADL/IADL for all three training conditions, suggesting that trained individuals experienced fewer impairments in their independent functioning in everyday life. Interestingly, at the long term follow up and an average age of 82 years, 60% of the trained participants were at or above their baseline everyday competence level, which was true for only 50% of the passive control participants. Taken together, the findings suggest that transfer effects on the ability to live independently may be detectable only in the long run rather than immediately following the training intervention. Outcome measures assessing everyday performance above the impairment level or everyday cognitive activities instead of abilities have hardly been used in the literature so far, but may be more promising to detect transfer to real life.

Evidence for Effects on Brain Structure and Function

Normal aging is accompanied by brain tissue loss and neurophysiological changes (Raz and Rodrigue 2006). While the loss of grey matter manifests itself as general volume decline and cortical thinning (Fjell and Walhovd 2010), the degradation of white matter is reflected in reduced integrity and the incidence of so-called white matter hyperintensities. With respect to brain function, aging has been linked with a complex pattern of local over- and under-recruitment of neural resources.

Effects on Brain Structure A growing number of structural neuroimaging studies in healthy older adults provide evidence for beneficial effects of cognitive training on brain structure, especially for the domains of memory and working memory, where most of the work has been carried out (see Lustig et al. 2009 for a review, see also Oschwald et al. 2019). These effects (compared to a control group) comprise reduced decreases, maintenance or even increases in volume or cortical thickness of brain structures relevant for the trained function (e.g., Raz et al. 2013; Lövdén et al. 2012). The integrity of white matter, which can be qualified by different measures of water diffusion (e.g., Fractional Anisotropy, FA) on the basis of diffusion tensor imaging (DTI), can also be maintained or even increased by cognitive training interventions (e.g., Engvig et al. 2012). The reported effects reflect processes of structural neuroplasticity, which (partly) counteract the tissue degradation normally observed with aging. However, as most of the previous studies used passive control groups only, future studies including active control groups need to confirm the specificity of such effects.

Effects on Brain Function The evidence emerging from studies investigating the effects of cognitive training interventions on brain function is less conclusive. On the one hand, studies adopting strategy-based training interventions report increased brain activity during posttraining task performance (e.g., Nyberg et al. 2003). Based on the observed correlations between neurophysiological and behavioral changes, the activation increase has been attributed to an enhanced recruitment of task-specific regions that enables behavioral gains. On the other hand, process-based training studies, particularly in the domains of working memory or executive functioning, showed decreased brain activity at the post- compared to pretraining assessment, indicating improved neural efficiency during posttraining task performance (e.g., Brehmer et al. 2011). This discrepancy in the pattern of activity might be due to the different neural mechanisms initiated by the different training types. However, there is evidence in younger adults, that the activity decrease seen at later phases of process-based training interventions is actually preceded by an increase of activity in early training phases (Hempel et al. 2004). Future studies need to confirm whether this trajectory holds for older adults and whether strategy-based training interventions would also lead to increased neural efficiency after an extended period of implementing the acquired strategies.

Using electroencephalography (EEG), recent studies in older adults have demonstrated facilitative effects of cognitive training on early electrophysiological markers of the trained cognitive function with the extent of the ERP change predicting posttraining performance (e.g., Berry et al. 2010).

Future Directions

The central goal of cognitive training is to positively impact cognitive ability, cognitive plasticity, and ultimately functional ability in everyday life. Given the inconsistent findings with regard to training and transfer effects, as well as moderators thereof, we argue that future research should focus on three aspects (see also Guye 2018): (1) Investigate change in the cognitive processes involved in cognitive training to gain insights into the mechanism and boundaries of cognitive plasticity, (2) identify training contexts that maximize training and transfer effects, and (3) quantify the impact on meaningful real-life outcomes.

Cognitive Processes During Cognitive Training

The basic assumption underlying cognitive training is that repeated practice of cognitively challenging tasks expands the cognitive capacity (i.e., cognitive plasticity). However, the limited evidence for both near and far transfer after intensive cognitive training challenges this basic assumption. We argue that a better understanding of the changes occurring in the cognitive processes involved in training is needed to identify the potential mechanisms that could drive cognitive plasticity. Oberauer and Lewandowsky (2019) have recently presented measurement models for working memory tasks that assume two dimensions contributing to working memory performance: memory for individual elements (e.g., single digits) and memory for relations (e.g., the temporary binding of a digit to its position in a list of digits). The authors found evidence that, relative to younger adults, older adults showed specific deficits in memory for relations but not in memory for individual elements. Thus, directly targeting memory for relations as a cognitive process that deteriorates in old age could be a potential way to boost training and transfer gains.

Optimal Training Contexts

Understanding the parameters of the individually optimal training context (e.g., location, social setting, psychological state and training schedule) may direct us towards better tailoring cognitive training to the individual needs and preferences. Therefore, in addition to focusing on between-person factors associated with cognitive performance (see Katz et al., this volume; Guye et al. 2017; and for review see von Bastian and Oberauer 2014), we and others (e.g., Könen and Karbach 2015) argue that future research efforts should aim towards a better understanding of the influence of within-person covariates and environmental factors on cognitive training and transfer performance. Assessing those variables in naturalistic settings has become easier than ever given the technological advancements in recent years.

There is accumulating research showing within-person associations between cognitive performance and other factors including stress (Sliwinski et al. 2006; Stawski et al. 2011), both positive and negative affect, as well as motivation (Brose et al. 2012; Brose et al. 2014), sleep (Könen et al. 2015), and social activity (Bielak et al. 2019). The assessment of such contextual factors could be integrated in cognitive training studies to provide real-time feedback to participants on which factors are positively or negatively associated with the current training performance. Based on such just-in-time information (Nahum-Shani et al. 2018), participants would then be able to dynamically adapt their training regime to match the personally most beneficial training context, potentially maximizing the overall longer-term training benefit. Thus, a comprehensive assessment of daily personal and environmental factors, which are theoretically assumed to co-vary with daily cognitive performance, may further contribute to understand how daily training performance can be boosted.

Given the technological developments in the field of mobile sensing, ambulatory assessments to collect self-reported outcomes can easily be complemented with other tracking technologies to objectively assess further contextual parameters such as physical activity (e.g., GPS and accelerometer), physiological parameters (e.g., sleep and heart rate), or social interactions (e.g., Brose and Ebner-Priemer 2015, Mehl et al. 2001; see also Cochrane and Green, Colzato and Hommel, this volume).

Meaningful Real-Life Outcomes

One of the main goals of cognitive training interventions is to enhance cognitive performance in real-life settings and functional ability in everyday life. Although research has shown that specific aspects of cognitive ability, engaged lifestyle, and functional ability are associated (Guye et al. 2019), cognitive training studies have primarily focused on lab-based cognitive transfer tasks. Ecologically valid assessments of transfer to measure functional ability or everyday cognition are scarce (but see Ball et al. 2010; Cantarella et al. 2017; Willis et al. 2006 for exceptions).

Especially in older adults, it may come as a surprising observation that age-related decline in basic cognitive functions measured with lab-based cognitive tasks can go hand in hand with high levels of life satisfaction (e.g., Lachman et al. 2008; Scheibe and Carstensen 2010), and the ability to clearly manage tasks and activities in real-life successfully (Salthouse 2011). Thus, in order to understand whether cognitive training is beneficial for older adults beyond lab-based measures of cognition, it is crucial to embed the evaluation of cognitive training into real-life settings and measure everyday cognition (Bielak et al. 2017; Verhaeghen et al. 2012). Some studies have used self-reported measures (e.g., IADL; Lawton and Brody 1969; CFQ; Broadbent et al. 1982), and performance-based measures of everyday cognition in the lab (EPT; Willis and Marsiske 1993). However, in addition to the potential ceiling effects discussed above, these efforts do not capture the complexity and richness of real-life activities under natural circumstances. Thus, to quantify the real-life impact of cognitive training interventions, the development of objective, ecologically valid and comprehensive measures of cognitive functional ability are needed (see Mazurek et al. 2015 for an exception).

Acknowledgements Part of this work was supported by a grant and financial support awarded to the first author from the Suzanne and Hans Biäsch Foundation for Applied Psychology and the Jacobs Foundation. Moreover, this work was supported by the University Research Priority Program "Dynamics of Healthy Aging" of the University of Zurich.

References

Ball, K., Berch, D. B., Helmers, K. F., Jobe, J. B., Leveck, M. D., Marsiske, M., ... & Unverzagt, F. W. (2002). Effects of cognitive training interventions with older adults: A randomized controlled trial. *JAMA, 288*, 2271–2281.

Ball, K. K., Edwards, J. D., Ross, L. A., & McGwin, G. (2010). Cognitive training decreases motor vehicle collision involvement of older drivers. *Journal of the American Geriatrics Society, 58*, 2107–2113.

Baltes, P. B., & Kliegl, R. (1992). Further testing of limits of cognitive plasticity: Negative age differences in a mnemonic skill are robust. *Developmental Psychology, 28*, 121–125.

Berry, A. S., Zanto, T. P., Clapp, W. C., Hardy, J. L., Delahunt, P. B., Mahncke, H. W., & Gazzaley, A. (2010). The influence of perceptual training on working memory in older adults. *PLoS ONE, 5*, e11537.

Bielak, A. A. M., Hatt, C. R., & Diehl, M. (2017). Cognitive performance in adults' daily lives: Is there a lab-life gap? *Research in Human Development, 14*, 219–233.

Bielak, A. A., Mogle, J., & Sliwinski, M. J. (2019). What did you do today? Variability in daily activities is related to variability in daily cognitive performance. *The Journals of Gerontology: Series B, 74*, 764–771.

Binder, J. C., Zöllig, J., Eschen, A., Mérillat, S., Röcke, C., Schoch, S., ... & Martin, M. (2015). Multi-domain training in healthy old age: Hotel Plastisse as an iPad-based serious game to systematically compare multi-domain and single-domain training. *Frontiers in Aging Neuroscience, 7*, 137.

Brehmer, Y., Rieckmann, A., Bellander, M., Westerberg, H., Fischer, H., & Bäckman, L. (2011). Neural correlates of training-related working-memory gains in old age. *NeuroImage, 58*, 1110–1120.

Broadbent, D. E., Cooper, P. F., FitzGerald, P., & Parkes, K. R. (1982). The Cognitive Failures Questionnaire (CFQ) and its correlates. *British Journal of Clinical Psychology, 21*, 1–16.

Brose, A., & Ebner-Priemer, U. W. (2015). Ambulatory assessment in the research on aging: Contemporary and future applications. *Gerontology, 61*, 372–380.

Brose, A., Schmiedek, F., Lövdén, M., & Lindenberger, U. (2012). Daily variability in working memory is coupled with negative affect: The role of attention and motivation. *Emotion, 12*, 605–617.

Brose, A., Lövdén, M., & Schmiedek, F. (2014). Daily fluctuations in positive affect positively co-vary with working memory performance. *Emotion, 14*, 1–6.

Cantarella, A., Borella, E., Carretti, B., Kliegel, M., & de Beni, R. (2017). Benefits in tasks related to everyday life competences after a working memory training in older adults. *International Journal of Geriatric Psychiatry, 32*, 86–93.

Engvig, A., Fjell, A. M., Westlye, L. T., Moberget, T., Sundseth, Ø., Larsen, V. A., & Walhovd, K. B. (2012). Memory training impacts short-term changes in aging white matter: A longitudinal diffusion tensor imaging study. *Human Brain Mapping, 33*, 2390–2406.

Fjell, A. M., & Walhovd, K. B. (2010). Structural brain changes in aging: Courses, causes and cognitive consequences. *Reviews in the Neurosciences, 21*, 187–221.

Guye, S. (2018). *Healthy cognition in old age: Effects of an engaged lifestyle and cognitive training*. (Unpublished doctoral dissertation. Zurich, Switzerland: University of Zurich.

Guye, S., & von Bastian, C. C. (2017). Working memory training in older adults: Bayesian evidence supporting the absence of transfer. *Psychology and Aging, 32*, 732–746.

Guye, S., De Simoni, C., & von Bastian, C. C. (2017). Do individual differences predict change in cognitive training performance? A latent growth curve modeling approach. *Journal of Cognitive Enhancement, 1*, 374–393.

Guye, S., Röcke, C., Martin, M., & von Bastian, C. C. (2019). Functional ability in everyday life: Are associations with an engaged lifestyle mediated by working memory? *The Journal of Gerontology Series B: Psychological and Social Sciences*. Advanced online publication. https://opac.nebis.ch/ediss/20183405.pdf.

Hempel, A., Giesel, F. L., Garcia Caraballo, N. M., Amann, M., Meyer, H., Wüstenberg, T., ... & Schröder, J. (2004). Plasticity of cortical activation related to working memory during training. *American Journal of Psychiatry, 161*, 745–747.

Karbach, J., & Verhaeghen, P. (2014). Making working memory work: A meta-analysis of executive-control and working memory training in older adults. *Psychological Science, 25*, 2027–2037.

Kelly, M. E., Loughrey, D., Lawlor, B. A., Robertson, I. H., Walsh, C., & Brennan, S. (2014). The impact of cognitive training and mental stimulation on cognitive and everyday functioning of healthy older adults: A systematic review and meta-analysis. *Ageing Research Reviews, 15*, 28–43.

Könen, T., & Karbach, J. (2015). The benefits of looking at intraindividual dynamics in cognitive training data. *Frontiers in Psychology, 6*, 615.

Könen, T., Dirk, J., & Schmiedek, F. (2015). Cognitive benefits of last night's sleep: Daily variations in children's sleep behavior are related to working memory fluctuations. *Journal of Child Psychology and Psychiatry, and Allied Disciplines, 56*, 171–182.

Lachman, M. E., Röcke, C., Rosnick, C., & Ryff, C. D. (2008). Realism and illusion in Americans' temporal views of their life satisfaction: Age differences in reconstructing the past and anticipating the future. *Psychological Science, 19*, 889–897.

Lampit, A., Hallock, H., & Valenzuela, M. (2014). Computerized cognitive training in cognitively healthy older Adults: A systematic review and meta-analysis of effect modifiers. *PLoS Medicine, 11*, e1001756.

Lawton, M. P., & Brody, E. M. (1969). Assessment of older people: Self-maintaining and instrumental activities of daily living. *The Gerontologist, 9*, 179–186.

Lövdén, M., Schaefer, S., Noack, H., Bodammer, N. C., Kühn, S., Heinze, H. J., … & Lindenberger, U. (2012). Spatial navigation training protects the hippocampus against age-related changes during early and late adulthood. *Neurobiology of Aging, 33*, 620.e9.

Lustig, C., Shah, P., Seidler, R., & Reuter-Lorenz, P. A. (2009). Aging, training, and the brain: A review and future directions. *Neuropsychology Review, 19*, 504–522.
Martin, M., Clare, L., Altgassen, A. M., Cameron, M. H., & Zehnder, F. (2011). Cognition-based interventions for healthy older people and people with mild cognitive impairment. *Cochrane Database of Systematic Reviews, 1*, CD006220.
Mazurek, A., Bhoopathy, R. M., Read, J. C., Gallagher, P., & Smulders, T. V. (2015). Effects of age on a real-world What-Where-When memory task. *Frontiers in Aging Neuroscience, 7*, 74.
Mehl, M. R., Pennebaker, J. W., Crow, D. M., Dabbs, J., & Price, J. H. (2001). The electronically activated recorder (EAR): A device for sampling naturalistic daily activities and conversations. *Behavior Research Methods, Instruments, & Computers, 33*, 517–523.
Nahum-Shani, I., Smith, S. N., Spring, B. J., Collins, L. M., Witkiewitz, K., Tewari, A., & Murphy, S. A. (2018). Just-in-time adaptive interventions (JITAIs) in mobile health: Key components and design principles for ongoing health behavior support. *Annals of Behavioral Medicine, 52*, 446–462.
Noack, H., Lövdén, M., Schmiedek, F., & Lindenberger, U. (2009). Cognitive plasticity in adulthood and old age: Gauging the generality of cognitive intervention effects. *Restorative Neurology and Neuroscience, 27*, 435–453.
Nyberg, L., Sandblom, J., Jones, S., Stigsdotter Neely, A., Petersson, K. M., Ingvar, M., & Bäckman, L. (2003). Neural correlates of training-related memory improvement in adulthood and aging. *Proceedings of the National Academy of Sciences, 100*, 13728–13733.
Oberauer, K., & Lewandowsky, S. (2019). Simple measurement models for complex working-memory tasks. *Psychological Review, 126*, 880–932.
Oschwald, J., Guye, S., Liem, F., Rast, P., Willis, S., Röcke, C., Jäncke, L., Martin, M., & Mérillat, S. (2019). Brain structure and cognitive ability in healthy aging: A review on longitudinal correlated change. *Reviews in the Neurosciences*. https://doi.org/10.1515/revneuro-2018-0096.
Park, D. C., Lodi-Smith, J., Drew, L., Haber, S., Hebrank, A., Bischof, G. N., & Aamodt, W. (2014). The impact of sustained engagement on cognitive function in older adults: The synapse project. *Psychological Science, 25*, 103–112.
Raz, N., & Rodrigue, K. M. (2006). Differential aging of the brain: Patterns, cognitive correlates and modifiers. *Neuroscience & Biobehavioral Reviews, 30*, 730–748.
Raz, N., Schmiedek, F., Rodrigue, K. M., Kennedy, K. M., Lindenberger, U., & Lövdén, M. (2013). Differential brain shrinkage over 6 months shows limited association with cognitive practice. *Brain and Cognition, 82*, 171–180.
Rebok, G. W., Ball, K., Guey, L. T., Jones, R. N., Kim, H. Y., King, J. W., ... & Willis, S. L. (2014). Ten-year effects of the advanced cognitive training for independent and vital elderly cognitive training trial on cognition and everyday functioning in older adults. *Journal of the American Geriatrics Society, 62*, 16–24.
Salthouse, T. A. (2006). Mental exercise and mental aging. Evaluating the validity of the "use it or lose it" hypothesis. *Perspectives on Psychological Science, 1*, 68–87.
Salthouse, T. A. (2010). Selective review of cognitive aging. *Journal of the International Neuropsychological Society, 16*, 754–760.
Salthouse, T. A. (2011). Consequences of age-related cognitive declines. *Annual Review of Psychology, 63*, 1–26.
Scheibe, S., & Carstensen, L. L. (2010). Emotional aging: Recent findings and future trends. *The Journals of Gerontology: Psychological Sciences, 65B*, 135–144.
Shipstead, Z., Redick, T. S., & Engle, R. W. (2012). Is working memory training effective? *Psychological Bulletin, 138*, 628–654.
Sliwinski, M. J., Smyth, J. M., Hofer, S. M., & Stawski, R. S. (2006). Intraindividual coupling of daily stress and cognition. *Psychology and Aging, 21*, 545–557.
Stawski, R. S., Mogle, J., & Sliwinski, M. J. (2011). Intraindividual coupling of daily stressors and cognitive interference in old age. *The Journals of Gerontology Series B: Psychological Sciences and Social Sciences, 66*, 121–129.
Stine-Morrow, E. A. L., Parisi, J. M., Morrow, D. G., & Park, D. C. (2008). The effects of an engaged lifestyle on cognitive vitality: A field experiment. *Psychology and Aging, 23*, 778–786.

Verhaeghen, P. (Ed.). (2014). *The elements of cognitive aging: Meta- analyses of age-related differences in processing speed and their consequences*. New York: Oxford University Press.
Verhaeghen, P., & Marcoen, A. (1996). On the mechanisms of plasticity in young and older adults after instruction in the method of loci: Evidence for an amplification model. *Psychology and Aging, 11*, 164–178.
Verhaeghen, P., Martin, M., & Sędek, G. (2012). Reconnecting cognition in the lab and cognition in real life: The role of compensatory social and motivational factors in explaining how cognition ages in the wild. *Neuropsychology, Development, and Cognition. Section B. Aging, Neuropsychology, and Cognition, 19*, 1–12.
von Bastian, C. C., & Oberauer, K. (2014). Effects and mechanisms of working memory training: A review. *Psychological Research, 78*, 803–820.
von Bastian, C. C., Guye, S., & De Simoni, C. (2019). How strong is the evidence for the effectiveness of working memory training? In M. F. Bunting, J. M. Novick, M. R. Dougherty, & R. W. Engle (Eds.), *Cognitive and working memory training: Perspectives from psychology, neuroscience, and human development*. New York: Oxford University Press.
Willis, S. L., & Marsiske, M. (1993). Manual for the everyday problems test. University Park: Department of Human Development and family Studies, Pennsylvania State University.
Willis, S. L., Tennstedt, S. L., Marsiske, M., Ball, K., Elias, J., Koepke, K. M., ... & Wright, E. (2006). Long-term effects of cognitive training on everyday functional outcomes in older adults. *JAMA, 296*, 2805–2814.

Part III
Plasticity of Different Cognitive Domains

Working Memory Training

Tanja Könen, Tilo Strobach, and Julia Karbach

Contents

Abstract Working memory (WM) is a limited capacity system which is responsible for simultaneously maintaining and processing information. Reliable individual differences in this capacity place limiting constraints for performing other cognitive activities. Thus, WM training might even benefit a wide range of cognitive functions. This prospect makes WM training very prominent and also controversial. In the present chapter, we briefly illustrated common training regimes and reviewed the empirical evidence for training effects on the trained WM tasks, near transfer to nontrained WM tasks, and far transfer to different cognitive functions. Consistent evidence across different age groups from all over the lifespan and across several meta-analyses speaks in favor of significant average training effects and significant near transfer to nontrained WM tasks. However, evidence for far transfer to, for example, fluid intelligence, executive functions, and academic achievement, is mixed. We reviewed current topics of discussion in the field and concluded that a greater focus on variables possibly moderating transfer effects (e.g., individual differences and situational characteristics during training) is necessary to better understand conflicting findings. More research on far transfer effects is needed because even small effects could actually make a difference relevant to everyday life.

T. Könen (✉) · J. Karbach
Department of Psychology, University of Koblenz-Landau, Landau, Germany
e-mail: koenen@uni-landau.de

T. Strobach
Department of Psychology, Medical School Hamburg, Hamburg, Germany

T. Strobach, J. Karbach (eds.), *Cognitive Training*,
https://doi.org/10.1007/978-3-030-39292-5_11

Definition, Models, and the Significance of Working Memory

Working memory (WM) allows for simultaneously maintaining and processing information in a controlled manner (Baddeley and Hitch 1994). Several competitive theoretical models of WM are existing and are still vividly discussed (Baddeley 2012; Oberauer et al. 2018). Most WM models contributed substantially to our current understanding of WM and largely agree on the basic assumptions that WM capacity is limited and that reliable individual differences in this capacity exist, which place limiting constraints for performing a wide range of other cognitive activities (e.g., Baddeley 2012; Oberauer 2009). In other words, WM is a limited capacity system providing the temporary storage and manipulation of information that is necessary for higher cognitive functioning (e.g., for reasoning; Baddeley 2012). The WM models do, however, significantly differ in the assumptions about the structure of this limited capacity system. We will shortly introduce the main ideas of the models of Oberauer, Baddeley, Miyake and colleagues because they are particularly helpful for understanding well-known WM training paradigms.

Oberauer defines WM as the cognitive system that allows for building, maintaining, and updating structural representations via dynamic bindings (cf. Oberauer 2009; Wilhelm et al. 2013). This WM system consists of two parts: Bindings temporarily organize information such as words, objects, or events in a declarative part, and connect this information to allowed or inhibited responses in a procedural part (Oberauer 2009). Baddeley, however, defines WM as a cognitive system with at least three components: The central executive, which is responsible for focusing and dividing attention and for coordinating the information flow between at least two temporary storage systems, one for phonological and one for visuo-spatial information (Baddeley and Hitch 1994). Miyake emphasizes the special role of WM updating (i.e., monitoring and refreshing information held in WM) as an executive function (Friedman and Miyake 2017; Miyake et al. 2000; Karbach and Kray, this volume).

Taken together, these WM models differ in the assumed underlying structure of the WM system but agree that it allows for simultaneously maintaining and processing information. Because of this fundamental function, it is not surprising that WM has shown to be a central determinant of fluid intelligence (e.g., Fuhrmann et al. 2019; Kane et al. 2004), school achievements in various domains (e.g., Peng et al. 2016, 2018), and a large number of other cognitive tasks that are highly relevant in daily life (e.g., language comprehension, following directions, and writing; Barrett et al. 2004, for a review).

The Rationale Behind Working Memory Training

The idea that WM capacity is the main limiting factor for performing a wide range of cognitive activities (e.g., Baddeley 2012) has the implication that WM training could not only benefit WM functioning but a wide range of cognitive functions.

Thus, in addition to performance improvements on the trained WM tasks and near transfer to other nontrained WM tasks, one might even expect far transfer to a range of alternative functions. For example, given the close relation of WM capacity and fluid intelligence (e.g., Kane et al. 2004) one could assume that WM training might also benefit reasoning. Improving WM functioning even slightly might therefore have enormous practical implications relevant to everyday life, which is why this topic has raised so much attention in several areas of psychology.

Two general mechanisms could mediate transfer effects: Enhanced WM capacity and/or enhanced efficiency using the available WM capacity (cf. von Bastian and Oberauer 2014). Enhancing WM capacity is the traditional goal of WM training and a classic explanation for transfer effects (Klingberg 2010, for a review). Enhanced efficiency has long been considered to be largely material- or process specific, for example, through the acquisition of strategies suited for a specific task paradigm only. Although there is evidence that enhanced efficiency could also work on a more general level, such as faster visual encoding or faster attentional processes (von Bastian and Oberauer 2014), enhancing WM capacity remains the aim and focus of most training studies. WM training is assumed to enhance general WM capacity if there is evidence for transfer effects to multiple WM tasks varying in the type of material and mode of testing (Klingberg 2010).

Enhanced WM capacity can theoretically be explained with training-induced cognitive plasticity (Lövdén et al. 2010; see also Karbach and Kray, this volume). Plasticity denotes that a prolonged mismatch between cognitive resources and situational demands can foster reactive changes in the possible ranges of individual cognitive performance – such as changes in WM capacity (cf. Lövdén et al. 2010). To create a prolonged mismatch, WM training needs to be challenging but manageable with a high degree of effort. No mismatch arises if the WM tasks can either be solved with the existing WM capacity or if they are so frustrating that participants give up. Therefore, WM training groups are often assigned to adaptive task-difficulty conditions to foster plasticity by keeping WM demands perpetually at the individual limit, whereas active control groups are assigned to consistently low WM task-difficulty conditions or tasks tapping on functions alternative to WM (cf. Lövdén et al. 2010).

The *cognitive routine framework* suggested by Gathercole et al. (2019) follows a similar idea. WM task features which create unfamiliar and challenging cognitive demands require participants to develop novel cognitive routines, because the demands cannot be met by existing mechanisms. New cognitive routines can then be applied to untrained tasks sharing the same requirements, which is the basis for transfer effects. This principle is largely in line with the concept of plasticity, but the framework also focuses on specific predictions about which common features will likely generate transfer and which will not (cf. Gathercole et al. 2019). For example, a crucial feature of recall paradigms (the recall of lists) is the presence or absence of distractor interference (distraction during the encoding of lists). Distractor interference requires cognitive routines to reduce the impact of interference (e.g., the removal of distractor representations), which can only be transferred to tasks sharing this requirement. Notably, these routines are automated cognitive procedures and more general than task-specific strategies. The process of constructing a new cognitive routine follows conventional models of skill acquisition and draws on general cognitive resources such as intelligence (cf. Gathercole et al. 2019).

Selected Training Regimes

A basic distinction can be drawn between (1) single-paradigm training regimes, focusing on one WM paradigm, (2) multiparadigm regimes including multiple WM paradigms (both 1 and 2 are single-domain regimes focusing only on the domain WM), and (3) multidomain regimes including not only WM tasks but also tasks drawing on other abilities (e.g., on processing speed; von Bastian and Oberauer 2014). Naturally, single-paradigm regimes have the advantage that training and transfer effects can be attributed to specific mechanisms more easily. Multiparadigm or multidomain regimes could in theory be more effective because they require more heterogeneous cognitive processes, but the effects cannot be isolated. A recent meta-analysis provided the first evidence on the effectiveness in older adults: Single-domain training resulted in larger effect sizes on near-transfer outcomes (compared to far-transfer outcomes), whereas multidomain training obtained larger effect sizes on far-transfer outcomes (compared to near-transfer outcomes; Nguyen et al. 2019). This pattern directly corresponds to training contents (training specific vs. numerous cognitive processes), but needs further validation (e.g., in other age groups). Only a few studies directly compared different WM training regimes (Holmes et al. 2019; von Bastian and Oberauer 2013). Most studies investigate the effectiveness of a specific regime. We will briefly introduce a selection of well-known WM training regimes.

Simple Span Training In simple span tasks, participants have to recall a list of stimuli (e.g., digits or colors) after a brief retention interval. In case of successful recall, they are given a longer list of stimuli. Recall takes place in either the presented order (e.g., digit span forwards) or in reverse order (e.g., digit span backwards). Recall in the presented order requires temporary storage and thus draws on the storage systems assumed in Baddeley's WM model. Backward span tasks draw on central executive functioning. Therefore, training regimes based on Baddeley's WM model usually include both forward and backward span tasks to train all components of WM. The probably best known regime based on simple span tasks is *Cogmed* WM training (*www.cogmed.com*), which is very common, particularly for children with ADHD. *Cogmed* has been tested in a large number of studies and is the topic of several ongoing discussions and current reviews (e.g., Aksayli et al. 2019; Shinaver et al. 2014).

Complex Span Training Complex span tasks combine simple span tasks with a simultaneous and often unrelated secondary task, such as evaluating equations or pictures. Thus, they draw on both storage and processing, which particularly corresponds with Baddeley's WM model (which includes storage and processing units). Empirical evidence suggests that they are almost perfectly correlated with binding and updating tasks (e.g., Wilhelm et al. 2013) and can thus also be mapped to Oberauer's WM model. Complex span tasks are well-established and popular indicators of WM capacity (e.g., Kane et al. 2004), which are regularly used as training tasks in cognitive training. For example, they are implemented in the WM training battery *Braintwister* (Buschkuehl et al. 2008) and the WM tasks in *Tatool* (von Bastian et al. 2013).

N-Back Training In the *n*-back task, participants are presented with sequences of stimuli and must decide whether the current stimulus matches the one presented *n* items back in a given modality (e.g., visuo-spatial or auditory). Importantly, *n* is a variable number that can be adjusted to increase or decrease task difficulty. Dual *n*-back tasks combine two modalities and are considered to be more difficult and effective than single *n*-back tasks. The *n*-back task is a valid indicator of WM capacity (e.g., Wilhelm et al. 2013; but see Jaeggi et al. 2010) and particularly corresponds with the theoretical understanding of Oberauer and Miyake as it requires the updating of information in WM. Cognitive training with *n*-back tasks is common in various age groups and is implemented in, for example, the *Braintwister* WM training battery (Buschkuehl et al. 2008) and the *Lumosity* cognitive training battery (e.g., Hardy et al. 2015).

Training and Transfer Effects

To evaluate the effectiveness of WM training, one considers whether a training group (compared to a control group) showed (1) performance improvements on the trained WM tasks, (2) near transfer to nontrained WM tasks, and (3) far transfer to different cognitive functions.

Training Effects WM training studies ubiquitously report that trained participants significantly improve their performance on the trained WM task(s) over the course of training (cf. Morrison and Chein 2011). This applies to a wide variety of training regimes and age ranges of the participants. Even generally critical reviews acknowledge that participants typically advance considerably (e.g., Shipstead et al. 2012). One meta-analytical integration of 12 WM training effects derived from studies with older adults found a large average standardized increase between pre- and posttest of $d = 1.1$ compared to the control groups (Karbach and Verhaeghen 2014), which was confirmed recently by a different research group (Hedges's $g = 1.2$ across 15 effect sizes; Nguyen et al. 2019). While average comparisons of standardized pre- and posttest performances are a classical requirement in WM training studies, analyzing individual performance trajectories over the course of training sessions can even provide additional information. For example, growth modeling with $N = 190$ younger and older adults revealed that individual performance substantially increased across the training phase, with a steeper increase at the beginning (Guye et al. 2017), which is in line with the power law of practice (Heathcote et al. 2000). By comparing the individual performance growth of younger and older adults, Bürki et al. (2014) demonstrated that older adults showed on average a slower WM performance growth during training than younger adults.

However, improved performance on a training task does not necessarily imply an enhanced WM capacity (Shipstead et al. 2012). The conclusion of training-induced increases in WM capacity is only valid in comparison to an adequate control group (e.g., Green et al. 2014, for a review) and with evidence for near transfer effects to multiple WM tasks varying in the type of material and mode of testing (Klingberg 2010).

Near Transfer Effects A large number of meta-analyses and reviews agree that WM memory training produces near transfer to nontrained WM tasks in children, younger adults, and older adults (e.g., Karbach and Verhaeghen 2014; Melby-Lervåg and Hulme 2013; Nguyen et al. 2019; Sala et al. 2019; Schwaighofer et al. 2015). For example, in a meta-analytical integration of 18–21 near transfer effects derived from studies with children and adults, Melby-Lervåg and Hulme (2013) found moderate and large average standardized increases on visuo-spatial/verbal WM tasks of $d = 0.5/0.8$ between pre- and posttest compared to control groups. Age was a significant moderator of the effect on verbal WM, with children showing larger benefits than adolescents (Melby-Lervåg and Hulme 2013). However, the effects are found across the whole lifespan (e.g., children, younger adults, and older adults in Sala et al. 2019) and are also valid when only comparisons between trained groups and active control groups were considered for the analysis (Sala et al. 2019). Notably, near transfer effects are usually smaller than training effects. For example, with *Cogmed* Training for children, improvements in trained tasks were about 30–40%, whereas improvements in nontrained WM tasks were about 15% (cf. Klingberg 2010; see also Karbach and Verhaeghen 2014, for similar findings on older adults).

Despite this promising evidence, it is important to consider that not all studies have minimized task-specific overlaps between the training and near transfer tasks (cf. Shipstead et al. 2012). This is particularly relevant for n-back training, because some learning processes that occur during *n*-back are assumed to be paradigm specific and thus not directly transferable to other WM paradigms (Shipstead et al. 2012). A recent meta-analysis found that a substantial part of near transfer following *n*-back training was indeed paradigm specific (Soveri et al. 2017). This demonstrates why transfer should be evaluated on the latent ability level (see Könen and Auerswald, this volume for details). Evidence for near transfer on the latent ability level would be strong evidence for training-induced increases in WM capacity and thus an optimal foundation for the investigation of far transfer effects.

Far Transfer Effects The question whether valid far transfer effects to different cognitive functions exist is highly controversial. They would be a central determinant of the value of WM interventions because training outcomes need to generalize to other cognitive abilities to optimally support participants in their daily life. Most views on transfer suggest that the likelihood and strength of far transfer varies as a function of the similarity in processing demands between the training and transfer tasks (see Taatgen, this volume for details). Thus, one would expect transfer to abilities that are generally known to be strongly related to WM, such as, for example, fluid intelligence, executive functions, and academic achievement (e.g., Kane et al. 2004; Peng et al. 2016, 2018). The evidence for far transfer effects, however, is mixed. Meta-analyses on WM training differed in the conclusion on the presence (Au et al. 2015, 2016; Karbach and Verhaeghen 2014; Schwaighofer et al. 2015) or absence of far transfer effects (Melby-Lervåg and Hulme 2013, 2016; Sala et al. 2019).

For example, the meta-analysis of Au et al. (2015) focused on fluid intelligence as transfer outcome. They integrated 24 effect sizes of *n*-back training with healthy adults (18–50 years of age) and found small average standardized increases on fluid intelligence tasks of Hedges's $g = 0.2$ between pre- and posttest compared to control groups. The meta-analysis of Schwaighofer et al. (2015) came to a similar conclusion on this issue, whereas two others did not (Melby-Lervåg and Hulme 2013; Nguyen et al. 2019). This is not surprising because different selection criteria can result in different samples and findings. For instance, Melby-Lervåg and Hulme (2013) included studies investigating different age groups from all over the lifespan (up to 75 years of age) and they did not differentiate between healthy and cognitively impaired participants. Considering the large individual differences in the magnitude of transfer effects, it is not surprising that data averaged over these very diverse groups do not show any significant far transfer effects on the group level. However, more evidence is needed before a converging view on far transfer to fluid intelligence can evolve in the field. Interestingly, Bürki et al. (2014) analyzed the individual performance growth in WM training with younger and older adults and found that those who improved more during training showed higher gains in a fluid intelligence transfer task. This is a correlational and by no means a causal finding, but it can help to understand individual differences in transfer outcomes.

Further, recent evidence shows far transfer to executive functions (e.g., Melby-Lervåg and Hulme 2013; Nguyen et al. 2019; Salminen et al. 2012), but a complete picture with findings on all age groups and all executive functions is yet missing. The meta-analyses of Melby-Lervåg and Hulme (2013) including children and adults demonstrated small transfer effects to inhibition (Stroop task, $d = 0.3$, 10 effect sizes). There is further meta-analytical evidence for small transfer effects to executive functioning (inhibition and flexibility) in adults in general (Soveri et al. 2017) and specifically older adults (e.g., Hedges's $g = 0.2$, 15 effect sizes; Nguyen et al. 2019). One meta-analysis, however, tested transfer of WM training to executive control together with other measures (fluid intelligence, processing speed, and language) and found no evidence for transfer effects over all measures in children and adults (Sala et al. 2019). Given the close theoretical and empirical relations of WM and executive functions (Friedman and Miyake 2017, for a review), it is rather surprising that we are missing a more differentiated understanding on the transfer of WM training to executive functions.

Concerning far transfer to academic achievement, the present findings on children demonstrate converging evidence for positive effects on reading but not mathematics (Titz and Karbach 2014, for a review; see also Johann and Karbach, this volume). Findings of children and adults combined, however, do not show transfer effects to either reading or mathematical abilities (Melby-Lervåg and Hulme 2013; Schwaighofer et al. 2015). Future meta-analyses including only children have to decide whether this transfer effect might be only valid for children who are still developing their reading skills.

Moderating Variables The current controversy about the existence of far transfer effects demonstrates the importance of considering moderating variables in evaluating training and transfer effects. Possible moderating variables are training-specific features (e.g., type, intensity, and duration of training; von Bastian and Oberauer 2014, for a review), individual differences (e.g., baseline performance, age, and personality; see Katz et al., this volume, for a review), and within-person processes during training (e.g., the strength of the relation between daily motivation and WM performance; Könen and Karbach 2015). As elaborate reviews on these issues do already exist (see above), we do not repeat their empirical findings here. We are, however, strongly convinced that the failure to consider moderating variables – not only in meta-analyses but also in primary studies – could mask training and transfer effects.

Maintenance The longevity of training-induced benefits is a key aspect of the value of WM interventions. Near transfer effects appear to be mostly stable, which is even acknowledged by generally critical reviews (e.g., Shipstead et al. 2012). A meta-analysis on studies with children and adults provided valuable evidence as it included 42 immediate effect sizes of near transfer to verbal WM and eleven long-term effect sizes derived from follow-up tests conducted on average 8 months after the posttests. After the removal of outliers, immediate near transfer effect sizes were moderate (Hedges's $g = 0.3–0.6$) and long-term effect sizes were small to moderate (Hedges's $g = 0.2–0.4$). The meta-analyses further demonstrated comparable immediate and long-term effects for visuo-spatial WM, albeit based on fewer effect sizes (Schwaighofer et al. 2015). Thus, even several months after WM training, near transfer effects to other WM tasks are still valid.

The longevity of far transfer effects, however, is unclear. Important evidence comes from the COGITO study (Schmiedek et al. 2014), in which a sample of younger adults practiced 12 tests of perceptual speed, WM, and episodic memory for over 100 daily 1-hr sessions. The findings demonstrated a net far transfer effect of 0.23 to a *latent factor* of reasoning 2 years later (compared to a passive control group), which did not differ in size from the immediate effect 2 years earlier. This shows that intensive cognitive training interventions can have long-term broad transfer at the level of cognitive abilities. However, as this was a multidomain training, the contribution of the WM training component cannot be isolated. This is essential, since a meta-analysis on single-domain WM training studies provided no evidence for the longevity of far transfer effects (Schwaighofer et al. 2015).

Neuropsychological and Everyday Correlates Identifying correlates to both neural functions and behavior in everyday life is another key aspect when assessing the value of WM interventions. Neuroimaging studies provided the first evidence that training-induced increases of WM performance were related to changes within a network of brain regions generally known for its association with WM functioning (i.e., dorsolateral prefrontal cortex, posterior parietal cortex, and basal ganglia; Morrison and Chein 2011, for a review). They suggest that WM training leads to neuroplastic processes that represent a reduced demand for attentional control with increasing practice (e.g., Clark et al. 2017; Thompson et al. 2016). Training-induced transfer was related to changes within networks of brain regions associated with performance on both the training and transfer tasks (cf. Morrison and Chein 2011). This could indicate that far transfer is more likely if the training and transfer tasks engage specific overlapping neural processing mechanisms and brain regions (Dahlin et al. 2008; see also Wenger and Kühn, this volume).

Correlates to behavior in everyday life are mostly tested in the context of ADHD symptoms. A meta-analysis integrated 13 effect sizes of studies with children and adults and indicated a moderate training-induced decrease of inattention in daily life ($d = -0.5$). Seven effect sizes from follow-up tests conducted 2–8 months after the posttests suggested persisting training benefits for inattention ($d = -0.3$; Spencer-Smith and Klingberg 2015). Thus, benefits of WM training might generalize to improvements in everyday functioning.

Methodological Issues

As the review above indicated, there is a huge controversy on far transfer effects of WM training. Many arguments apply to cognitive training in general but are largely discussed in the context of WM training. We briefly review three main methodological issues that have been repeatedly discussed over the years (see Schmiedek, this volume, for more details).

Adequate Control Groups A major concern in the field of WM training is the appropriateness of the control condition(s). The field fundamentally agrees on the advantages of active control groups and the necessity of considering the type of control group in interpreting findings (passive control groups receive no treatment and active control groups receive a treatment that does not qualify as WM training or not as cognitively demanding WM training). The type of control group is a standard moderator tested in meta-analyses and topic of several reviews (e.g., Green et al. 2014). There is, however, disagreement on the potential benefit of passive control groups. Some emphasize the risks of overestimating training and transfer effects and false claims of causality in passive control designs (they cannot control for expectancy and other nonfocal effects; e.g., Melby-Lervåg and Hulme 2016). Others in turn emphasize the difficulty of finding an adequate active control condition, which produces the same nonfocal effects (e.g., which is motivating and challenging) but does not draw on WM (cf. Oberauer 2015). If the active control condition draws significantly on WM, an underestimation of training and transfer effects is likely. A self-evident consequence of all risks would be to include both passive and active control groups and assess motivation and expectancy in active control groups.

Underpowered Studies Underpowered studies with too few participants per training group are a common problem in the field. Naturally, null findings in underpowered studies should not be interpreted, but underpowered studies can theoretically produce spurious significant effects, too. Meta-analytic procedures typically adjust effect sizes for the sample sizes of the included studies but the estimates can still be affected. Given the currently large number of meta-analyses in the field summarizing mostly the same partly underpowered studies, we would strongly profit from carefully designed new studies and carefully conducted replications of known effects with adequate power (e.g., Brandt et al. 2014, for a tutorial). One solution for this issue is a more consequent peer-review system requesting power estimates. A couple of notable exceptions exist, for example, a study on a multidomain online training (including WM training) with $N = 4715$ participants. It demonstrated moderate transfer effects to several cognitive functions such as WM and reasoning compared to an active control condition (Hardy et al. 2015).

Research Bias It is obvious that the present research labs fundamentally differ in whether they have an optimistic or pessimistic view on WM training outcomes, particularly on far transfer. This could be very valuable because it could be the foundation of a fruitful discussion. However, the current debate is far too heated, which could – in the worst case – result in biased research. That is, it could result in a biased publication of one's own work and a biased reading of other work. In our view, four things are helpful to address this issue: (1) consideration of labs/authors as a moderating factor in meta-analyses (e.g., in Au et al. 2015), (2) reports of Bayesian analyses which allow for quantifying the strength of evidence in favor of both the null and the alternative hypothesis (e.g., in Gathercole et al. 2019), (3) preregistration of methods and hypotheses (e.g., Weicker et al. 2018, for a registered clinical trial), and (4) endorsement of a more differentiated perspective and language through senior researchers (e.g., Oberauer 2015) and peer review.

Taken together, the necessary tools to overcome research bias already exist and should be applied. A recently published consensus of 48 scientists discusses further aspects of methodological standards in cognitive training research (Green et al. 2019; see also Cochrane & Green, this volume (Chap. 3)).

Conclusion

In summary, consistent evidence suggests significant average training effects and significant near transfer to nontrained WM tasks. However, evidence for far transfer to other cognitive functions is mixed, which caused a vivid controversy in the field. Still, the prospect of successful WM training has so many significant theoretical and practical outcomes that we should be more than motivated to investigate conflicting findings. If the existing evidence for transfer could be further validated, it would significantly impact our theoretical understanding of both WM and the transfer constructs (e.g., in terms of plasticity). It could also positively impact intervention programs, where even small gains in WM capacity and transfer constructs could actually make a difference relevant to everyday life (e.g., for school children relying on WM capacity to improve learning processes). Further, the large individual differences in training outcomes (Katz et al., this volume) should also motivate us to understand these differences. We agree with Colzato and Hommel (this volume) that the current controversy about the effectiveness of training is likely partly due to the failure to consider individual differences. Not considering the personality of the trained participants, their experiences, and life contexts during training could mask training effects. We should not only ask whether WM training works on average but also for whom it works and in which contexts and situations it works.

References

Aksayli, N. D., Sala, G., & Gobet, F. (2019). The cognitive and academic benefits of Cogmed: A meta-analysis. *Educational Research Review, 27*, 229–243.

Au, J., Sheehan, E., Tsai, N., Duncan, G. J., Buschkuehl, M., & Jaeggi, S. M. (2015). Improving fluid intelligence with training on working memory: A meta-analysis. *Psychonomic Bulletin & Review, 22*, 366–377.

Au, J., Buschkuehl, M., Duncan, G. J., & Jaeggi, S. M. (2016). There is no convincing evidence that working memory training is NOT effective: A reply to Melby-Lervåg and Hulme (2015). *Psychonomic Bulletin and Review, 23*, 331–337.

Baddeley, A. (2012). Working memory: Theories, models, and controversies. *Annual Review of Psychology, 63*, 1–29.

Baddeley, A. D., & Hitch, G. J. (1994). Developments in the concept of working memory. *Neuropsychology, 8*, 485–493.

Barrett, L. F., Tugade, M., & Engle, R. W. (2004). Individual differences in working memory capacity and dual-process theories of the mind. *Psychological Bulletin, 130*, 553–573.

Brandt, M. J., IJzerman, H., Dijksterhuis, A., Farach, F. J., Geller, J., Giner-Sorolla, R., … & Van't Veer, A. (2014). The replication recipe: What makes for a convincing replication?. *Journal of Experimental Social Psychology, 50*, 217–224.

Bürki, C., Ludwig, C., Chicherio, C., & de Ribaupierre, A. (2014). Individual differences in cognitive plasticity: An investigation of training curves in younger and older adults. *Psychological Research, 78*, 821–835.

Buschkuehl, M., Jaeggi, S., Kobel, A., & Perrig, W. J. (2008). *Braintwister: Aufgabensammlung für kognitives training [Braintwister: Tasks for cognitive training]*. Bern: Universität Bern, Institut für Psychologie.

Clark, C. M., Lawlor-Savage, L., & Goghari, V. M. (2017). Functional brain activation associated with working memory training and transfer. *Behavioral Brain Research, 334*, 34–49.

Dahlin, E., Neely, A., Larsson, A., Bäckman, L., & Nyberg, L. (2008). Transfer of learning after updating training mediated by the striatum. *Science, 320*, 1510–1512.

Friedman, N. P., & Miyake, A. (2017). Unity and diversity of executive functions: Individual differences as a window on cognitive structure. *Cortex, 86*, 186–204.

Fuhrmann, D., Simpson-Kent, I. L., Bathelt, J., The CALM Team, & Kievit, R. A (2019). A hierarchical watershed model of fluid intelligence in childhood and adolescence. *Cerebral Cortex.* Advance online publication. https://doi.org/10.1093/cercor/bhz091

Gathercole, S. E., Dunning, D. L., Holmes, J., & Norris, D. (2019). Working memory training involves learning new skills. *Journal of Memory and Language, 105*, 19–42.

Green, C. S., Strobach, T., & Schubert, T. (2014). On methodological standards in training and transfer experiments. *Psychological Research, 78*, 756–772.

Green, C. S., Bavelier, D., Kramer, A. F., Vinogradov, S., Ansorge, U., Ball, K. K., ... & Facoetti, A. (2019). Improving methodological standards in behavioral interventions for cognitive enhancement. *Journal of Cognitive Enhancement, 3*, 2–29.

Guye, S., De Simoni, C., & von Bastian, C. C. (2017). Do individual differences predict change in cognitive training performance? A latent growth curve modeling approach. *Journal of Cognitive Enhancement, 1*, 374–393.

Hardy, J. L., Nelson, R. A., Thomason, M. E., Sternberg, D. A., Katovich, K., Farzin, F., & Scanlon, M. (2015). Enhancing cognitive abilities with comprehensive training: A large, online, randomized, active-controlled trial. *PLoS ONE, 10*, e0134467.

Heathcote, A., Brown, S., & Mewhort, D. J. (2000). The power law repealed: The case for an exponential law of practice. *Psychonomic Bulletin & Review, 7*, 185–207.

Holmes, J., Woolgar, F., Hampshire, A., & Gathercole, S. E. (2019). Are working memory training effects paradigm-specific? *Frontiers in Psychology, 10*, 1103.

Jaeggi, S. M., Buschkuehl, M., Perrig, W. J., & Meier, B. (2010). The concurrent validity of the *N*-back task as a working memory measure. *Memory, 18*, 394–412.

Kane, M. J., Hambrick, D. Z., Tuholski, S. W., Wilhelm, O., Payne, T. W., & Engle, R. W. (2004). The generality of working-memory capacity: A latent-variable approach to verbal and visuospatial memory span and reasoning. *Journal of Experimental Psychology: General, 133*, 189–217.

Karbach, J., & Verhaeghen, P. (2014). Making working memory work: A meta-analysis of executive-control and working memory training in older adults. *Psychological Science, 25*, 2027–2037.

Klingberg, T. (2010). Training and plasticity of working memory. *Trends in Cognitive Sciences, 14*, 317–324.

Könen, T., & Karbach, J. (2015). The benefits of looking at intraindividual dynamics in cognitive training data. *Frontiers in Psychology, 6*, 615.

Lövdén, M., Bäckman, L., Lindenberger, U., Schaefer, S., & Schmiedek, F. (2010). A theoretical framework for the study of adult cognitive plasticity. *Psychological Bulletin, 136*, 659–676.

Melby-Lervåg, M., & Hulme, C. (2013). Is working-memory training effective? A meta-analytic review. *Developmental Psychology, 49*, 270–291.

Melby-Lervåg, M., & Hulme, C. (2016). There is no convincing evidence that working memory training is effective: A reply to Au et al. (2014) and Karbach and Verhaeghen (2014). *Psychonomic Bulletin and Review, 23*, 324–330.

Miyake, A., Friedman, N. P., Emerson, M. J., Witzki, A. H., Howerter, A., & Wagner, T. (2000). The unity and diversity of executive functions and their contributions to complex "frontal lobe" tasks: A latent variable analysis. *Cognitive Psychology, 41*, 49–100.

Morrison, A., & Chein, J. (2011). Does working memory training work? The promise and challenges of enhancing cognition by training working memory. *Psychonomic Bulletin & Review, 18*, 46–60.

Nguyen, L., Murphy, K., & Andrews, G. (2019). Immediate and long-term efficacy of executive functions cognitive training in older adults: A systematic review and meta-analysis. *Psychological Bulletin, 145*, 698–733.

Oberauer, K. (2009). Design for a working memory. *Psychology of Learning and Motivation, 51*, 45–100.

Oberauer, K. (2015). *Training of intelligence: A question of intelligent training*. Retrieved from: http://www.psychonomic.org/featured-content-detail/training-of-intelligence-question-of-intelligent-t

Oberauer, K., Lewandowsky, S., Awh, E., Brown, G. D., Conway, A., Cowan, N., ... & Ma, W. J. (2018). Benchmarks for models of short-term and working memory. *Psychological Bulletin, 144*, 885–958.

Peng, P., Namkung, J., Barnes, M., & Sun, C. (2016). A meta-analysis of mathematics and working memory: Moderating effects of working memory domain, type of mathematics skill, and sample characteristics. *Journal of Educational Psychology, 108*, 455–473.

Peng, P., Barnes, M., Wang, C., Wang, W., Li, S., Swanson, H. L., ... & Tao, S. (2018). A meta-analysis on the relation between reading and working memory. *Psychological Bulletin, 144*, 48–76.

Sala, G., Aksayli, N. D., Tatlidil, K., Tatsumi, T., Gondo, Y., & Gobet, F. (2019). Near and far transfer in cognitive training: A second-order meta-analysis. *Collabra: Psychology, 5*, 18.

Salminen, T., Strobach, T., & Schubert, T. (2012). On the impacts of working memory training on executive functioning. *Frontiers in Human Neuroscience, 6*, 166.

Schmiedek, F., Lövdén, M., & Lindenberger, U. (2014). Younger adults show long-term effects of cognitive training on broad cognitive abilities over 2 years. *Developmental Psychology, 50*, 2304–2310.

Schwaighofer, M., Fischer, F., & Bühner, M. (2015). Does working memory training transfer? A meta-analysis including training conditions as moderators. *Educational Psychologist, 50*, 138–166.

Shinaver, C. S., Entwistle, P. C., & Söderqvist, S. (2014). Cogmed WM training: Reviewing the reviews. *Applied Neuropsychology: Child, 3*, 163–117.

Shipstead, Z., Redick, T. S., & Engle, R. W. (2012). Is working memory training effective? *Psychological Bulletin, 138*, 628–654.

Soveri, A., Antfolk, J., Karlsson, L., Salo, B., & Laine, M. (2017). Working memory training revisited: A multi-level meta-analysis of n-back training studies. *Psychonomic Bulletin & Review, 24*, 1077–1096.

Spencer-Smith, M., & Klingberg, T. (2015). Benefits of a working memory training program for inattention in daily life: A systematic review and meta-analysis. *PLoS ONE, 10*, e0119522.

Thompson, T. W., Waskom, M., & Gabrieli, J. D. (2016). Intensive working memory training produces functional changes in large-scale frontoparietal networks. *Journal of Cognitive Neuroscience, 28*, 575–588.

Titz, C., & Karbach, J. (2014). Working memory and executive functions: Effects of training on academic achievement. *Psychological Research, 78*, 852–868.

von Bastian, C. C., & Oberauer, K. (2013). Distinct transfer effects of training different facets of working memory capacity. *Journal of Memory and Language, 69*, 36–58.

von Bastian, C. C., & Oberauer, K. (2014). Effects and mechanisms of working memory training: A review. *Psychological Research, 78*, 803–820.

von Bastian, C. C., Locher, A., & Ruflin, M. (2013). Tatool: A Java-based open-source programming framework for psychological studies. *Behavior Research Methods, 45*, 108–115.

Weicker, J., Hudl, N., Frisch, S., Lepsien, J., Mueller, K., Villringer, A., & Thöne-Otto, A. (2018). WOME: Theory-based working memory training –A placebo-controlled, double-blind evaluation in older adults. *Frontiers in Aging Neuroscience, 10*, 247.

Wilhelm, O., Hildebrandt, A., & Oberauer, K. (2013). What is working memory, and how can we measure it? *Frontiers in Psychology, 4*, 433.

Episodic Memory Training

Elisabeth Wenger, Yana Fandakova, and Yee Lee Shing

Contents

Abstract Episodic memory circumscribes the ability to remember events, occurrences, and situations embedded in their temporal and spatial context – in other words, the memory of "what," "where," and "when." Throughout the lifespan, episodic memory functioning continuously undergoes extensive change, with rapid increases during childhood, some decreases in adulthood, and accelerated decline in very old age. Given the important role that episodic memory plays in our daily lives, the prospect of potential trainability of episodic memory is a highly attractive idea. This applies to educational settings that aim to facilitate children's memory, to younger adults hoping to optimize their episodic memory ability, but probably even more to older adults, who generally experience a profound decline in episodic memory functioning that can seriously affect their well-being and life quality. In this chapter, we first provide a brief definition and account of the processes that are involved

E. Wenger (✉) · Y. Fandakova
Center for Lifespan Psychology, Max Planck Institute for Human Development, Berlin, Germany
e-mail: wenger@mpib-berlin.mpg.de

Y. L. Shing
Center for Lifespan Psychology, Max Planck Institute for Human Development, Berlin, Germany

Institute of Psychology, Goethe University Frankfurt, Frankfurt am Main, Germany

T. Strobach, J. Karbach (eds.), *Cognitive Training*,
https://doi.org/10.1007/978-3-030-39292-5_12

in episodic memory. We then discuss two theoretical frameworks, one concerning the components of episodic memory across the lifespan, the other concerning the conception of plasticity. These theoretical frameworks help to guide us through the large body of memory training literature. We then summarize and integrate the behavioral and neuroimaging literature on memory training. Building upon some key points extracted from these sets of literature, we finally discuss the utility of multifactorial types of memory training and potential future work in this direction.

Introduction

Episodic memory circumscribes the ability to remember events, occurrences, and situations embedded in their temporal and spatial context – in other words, the memory of "what", "where," and "when" (Tulving 2002). It is the unique ability of humans to travel back in time and re-experience past events. To achieve this, elements belonging to the same event need to be associated with each other while being separated from other elements belonging to other events (Tulving 2002).

Throughout the lifespan, episodic memory functioning continuously undergoes extensive change, with rapid increases during childhood, some decreases in adulthood, and accelerated decline in very old age (Shing et al. 2010). Given the important role that episodic memory plays in our daily lives, the prospect of potential trainability of episodic memory is a highly attractive idea. This applies to educational settings that aim to facilitate children's memory (see Johann and Karbach, this volume), to younger adults hoping to optimize their episodic memory ability, but probably even more to older adults, who generally experience a profound decline in episodic memory functioning that can seriously affect their well-being and quality of life (see also Boller et al., Guye et al., this volume). In the following, we first provide a brief definition of the processes that are involved in episodic memory and lay out the conception of plasticity we subscribe to. We then summarize and integrate the behavioral and neuroimaging literature on memory training and aging. Building upon some key points extracted from these sets of literature, we finally discuss the utility of multifactorial types of memory training and potential future work in this direction.

Episodic Memory: Definition and Processes

Episodic memories – in contrast to semantic memories that are encyclopedic and not tied to a time or place – refer to specific episodes or events in a person's life. These memories are tied to the time and place in which the information was acquired. It follows that episodic memory as a form of explicit memory involves encoding, consolidation, and retrieval of events. When new declarative information is processed by the brain, it is presumed to be *encoded* by the medial temporal lobe (MTL) and then preserved in different cortical parts in the brain (Paller and Wagner 2002). The *consolidation* of memory traces is a process that stabilizes memory traces so they are preserved, and typically takes place during postlearning periods when the brain is not consciously encoding or retrieving a certain memory (McGaugh 2000; Dudai et al. 2015). *Memory retrieval* is assumed to approximate encoding processes in terms of activated brain regions (Nyberg et al. 2000) and is generally found to be dependent on the MTL as well as prefrontal cortex (PFC) and posterior parietal cortex (PPC).

Two-Component Framework of Episodic Memory

In an effort to combine and integrate neuronal and behavioral evidence, it has been proposed that episodic memory embodies two interacting components (Moscovitch 1992; Shing et al. 2010).

1. The *strategic* component refers to control processes that assist and coordinate memory processes at both encoding and retrieval. These processes may include elaboration and organization of memory content at encoding, and specification, verification, monitoring, and evaluation of relevant information at retrieval (e.g., Simons and Spiers 2003). On the neural level, the strategic component relies mostly on regions in PFC and PPC.
2. The *associative* component, on the other hand, refers to mechanisms that bind together different features of a memory item, different memory items, or a given memory episode and its context, into coherent representations, and is mediated by areas of the MTL.

Several behavioral experiments have indicated that these two components show distinct developmental trajectories across the lifespan (Brehmer et al. 2007). In short, the associative component has been found to be relatively functional by middle childhood, but exhibits age-related decline in older adults. These changes are thought to reflect the relatively earlier maturation of MTL during childhood, along with pronounced MTL declines in later adulthood. In contrast, the strategic component has been found to function at a level below that of young adults in children and older adults, most likely due to protracted maturation of PFC regions (and to some extent, of parietal regions as well) across childhood and early age-related decline in PFC regions across later parts of adulthood (Shing et al. 2010).

The Concept of Plasticity

Research has shown that the brain is malleable by experience – that is, plastic – from childhood to young adulthood and even into old age and can therefore adjust to new challenges, albeit to varying degrees (Lövdén et al. 2013; Kühn and Lindenberger 2016). In the conception that we subscribe to, plasticity denotes the capacity for change in brain structure induced by a mismatch between the demands of the environment and the current functional supply the brain can momentarily offer (Lövdén et al. 2010; see also Wenger and Kühn, this volume). In the majority of cases this can be accomplished through neuronal and behavioral variability and flexibility within a given equilibrium, that is, optimizing the use of already existent resources. If the system is capable of a response to altered requirements through this flexibility then no mismatch is experienced and no plastic (structural) change is necessary. However, if these processes do not suffice in fulfilling environmental demands, either due to dramatic changes in requirements or due to damaged functionality of the brain following brain injury, then change is demanded and can manifest in the form of plasticity. If the mismatch is too large, though, and new requirements are far too high for the momentary functional level of the brain, the system will not be able to assimilate in any way and plastic changes will not evolve. In other words, this model emphasizes that the system needs to experience mismatch, which means that the new environmental requirements need to lie between certain boundaries of task difficulty being not too high or too low in order to evince experience-dependent plastic changes. Such changes can then help the system to adapt to new circumstances.

Plasticity in Episodic Memory

Training programs designed to enhance memory performance have proliferated over the past decades and meta-analytic reviews support the efficacy of at least some of these types of memory training across a broad array of memory tasks (Verhaeghen et al. 1992; Lustig et al. 2009). Episodic memory can be trained by instructing people to use a specific strategy such as the Method of Loci,[1] name–face mnemonics, number mnemonics, story and sentence mnemonics, strategies altering the organization of material (categorization, chunking, associations, imagery) or optimizing basic processes like rehearsal or concentration, or even strategies making the best use of external memory cues (Gross et al. 2012). While attempts to train episodic memory via strategy instruction have dominated research on memory plasticity, there have

[1] In the Method of Loci, participants are presented with lists of words, which are learned by forming visual associations between the *n*th word and the *n*th place (locus) of a fixed trajectory of places (loci) scanned mentally by the participant. Retrieval occurs by taking a mental walk along the trajectory, retrieving the associated image at each locus, and deriving the original word from it.

also been a few attempts to target memory without strategy instruction. For example, in the repetition-lag training procedure (Jennings and Jacoby 2003) participants are given several trials of a continuous recognition task in which they have to use recollection to identify repeated items. After each trial, the number of intervening items between repetitions increases gradually. This incremented-difficulty approach has been shown to enhance the ability to recollect information across increasing delay intervals and has also been replicated and shown to generalize to at least some working memory tasks (e.g., Jennings et al. 2005; Bailey et al. 2011; Boller et al. 2012; Stamenova et al. 2014). Thus far, a variety of training routes have been shown to improve episodic memory performance in younger as well as older adults.

Age-Related Differences in the Efficacy of Memory Training

Given the pronounced developmental changes in the associative and strategic components of episodic memory across the lifespan, memory training benefits may vary across the lifespan depending on the degree to which different training regimes put different requirements on the two components. Turning to aging, cognitively healthy older adults are able to acquire and utilize memory strategies, even up to their 80s and can indeed improve their memory performance through this form of contextual support (Brehmer et al. 2014; see also Guye et al., this volume). Older adults often show much improvement in memory performance after strategy instruction, bringing them to the initial level of performance of younger adults before training. The benefits of strategy training in older adults can also be long-lasting – in the large ACTIVE trial participants in the memory training group showed increases in memory performance that were maintained up to 5 years after training (Willis et al. 2006). However, in terms of plasticity, younger adults seem to profit more from strategy-based memory-enhancing interventions than older adults do (Brehmer et al. 2007). This is clearly visible in the so-called testing-the-limits approach: after extensive training in serial recall of word lists with the Method of Loci (i.e., after 17 training sessions distributed over the course of more than 1 year), there was an almost perfect separation of age groups – a magnification of age differences in performance after training (Baltes and Kliegl 1992). Thus, while older adults can clearly benefit from strategy-based memory training, sometimes approaching or even reaching the initial performance level of younger adults, they do not benefit as much as younger adults do, leading to a magnification of age differences after training (see Fig. 1).

There may be several reasons for the above findings. For one, older adults may have difficulties in forming novel associations between landmarks and the to-be remembered information, for example due to age-related decline in MTL regions, which are crucial for the associative component of episodic memory. They may also have difficulties in the use of mental imagination for memorization and find it difficult to form bizarre or unnatural images, which is important for the efficacy of the Method of Loci. Both of these explanations would add up to being a barrier

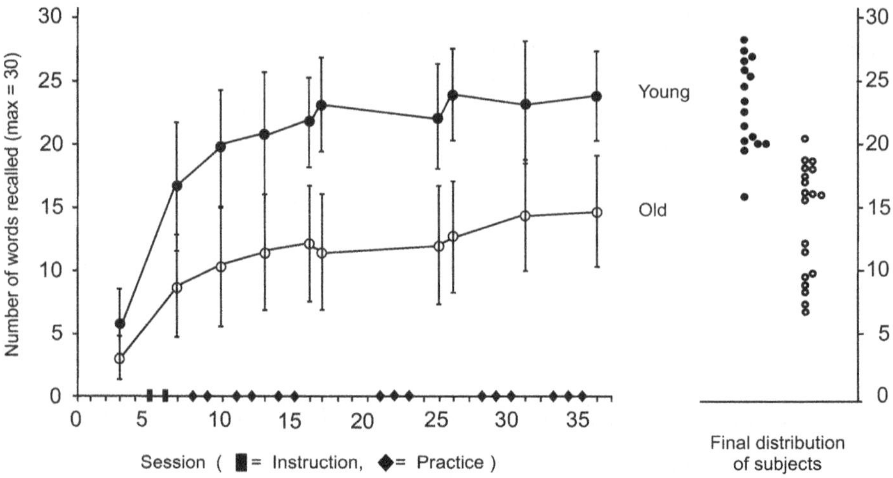

Fig. 1 Training effects and magnification of age differences. Older adults can benefit from memory training, thereby reaching initial performance levels of younger adults. However, younger adults benefit even more, leading to a magnification of age differences after training. (Adapted from Baltes and Kliegl 1992)

when using rather than acquiring the mnemonic. Age differences in compliance regarding the use of the taught strategy may also play a role. In addition, critical variables associated with learning in old age (e.g., the speed of mental operations) are characterized by large age differences favoring the young. Age differences in episodic memory performance could then be magnified by training (i.e., an amplification model) partly because the abilities known as fluid intelligence (working memory, executive control, perceptual speed) are reduced with advancing age (Verhaeghen and Marcoen 1996). While between-person and therefore age differences can be reduced after initial mnemonic instruction (i.e., compensation), age differences are magnified after extensive adaptive practice because baseline performance and general cognitive resources correlate positively with training gains, leading to reduced memory plasticity in older adults (Lövdén et al., 2012).

The repetition-lag training procedure has been consistently shown to improve older adult's recollection (Jennings and Jacoby 2003) with effects maintained up to 3 months after training (Anderson et al. 2018). However, evidence for generalizable benefits beyond the trained verbal task has been mixed. Other approaches focusing on perceptual processing have shown that training auditory perception tasks increased older adults' memory for auditory information (Mahncke et al. 2006). Finally, one might expect that training executive control processes, which are central to the strategic component, may help improve episodic memory (Ranganath et al. 2011). And indeed adaptive working memory updating training has been shown to transfer to an untrained episodic memory task (Flegal et al. 2019; see also Könen et al., this volume). Such a possibility is particularly interesting with respect to aging as older adults show an increased tendency to falsely remember details of events that did not happen in the past due to age-related declines in monitoring and control in the PFC (Fandakova et al. 2012). Thus, programs that aim at improving executive control may help reduce memory errors across the lifespan, and in particular in later adulthood (see also Karbach and Kray, this volume).

Children's episodic memory performance can also be improved through instruction and practice, and even more so than older adults': Children can advance to the trained level of young adults when they have the chance to practice the newly learnt strategy extensively (Brehmer et al. 2007; see also Rueda et al., this volume). In this case, instruction gains may primarily reflect developmental and individual differences in the strategic component of memory – that is, the current ability of individuals to make use of the newly learnt mnemonic strategy to actively organize (or categorize) the to-be-remembered material. Practice gains on the other hand, which are much stronger in children than in older adults, may then reflect developmental and individual differences in the associative component of episodic memory more than differences in the strategic component – that is, individuals' latent potential in fine-tuning mechanisms involved in the execution of the mnemonic strategy to optimize the formation and retrieval of new associations.

Taken together, this evidence reveals that while both children and older adults benefit from memory strategy instruction, only children can improve significantly more through extensive training and practice because they can rely upon the associative component of memory, which is relatively mature. Older adults, on the other hand, show deterioration in the associative component such that even after the strategic deficit has been reduced by strategy instruction, they are limited in their memory improvement.

Training-Related Changes on the Neural Level

Given the improvement in behavioral performance, it is unsurprising that some studies (mostly employing the Method of Loci training) have found associated change in brain activation. A comparison of encoding before and after instruction revealed increased activity in frontal areas and fusiform gyrus, and recall after instruction additionally showed significant activation in parahippocampal gyrus and parietal regions as compared to recall before instruction (Kondo et al. 2005). Maguire and colleagues investigated superior memorizers in contrast to control subjects and found increased activation during encoding in very similar regions: namely, medial parietal cortex, retrosplenial cortex, and right posterior hippocampus (Maguire et al. 2003). Importantly, nearly all of the superior memorizers in this study happened to use a spatial learning strategy like the Method of Loci. In general, the activation of frontal regions in these studies underlines the increased engagement of control processes and thus the strategic component, with more posterior, parietal activation pointing to the specific involvement of imagery due to the nature of the training, while activation in temporal lobe indicates increased engagement of the associative component.

In the context of an aging study, Nyberg and colleagues demonstrated increased activity during memory encoding in occipital-parietal and frontal brain regions after learning the Method of Loci in young adults. Older adults did not show increased frontal activity, and only those older participants who had benefited from the mnemonic exhibited increased occipital-parietal activity (Nyberg et al. 2003). Focusing on memory retrieval, a semantic strategy training was found to improve older adults' word recollection along with increased hippocampal/MTL activity during retrieval (Kirchhoff et al. 2012). Notably, older adults who showed greater training-related changes in MTL activity also showed greater training-related increases in PFC during semantic elaboration at encoding. Together, these findings suggest that age-related differences in memory plasticity may reflect both diminished processing resources along with failure to engage those resources appropriately in crucial task-relevant processing. Interestingly, a study focusing on encoding success (i.e., successful memory formation) instead of encoding processes generally, regardless of outcome (as in the case of Nyberg et al. 2003), found no age differences in neural activation, but rather comparable training-induced activation changes across the lifespan (Brehmer et al. 2016). This might speak to the proposition that brain areas supporting successful memory encoding following strategy instruction and practice remain quite stable across the lifespan, particularly in those older adults that have more youth-like brains, such as the positively selected study sample in the Brehmer et al. study (brain maintenance hypothesis of cognitive aging, Nyberg et al. 2012).

A few studies have also investigated changes in brain structure evoked by memory training. Eight weeks of training with the Method of Loci have been shown to result in improved memory performance along with increases in cortical thickness in the right insula, left and right orbitofrontal cortex, and right fusiform cortex (Engvig et al. 2010). This memory training additionally led to maintenance of frontal fractional anisotropy, a measure of white matter integrity, as compared to a control group that showed decreases over the course of training. Older adults who showed maintenance or increase in frontal white matter also showed greater improvement in memory performance (Engvig et al. 2012b). Another training paradigm for episodic memory has been implemented in the form of vocabulary learning. As the matching of an unknown word with a semantic meaning comes close to the concept of associative memory (Davis and Gaskell 2009) and language learning holds a high motivational aspect, it provides ideal grounds to investigate episodic memory plasticity. Mårtensson et al. (2012) studied changes in brain structure following 3 months of intense foreign-language acquisition in Swedish interpreters. Results showed increases in hippocampal volume and in cortical thickness in the left middle frontal gyrus, inferior frontal gyrus, and superior temporal gyrus for interpreters compared to a control group, whereby some of these regions showed a correlation with behavioral measures of proficiency or struggling (Mårtensson et al. 2012). But even in less intensive regimes, for example when learning Italian vocabulary at a more normal pace, improvements in memory performance were associated with increases in hippocampal volume, independent of time devoted to the studies and amount of acquired vocabulary (Bellander et al. 2016).

To date, memory training studies have focused primarily on memory at short intervals after studying the to-be-remembered information. At the same time, using an effective strategy or creating a well-bound representation of different events can positively affect the longevity of memory traces, making them more resistant to forgetting. Thus, it is possible that memory training reduces forgetting at longer delays via improved strategy use and associative binding. On the other hand, in animal research increased neurogenesis in the hippocampus has been associated with forgetting due to weakening of existing memories while at the same time facilitating encoding of new memories (Akers et al. 2014; Epp et al. 2016). This research suggests that to the extent to which structural changes in the hippocampus may reflect neurogenesis (see Wenger and Kühn, this volume), interventions that promote hippocampal growth may increase rates of forgetting of existing memories while at the same improving new learning. To test these predictions, one would need to extend existing measures to include delayed memory and measures of learning as opposed to pure performance at a given point in time.

Strategy Instruction – Manifestations of Plasticity or Not?

Instructions for the use of a new strategy to improve episodic memory performance can be viewed as a case study for the plasticity model introduced above (Lövdén et al. 2010, 2013). Can we regard functional changes as evidence for plasticity when participants show improved performance after instruction for strategy use? Following the theoretical definition laid out above, a more or less immediate change in behavioral performance and its accompanying change in functional activation due to strategy instruction would not be considered as manifested plasticity but rather as flexibility. In the words of Paul Baltes, this improvement following instruction could be termed *baseline reserve capacity*, namely what an individual is capable of when the conditions of assessment are optimized, that is, providing for an extended range of possible performances with additional resources (Baltes 1987). *Developmental reserve capacity*, on the other hand, would then be the plasticity as defined in our theoretical model above, namely a further extension of performance range after conditions have been altered, with the aim of full activation, and possibly expansion, of an individual's task-relevant cognitive or neural resources. The strongest evidence for such developmental reserve capacity or plasticity would then be given if memory performance as such *generally* improved after strategy training, even if the newly acquired strategy was not used at this specific moment. Theoretically, the extensive use of such a mnemonic technique as the Method of Loci may enable an aged individual to re-challenge brain regions important for episodic memory tasks that have become under-challenged due to age-related decline. The heightened recruitment and engagement of these brain regions may then evoke macroscopic changes in brain structure – hence manifestations of plasticity.

Who Benefits the Most from Memory Training?

As highlighted above, training benefits vary considerably across individuals and age groups (see also Cochrane and Green, Karbach and Kray, Katz et al., Könen et al., this volume). With strategy instruction of the Method of Loci, children and younger and older adults showed a similar pattern such that participants who started out with the lowest performance showed the greatest benefit from strategy instruction, consistent with the idea of flexibility and baseline reserve capacity (Baltes 1987; Lövdén et al. 2012). At the same time, among children, those who had higher baseline ability showed greater benefit from adaptive practice of the strategy.

In addition, for older adults with a relatively spared strategic component, it may be sufficient to arrange the environment in a way that promotes the use of an effective strategy, whereas for older adults with more pronounced declines a more directed instruction of a strategy may yield the largest memory benefit (Fandakova et al. 2012). Similar effects have been reported using the repetition-lag procedure such that older adults who spent more time encoding an item in a proactive controlled manner were also the ones who showed the largest training benefits (Bissig and Lustig 2007). On the neural level, among older adults with memory complaints, individuals with larger hippocampal volumes showed larger improvements with memory training, possibly reflecting greater potential for change with an intervention (Engvig et al. 2012a).

Together, these examples suggest that successful boosting of memory performance may be achieved through different training manipulations, depending on the functional status of the associative and strategic memory components. While research on individual differences has focused primarily on memory encoding and/or retrieval, consolidation processes, especially in relation to sleep constitute another potentially important predictor of memory training gains that shows considerable heterogeneity across the lifespan (e.g., Muehlroth et al. 2019).

Combination of Training Types to Enhance Generalizability and Maintenance

In general, it seems to be beneficial, if not necessary, for the enhanced magnitude and preservation of behavioral effects to combine training of mnemonic techniques with other important factors affecting memory performance. A crucial limitation of targeted training interventions has been the widespread inability to sustain and generalize (i.e., transfer) the benefits of training in a specific strategy beyond the tasks actually used for training (Noack et al. 2014). The most promising results have been provided by multifactorial interventions, in which different memory enhancing techniques were combined with training of other skills (e.g., attention and relaxation). Under these circumstances, memory performance can improve and be sustained for up to 3.5 years (Stigsdotter Neely and Bäckman 1993). Stigsdotter Neely and Bäckman provide well-founded arguments for the benefit of involving several critical aspects of memory functioning in memory training programs if they are to be maximally effective. Age-related deficits in episodic memory have an array of different sources (Bäckman 1989). Deficient retrieval mechanisms alone, or impaired encoding and retrieval mechanisms could just as well play a role as attentional deficits. Older adults also seem to be disadvantaged with respect to a number of noncognitive factors, such as laboratory anxiety and level of arousal. As memory deficits accompanying the aging process have several origins, efforts to alleviate these deficits should ideally be multifactorial as well, to best target the problems. Training of encoding operations to provide effective strategies for organization and visualization of the material could then be combined with training of attentional skills – to improve concentration, focusing of attention, and vigilance, all of which are necessary to meet the attentional demands of remembering – and should additionally be combined with training to reduce levels of situational anxiety. Specific pretraining techniques focusing on image elaboration, verbal judgement, and relaxation have also been shown to enhance the application of a mnemonic technique and helped to maintain its efficacy (Sheikh et al. 1986).

Boosting Memory Training: A Promising Future Training Paradigm

Furthermore, reaching beyond the rationale for multifactorial combined training, we would like to emphasize that physical exercise intervention also needs to be taken into consideration (see also Bherer and Pothier, this volume). In particular, this applies to children and older adults whose bodily functioning is also undergoing pronounced changes that may have strong implications for cognition. Observational studies continue to suggest that adults who engage in physical activity have a reduced risk of cognitive decline and dementia (Düzel et al. 2016). Exercise can exert a protective effect, even if initiated in later life. Although the mechanisms through which physical exercise affects cognition and especially episodic memory are not yet fully understood, there is growing evidence that selected aspects of cognition are responsive to increases in physical exercise (Cotman and Berchtold 2002). This association is obviously particularly relevant in children, where numerous studies have now shown that children's aerobic fitness is associated with higher levels of cognition and differences in regional brain structure and function and that aerobic fitness levels can predict cognition over time (Chaddock et al. 2011). Also in adulthood the powerful influence of exercise training has been shown repeatedly. For example, Erickson and colleagues observed that the hippocampus increased in size after 1 year of moderate exercise, and this structural change was correlated with changes in spatial memory performance (Erickson et al. 2011; but note that memory changes did not differ between experimental and control groups). Another study reported selective increases in cerebral blood volume in dentate gyrus – possibly an indicator for exercise-induced neurogenesis – after 3 months of exercising, which correlated with changes in cognitive performance (Pereira et al. 2007). In an earlier study, a combination of mental and physical training led to greater effects on a memory score than either activity alone (Fabre et al. 2002). The mental training program was multifactorial and comprised tasks involving perception, attention, association, and imagination.

Taken together, we propose that future studies should focus on such multidomain training approaches based on findings from the animal literature. Researchers examining rodents have emphasized both cognitive enrichment and enhanced physical activity as the driving forces behind plastic changes (Kempermann et al. 2010). One can speculate that physical activity may not only enhance cognition directly but also improve plasticity as the capacity for change per se. Physical activity may therefore boost the effects of cognitive enrichment or training on both the behavioral and the neural level. Such an additive effect of physical exercise and environmental enrichment has been shown before in the mouse hippocampus (Fabel et al. 2009). Voluntary physical exercise and environmental enrichment both stimulate adult hippocampal neurogenesis in mice, but via different mechanisms. That is, running in a wheel induces precursor cell proliferation, whereas environmental enrichment exerts a survival-promoting effect on newborn cells. Fabel and colleagues reported an increased potential for neurogenesis in that proliferating precursor cells were activated by running and then received a survival-promoting stimulus due to environmental enrichment following the exercise. Ten days of running followed by 35 days of environmental enrichment were additive such that the combined stimulation resulted in a 30% greater increase in new neurons as compared to either paradigm alone (see Fig. 2; Fabel et al. 2009). Translated to the human hippocampus, this may mean that physical exercise could stimulate proliferating precursor cells that would then be more likely to survive if challenged by appropriate cognitive enrichement relying on the hippocampal structure, as for example memory demands. In this way, physical exercise could first "prepare" the aged MTL for increased usage. Hypothetically, any ensuing strategy instruction and specifically the practice of memory strategies could then – and perhaps only then – be successfully and fully exploited. As the associative and strategic components of memory function in intricate ways and are critically important for episodic memory performance, it seems to be a promising route to target both components and the neural regions underlying their functioning, namely MTL and frontal lobe, in a combined multidomain training paradigm.

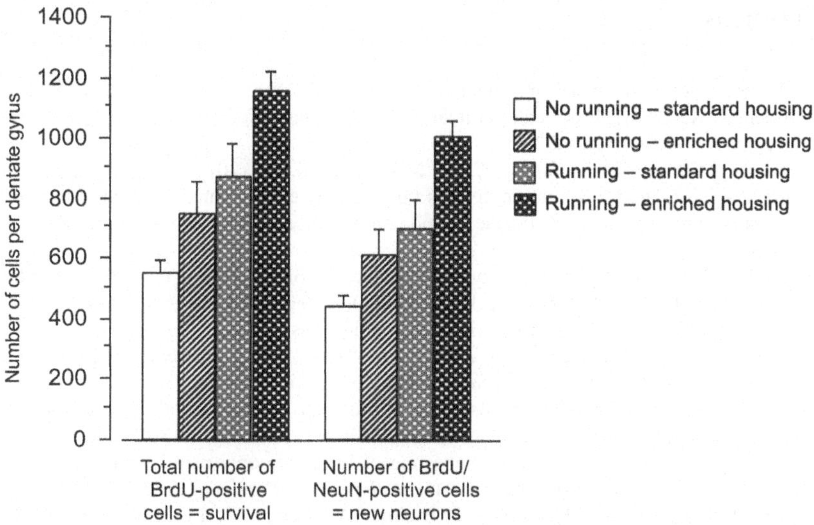

Fig. 2 Additive effect of physical exercise and environmental enrichment in the mouse hippocampus. Voluntary wheel running and enriched housing have each been shown to result in an increased number of cell labels with Bromodeoxyuridine (BrdU) and new neurons compared to no running in standard housing. Combined running and enriched housing results in an even greater increase of BrdU-positive cells and newborn neurons. BrdU is commonly used to detect proliferating cells in living tissue. (Adapted from Fabel et al. 2009)

Concluding Remarks

Put simply, episodic memory can be trained. Children as well as younger and older adults profit from training, most often shown using strategy instruction, and it is encouraging to see that older adults can reach initial performance levels of younger adults after strategy instruction. Importantly, such performance gains most likely reflect manifestations of flexibility – defined as the adaptive reconfiguration of the existing functional and structural repertoire, and if implemented correctly, rely most heavily on the strategic component of memory, that is, on prefrontal regions of the brain. Further performance gains following extensive practice are then most likely to be manifestations of plasticity. Unlike flexibility, plasticity does not only make use of preexisting neural resources, but also changes them fundamentally. Here, older adults show reduced levels of plasticity compared to children and younger adults, as indicated by their lower performance gains following practice. In our view, one promising route for intervention is to provide older adults with memory training in combination with physical exercise to revitalize plasticity and thereby boost training effectiveness. Strategy training alone may be too narrow an intervention to result in substantive transfer and lasting maintenance of acquired skills. Currently, combined memory training types, most promisingly in concert with physical exercise, seem to be the best bet to not only target the strategic, but also the associative component of memory, thereby hopefully having a widespread and lasting effect on memory functioning.

References

Akers, K. G., Martinez-Canabal, A., Restivo, L., Yiu, A. P., De Cristofaro, A., Hsiang, H. L. L., ... & Ohira, K. (2014). Hippocampal neurogenesis regulates forgetting during adulthood and infancy. *Science, 344*, 598–602.

Anderson, N. D., Ebert, P. L., Grady, C. L., & Jennings, J. M. (2018). Repetition lag training eliminates age-related recollection deficits (and gains are maintained after three months) but does not transfer: Implications for the fractionation of recollection. *Psychology and Aging, 33*, 93–108.

Bäckman, L. (1989). Varieties of memory compensation by older adults in episodic remembering. In L. W. Poon, D. C. Rubin, & B. A. Wilson (Eds.), *Everyday cognition in adulthood and late life* (pp. 509–544). New York: Cambridge University Press.

Bailey, H., Dagenbach, D., & Jennings, J. M. (2011). The locus of the benefits of repetition-lag memory training. *Aging, Neuropsychology, and Cognition, 18*, 577–593.

Baltes, P. B. (1987). Theoretical propositions of life-span developmental psychology: On the dynamics between growth and decline. *Developmental Psychology, 23*, 611–626.

Baltes, P. B., & Kliegl, R. (1992). Further testing of limits of cognitive plasticity: Negative age differences in a mnemonic skill are robust. *Developmental Psychology, 28*, 121–125.

Bellander, M., Berggren, R., Mårtensson, J., Brehmer, Y., Wenger, E., Li, T. Q., ... & Lövdén, M. (2016). Behavioral correlates of changes in hippocampal gray matter structure during acquisition of foreign vocabulary. *NeuroImage, 131*, 205–213.

Bissig, D., & Lustig, C. (2007). Who benefits from memory training? *Psychological Science, 18*, 720–726.

Boller, B., Jennings, J. M., Dieudonné, B., Verny, M., & Ergis, A. M. (2012). Recollection training and transfer effects in Alzheimer's disease: Effectiveness of the repetition-lag procedure. *Brain and Cognition, 78*, 169–177.

Brehmer, Y., Li, S. C., Müller, V., von Oertzen, T., & Lindenberger, U. (2007). Memory plasticity across the life span: uncovering children's latent potential. *Developmental Psychology, 43*, 465–478.

Brehmer, Y., Kalpouzos, G., Wenger, E., & Lövdén, M. (2014). Plasticity of brain and cognition in older adults. *Psychological Research, 78*, 790–802.

Brehmer, Y., Shing, Y. L., Heekeren, H. R., Lindenberger, U., & Bäckman, L. (2016). Training-induced changes in subsequent-memory effects: No major differences among children, younger adults, and older adults. *NeuroImage, 131*, 214–225.

Chaddock, L., Pontifex, M. B., Hillman, C. H., & Kramer, A. F. (2011). A review of the relation of aerobic fitness and physical activity to brain structure and function in children. *Journal of the International Neuropsychological Society, 17*, 975.

Cotman, C. W., & Berchtold, N. C. (2002). Exercise: A behavioral intervention to enhance brain health and plasticity. *Trends in Neurosciences, 25*, 295–301.

Davis, M. H., & Gaskell, M. G. (2009). A complementary systems account of word learning: Neural and behavioural evidence. *Philosophical Transactions of the Royal Society, B: Biological Sciences, 364*, 3773–3800.

Dudai, Y., Karni, A., & Born, J. (2015). The consolidation and transformation of memory. *Neuron, 88*, 20–32.

Düzel, E., Van Praag, H., & Sendtner, M. (2016). Can physical exercise in old age improve memory and hippocampal function? *Brain, 139*, 662–673.

Engvig, A., Fjell, A. M., Westlye, L. T., Moberget, T., Sundseth, Ø., Larsen, V. A., & Walhovd, K. B. (2010). Effects of memory training on cortical thickness in the elderly. *NeuroImage, 52*, 1667–1676.

Engvig, A., Fjell, A. M., Westlye, L. T., Skaane, N. V., Sundseth, Ø., & Walhovd, K. B. (2012a). Hippocampal subfield volumes correlate with memory training benefit in subjective memory impairment. *NeuroImage, 61*, 188–194.

Engvig, A., Fjell, A. M., Westlye, L. T., Moberget, T., Sundseth, Ø., Larsen, V. A., & Walhovd, K. B. (2012b). Memory training impacts short-term changes in aging white matter: A longitudinal diffusion tensor imaging study. *Human Brain Mapping, 33*, 2390–2406.

Epp, J. R., Mera, R. S., Köhler, S., Josselyn, S. A., & Frankland, P. W. (2016). Neurogenesis-mediated forgetting minimizes proactive interference. *Nature Communications, 7*, 1–8.

Erickson, K. I., Voss, M. W., Prakash, R. S., Basak, C., Szabo, A., Chaddock, L., … & Wojcicki, T. R. (2011). Exercise training increases size of hippocampus and improves memory. *Proceedings of the National Academy of Sciences, 108*, 3017–3022.

Fabel, K., Wolf, S., Ehninger, D., Babu, H., Galicia, P., & Kempermann, G. (2009). Additive effects of physical exercise and environmental enrichment on adult hippocampal neurogenesis in mice. *Frontiers in Neuroscience, 3*, 2.

Fabre, C., Chamari, K., Mucci, P., Masse-Biron, J., & Prefaut, C. (2002). Improvement of cognitive function by mental and/or individualized aerobic training in healthy elderly subjects. *International Journal of Sports Medicine, 23*, 415–421.

Fandakova, Y., Shing, Y. L., & Lindenberger, U. (2012). Heterogeneity in memory training improvement among older adults: A latent class analysis. *Memory, 20*, 554–567.

Flegal, K. E., Ragland, J. D., & Ranganath, C. (2019). Adaptive task difficulty influences neural plasticity and transfer of training. *NeuroImage, 188*, 111–121.

Gross, A. L., Parisi, J. M., Spira, A. P., Kueider, A. M., Ko, J. Y., Saczynski, J. S., ... & Rebok, G. W. (2012). Memory training interventions for older adults: A meta-analysis. *Aging & Mental Health, 16*, 722–734.

Jennings, J. M., & Jacoby, L. L. (2003). Improving memory in older adults. *Neuropsychological Rehabilitation, 13*, 417–440.

Jennings, J. M., Webster, L. M., Kleykamp, B. A., & Dagenbach, D. (2005). Recollection training and transfer effects in older adults: Successful use of a repetition-lag procedure. *Aging, Neuropsychology, and Cognition, 12*, 278–298.

Kempermann, G., Fabel, K., Ehninger, D., Babu, H., Leal-Galicia, P., Garthe, A., & Wolf, S. (2010). Why and how physical activity promotes experience-induced brain plasticity. *Frontiers in Neuroscience, 4*, 189.

Kirchhoff, B. A., Anderson, B. A., Barch, D. M., & Jacoby, L. L. (2012). Cognitive and neural effects of semantic encoding strategy training in older adults. *Cerebral Cortex, 22*, 788–799.

Kondo, Y., Suzuki, M., Mugikura, S., Abe, N., Takahashi, S., Iijima, T., & Fujii, T. (2005). Changes in brain activation associated with use of a memory strategy: A functional MRI study. *NeuroImage, 24*, 1154–1163.

Kühn, S., & Lindenberger, U. (2016). Research on human plasticity in adulthood: A lifespan agenda. In K. W. Schaie & S. L. Willis (Eds.), *Handbook on the psychology of aging* (8th ed., pp. 105–123). Amsterdam: Academic Press. https://doi.org/10.1016/B978-0-12-411469-2.00006-6.

Lövdén, M., Bäckman, L., Lindenberger, U., Schaefer, S., & Schmiedek, F. (2010). A theoretical framework for the study of adult cognitive plasticity. *Psychological Bulletin, 136*, 659–676.

Lövdén, M., Brehmer, Y., Li, S. C., & Lindenberger, U. (2012). Training-induced compensation versus magnification of individual differences in memory performance. *Frontiers in Human Neuroscience, 6*, 141.

Lövdén, M., Wenger, E., Mårtensson, J., Lindenberger, U., & Bäckman, L. (2013). Structural brain plasticity in adult learning and development. *Neuroscience & Biobehavioral Reviews, 37*, 2296–2310.

Lustig, C., Shah, P., Seidler, R., & Reuter-Lorenz, P. A. (2009). Aging, training, and the brain: A review and future directions. *Neuropsychology Review, 19*, 504–522.

Maguire, E. A., Valentine, E. R., Wilding, J. M., & Kapur, N. (2003). Routes to remembering: The brains behind superior memory. *Nature Neuroscience, 6*, 90–95.

Mahncke, H. W., Connor, B. B., Appelman, J., Ahsanuddin, O. N., Hardy, J. L., Wood, R. A., ... & Merzenich, M. M. (2006). Memory enhancement in healthy older adults using a brain plasticity-based training program: A randomized, controlled study. *Proceedings of the National Academy of Sciences, 103*, 12523–12528.

Mårtensson, J., Eriksson, J., Bodammer, N. C., Lindgren, M., Johansson, M., Nyberg, L., & Lövdén, M. (2012). Growth of language-related brain areas after foreign language learning. *NeuroImage, 63*, 240–244.

McGaugh, J. L. (2000). Memory: A century of consolidation. *Science, 287*, 248–251.

Moscovitch, M. (1992). Memory and working-with-memory: A component process model based on modules and central systems. *Journal of Cognitive Neuroscience, 4*, 257–267.

Muehlroth, B. E., Sander, M. C., Fandakova, Y., Grandy, T. H., Rasch, B., Shing, Y. L., & Werkle-Bergner, M. (2019). Precise slow oscillation–Spindle coupling promotes memory consolidation in Younger and older adults. *Scientific Reports, 9*, 1–15.

Noack, H., Lövdén, M., & Schmiedek, F. (2014). On the validity and generality of transfer effects in cognitive training research. *Psychological Research, 78*, 773–789.

Nyberg, L., Habib, R., McIntosh, A. R., & Tulving, E. (2000). Reactivation of encoding-related brain activity during memory retrieval. *Proceedings of the National Academy of Sciences, 97*, 11120–11124.

Nyberg, L., Sandblom, J., Jones, S., Neely, A. S., Petersson, K. M., Ingvar, M., & Bäckman, L. (2003). Neural correlates of training-related memory improvement in adulthood and aging. *Proceedings of the National Academy of Sciences, 100*, 13728–13733.

Nyberg, L., Lövdén, M., Riklund, K., Lindenberger, U., & Bäckman, L. (2012). Memory aging and brain maintenance. *Trends in Cognitive Sciences, 16*, 292–305.

Paller, K. A., & Wagner, A. D. (2002). Observing the transformation of experience into memory. *Trends in Cognitive Sciences, 6*, 93–102.

Pereira, A. C., Huddleston, D. E., Brickman, A. M., Sosunov, A. A., Hen, R., McKhann, G. M., ... & Small, S. A. (2007). An in vivo correlate of exercise-induced neurogenesis in the adult dentate gyrus. *Proceedings of the National Academy of Sciences, 104*, 5638–5643.

Ranganath, C., Flegal, K. E., & Kelly, L. L. (2011). Can cognitive training improve episodic memory? *Neuron, 72*, 688–691.

Sheikh, J. I., Hill, R. D., & Yesavage, J. A. (1986). Long-term efficacy of cognitive training for age-associated memory impairment: A six-month follow-up study. *Developmental Neuropsychology, 2*, 413–421.

Shing, Y. L., Werkle-Bergner, M., Brehmer, Y., Müller, V., Li, S. C., & Lindenberger, U. (2010). Episodic memory across the lifespan: The contributions of associative and strategic components. *Neuroscience & Biobehavioral Reviews, 34*, 1080–1091.

Simons, J. S., & Spiers, H. J. (2003). Prefrontal and medial temporal lobe interactions in long-term memory. *Nature Reviews Neuroscience, 4*, 637–648.

Stamenova, V., Jennings, J. M., Cook, S. P., Walker, L. A., Smith, A. M., & Davidson, P. S. (2014). Training recollection in healthy older adults: Clear improvements on the training task, but little evidence of transfer. *Frontiers in Human Neuroscience, 8*, 898.

Stigsdotter Neely, A., & Bäckman, L. (1993). Long-term maintenance of gains from memory training in older adults: Two 3 1/2-year follow-up studies. *Journal of Gerontology, 48*, P233–P237.

Tulving, E. (2002). Episodic memory: From mind to brain. *Annual Review of Psychology, 53*, 1–25.

Verhaeghen, P., & Marcoen, A. (1996). On the mechanisms of plasticity in young and older adults after instruction in the method of loci: Evidence for an amplification model. *Psychology and Aging, 11*, 164–178.

Verhaeghen, P., Marcoen, A., & Goossens, L. (1992). Improving memory performance in the aged through mnemonic training: A meta-analytic study. *Psychology and Aging, 7*, 242–251.

Willis, S. L., Tennstedt, S. L., Marsiske, M., Ball, K., Elias, J., Koepke, K. M., ... & Wright, E. (2006). Long-term effects of cognitive training on everyday functional outcomes in older adults. *JAMA, 296*, 2805–2814.

Prospective Memory Training

Sharda Umanath, Joan Toglia, Francis Anderson, Erin Foster, and Mark A. McDaniel

Contents

Abstract Prospective memory (PM) involves remembering to perform intended actions in the future. PM failures are especially problematic for older adults, both in terms of frequency of occurrence and severity of consequences. As such, we tackle the challenge of developing a cognitive training program for PM specifically geared towards older adults. Departing from other popular cognitive training, our focus has been and continues to be on teaching effective and efficient strategies with the intention of promoting transfer to real-world PM challenges. We discuss several considerations in cognitive training including matching the type of PM task (focal or nonfocal) with effective strategies, variability and characteristics of training materials, and differences in methods used to train strategies. For example, training can involve explicit direct instruction or guided instruction aimed at helping a person self-generate and self-evaluate strategy effectiveness. Existing data and ongoing work aimed at identifying the key intervention components that enhance successful outcomes are presented. We report a new study with healthy older adults that includes these components and develops a metacognitive-strategy intervention for

S. Umanath
Claremont McKenna College, Claremont, CA, USA

J. Toglia
Mercy College, Palakkad, India

F. Anderson · E. Foster · M. A. McDaniel (✉)
Washington University in St. Louis, St. Louis, MO, USA
e-mail: markmcdaniel@wustl.edu

T. Strobach, J. Karbach (eds.), *Cognitive Training*,
https://doi.org/10.1007/978-3-030-39292-5_13

prospective memory that guides participants in analysis of task demands and self-generation of strategies. We also describe some initial prospective-memory training work with Parkinson's disease patients.

Prospective Memory

Prospective memory (PM) refers to tasks in which one must remember to carry out an intended action at some point in the future. Good PM is vital in everyday life (McDaniel and Einstein 2007), whether remembering an appointment, paying a bill, or taking a prescription. While PM is important for everyone, the consequences of failure can be much greater for older adults. A missed doctor's appointment or a forgotten pill can have dire repercussions. In addition, older adults complain most about PM failures compared to other memory issues (McDaniel and Einstein), and PM ability declines with age, at least for some types of PM (for a review, see Henry et al. 2004). Given the potential beneficial impact, PM is an ideal target for training, especially in older adults. Yet, very few cognitive training programs in general, or specifically for older adults, have attempted to train PM (see Waldum et al. 2016, for review). Here, we first discuss the theoretical approach—including *what* to train and *how* to train it—that has guided our attempts to train PM. We then provide evidence from existing data and current preliminary work supporting and informing this approach.

Theoretical Approach

The fundamental goal in developing a training protocol for PM and a standard goal in cognitive training is to promote transfer or generalization beyond the context of learning. That is, training that learners undergo should be helpful beyond the laboratory and applicable in the real world (see also Guye et al., Karbach and Kray, Könen et al., Rueda et al., Strobach and Schubert, Swaminathan and Schellenberg, this volume). However, transfer following cognitive training has been elusive (see Hertzog et al. 2009; McDaniel and Bugg 2012). With this challenge in mind, our broad approach is to look at existing literature and focus on identifying effective PM strategies that learners can be explicitly taught to apply and generalize more broadly. This is a somewhat innovative approach as other cognitive training protocols have embraced different underlying assumptions. For example, some cognitive training has taken a *restorative* approach, attempting to enhance the underlying neural physiology to improve cognition (see Lustig et al. 2009, for review; Taatgen, Wenger et al., this volume). Other cognitive training programs include only practice of relevant tasks rather than explicit instruction on how to approach them (e.g., for attentional control: Karbach and Kray 2009; Kramer et al. 1995; Mackay-Brandt 2011; for retrospective-memory control: Jennings and Jacoby 2003; for working memory: Harrison et al. 2013; Redick et al. 2013; see Könen et al., this volume). Even one of the very few training programs aimed at improving PM used only practice and was only somewhat successful in producing transfer (Rose et al. 2015). In contrast, rather than attempting to modify the nervous system or rely on learners gaining spontaneous insights into how to best handle PM tasks through repetitive practice, our approach is to teach effective, efficient strategies with which learners can tackle PM tasks.

We adopted such an approach for several interrelated reasons. First, the PM literature has revealed that dissociated processes underlie different PM tasks (described below), as opposed to perhaps more unitary skills (tasks) that seem to submit to restorative or practice-alone regimens (e.g., attentional control and working memory). Second, PM strategies have been identified that we assume are directly useful in everyday PM tasks (unlike some trained retrospective memory strategies; cf. McDaniel and Bugg 2012). Of note is that PM in the laboratory is quite different than PM in the real world. PM tasks that are encountered in everyday life are widely variable and occur in a myriad of contexts; for example, they include remembering to put a rent check in the mail every month, remembering to pick up a friend at the airport, and remembering to give a housemate a message. By contrast, laboratory PM tasks involve remembering to press a particular key when a given target appears (e.g., the word *president* or the syllable *tor*) during an ongoing task (e.g., answering trivia questions; Einstein et al. 1995). Thus, a challenge for a PM training program is creating strong connections between the laboratory training context and the situations learners are faced with in their daily lives (see also Guye et al., this volume). Because practice alone can produce brittle skills that are tightly tied to training (e.g., Healy et al. 2005), we felt that appropriately selected strategies and training could better allow learners to link the laboratory context to everyday PM situations. In fact, Bottiroli et al. (2013) found benefits of a strategy approach for promoting transfer—on retrospective memory tasks—specifically with older adults (see also Wenger et al., this volume; PRIMs Theory in Taatgen, this volume). Third, available evidence suggested that these strategies might help override age-related cognitive limitations that attenuate PM performance for older adults (e.g., Liu and Park 2004). In sum, for PM our aim has been to create and test a cognitive training intervention that is applicable for improving PM in the real world and teaches learners effective practical strategies informed by the basic PM literature.

Despite little work on *training* PM, the broader PM literature indicates a number of strategies that learners could use to improve their PM. As just noted, there are different types of PM that rely on different processes (McDaniel and Einstein 2007), and accordingly are associated with different effective strategies. Focal PM tasks involve cues that are presented in the focus of attention and thus are easy to recognize as a cue for performing the related task. For example, seeing a coworker in the hallway can act as a focal cue to give that person a message. In other words, simply seeing that coworker might automatically bring to mind the PM task of relaying the message. Because PM intentions like this are associated with focal cues that can stimulate spontaneous retrieval of the intention, they can be performed without actively looking for the cue. Previous research indicates that creating a strong association between the anticipated cue and the PM intention (an implementation intention strategy taking the general form, "When X occurs, I will remember to perform Y") can improve performance on focal tasks (e.g., McDaniel and Scullin 2010). This strategy has been explored more broadly and shows effectiveness beyond healthy aging: In a subsequent section, we report recent research with Parkinson's disease (PD) patients that train an implementation-intention strategy.

In contrast, nonfocal tasks involve cues that occur outside the focus of attention and are therefore more difficult to notice. For instance, one may need to stop at the grocery store after work, but the store itself is not easy to notice in the midst of a routine drive home where one must pay attention to traffic, etc. Here, actively monitoring for the cue is needed in order to successfully notice (Einstein et al. 2005), otherwise one might drive right by the store. The implementation intention strategy that is effective for focal tasks would not be as helpful in nonfocal PM tasks since the key is to notice the cue in the first place (Breneiser 2007). Thus, the best strategy for nonfocal tasks may be to simply check for the cue frequently and actively attend to that intention (an event monitoring strategy; see also Wenger et al., this volume).

Similarly, time-based PM tasks, wherein an intended action must occur at a particular time, require this type of active monitoring. Furthermore, the only cue is the time itself, whereas in focal and nonfocal tasks, events are the cues. This type of task is especially challenging for older adults (Einstein et al. 1995). Prior work indicates that learners who check the clock more often as the target time nears perform intended actions more frequently (Einstein et al.). Consistent with this finding, older adults are less likely than younger adults to ramp up their monitoring as the target time approaches (Einstein et al.; Park et al. 1997). Teaching older adults to use this strategic clock-checking may be the most effective strategy for improving their performance on time-based tasks.

Beyond the specific strategies to teach older adults, an important question is how to implement the training. In what form should these strategies be taught such that older adults learn them well and learn to apply them outside the context of learning? Several key factors may be critical for designing the most beneficial training program.

Key Factors for Training: The EXACT Study

As part of a larger cognitive training and aerobic exercise program (EXACT; McDaniel et al. 2014), McDaniel and colleagues developed a protocol specifically aimed at improving PM through strategy use (Waldum et al. 2016 describe this protocol in detail; see also Pothier and Bherer, this volume). Five main components were implemented in an 8-week intensive intervention. First, learners were given explicit instructions about effective strategies to use in PM tasks, specifically tailored for each type of task. Second, both to increase the generalizability of training and capitalize on previous memory research, the training context varied greatly. In terms of generalizability, as mentioned above, PM tasks are widely variable, both in task type (focal, non-focal, and time-based) and in context. Accordingly, learners were trained using several on-going tasks that tapped different types of PM. Encountering various scenarios during training might make learners' approach more flexible and resilient in the face of new PM challenges. Additionally, learners may start to be able to identify the different types of PM tasks and then transfer the appropriate strategies accordingly. This line of reasoning is also consistent with memory research on encoding variability wherein multiple contexts at the learning stage can improve later memory for the to-be-remembered material (Hintzman and Stern 1978).

Third, combined with the wide variety of laboratory tasks, homework was added to the program. That is, learners were given assignments to complete outside the laboratory regarding PM situations they faced in daily life. Explicit practice applying the training they received in the lab to their regular lives is likely to be beneficial for later transfer (e.g., Schmidt et al. 2001). Fourth, as the training program went on, the difficulty of the tasks increased. Learners were asked to keep in mind more PM objectives, and the nature of the tasks also became more challenging. Simultaneously, the trainer's involvement decreased from initially providing explicit strategy instruction prior to each training task to later expecting the learners to use the relevant strategies without prompting. This idea of increasing the difficulty across the training program is consistent with the broader literature on cognitive training. In the restorative approach, the demands of the task are incrementally increased to push the ultimate level of acquisition of the trained skill (e.g., retrospective memory training: Jennings and Jacoby 2003; attentional training: Mackay-Brandt 2011). Additionally, in the occupational therapy domain, strategies are trained such that learners are required to initiate and apply the strategies across activities that systematically differ in physical similarity and context but remain at the same level of complexity. In this sideways approach, task difficulty is only increased after strategy transfer has been observed (Toglia 2011). Again, intervention is designed to encourage transfer and generalize the training to learners' everyday lives.

Fifth, a key component of the EXACT project was to evaluate the training effects with computer simulations of cognitively challenging real-world tasks (e.g., cooking breakfast, Craik and Bialystok 2006; remembering health-related information and the sources of that information). To evaluate the PM training effects, older adults completed (pre- and posttraining) a simulation of going through the course of a day for three successive days (the Virtual Week task; Rose et al. 2010). During the course of each day, the older adults have to remember a number of prospective tasks, such as "remember to drop off dry cleaning when you go shopping" and "remember to take asthma medication at 11 a.m. and 9 p.m." (in the game, a person's token passes squares that indicate the virtual time for the day).

The results of the EXACT study were especially encouraging with regard to training PM (see McDaniel et al. 2014). Eight weeks of cognitive training on laboratory PM tasks with the components discussed above produced significant gains (from pre to posttests) in remembering to perform the real-world Virtual Week PM tasks relative to a control that did not receive PM training or an aerobic exercise control (a real clock, time-based task did not show training effects). By contrast, cognitive training did not produce significant gains for cooking breakfast or memory for health information tasks. However, the EXACT study was not designed to isolate the impact of particular training components to the success of the training protocol for improving PM; accordingly, many basic issues remain unanswered (see Waldum et al. 2016, for detailed discussion).

Briefly, the cognitive training included attentional control training tasks and retrospective training tasks in addition to PM training; thus, though plausible, it remains uncertain that the PM training alone would be sufficient to produce transfer to the ecologically valid VW tasks. Also, the PM training protocol included a number of components—including using a different laboratory task each week (variable training) and explicit strategy instruction—either or both theoretically could have been instrumental in promoting transfer. Initial support for the value of these components comes from noting that in the EXACT protocol, the attentional control training, and the retrospective memory training, following the precedent from the literature, generally did not include explicit strategy instruction and repeatedly used the same training task over the course of 8 weeks. As just mentioned, there was no significant transfer of training to the real-world attentional control task (cooking breakfast) or to the real-world retrospective memory task (memory for health information). Clearly, experiments that directly compare variable training (varying the parameters of the practice task, rather than keeping it constant; e.g., Kerr and Booth 1978; Goode et al. 2008) to single-task training and directly compare explicit strategy instruction with a typical practice-only procedure (e.g., Kramer et al. 1995; Jennings and Jacoby 2003) would provide valuable insights as to the importance of these factors in promoting the generalizability of cognitive training.

Finally, a feature of the EXACT project that poses practical limitations is that the cognitive training was a huge undertaking, requiring a great deal of commitment and investment from the trainers and the learners. A major practical issue is whether a more efficient training program focusing on PM per se and restricting training to one session (rather than multiple sessions as in EXACT) could support transfer of strategies to real-world PM tasks. Initial studies have reported significant improvements with older adults in everyday-like PM tasks using a brief implementation-intention instruction for the target PM task (Liu and Park 2004, with healthy adults; Shelton et al. 2016, with older adults with mild Alzheimer's disease; see also, Lee et al. 2016, for effective implementation-intention use with AD patients). Accordingly, it seemed possible that a single PM strategy training session could support transfer, and if so, then an efficient and nondemanding training protocol could be provided to older adults to improve their everyday PM success. We tested this possibility in a new experiment, reported next.

An Initial Experiment with Healthy Older Adults

The focus in our new experiment was to evaluate the success of PM training for a single 60–90 minute training session that compared the success of explicit prospective-memory strategy training relative to a practice-only condition and a test-retest control condition. We also included a new prospective-memory strategy training condition that we developed: *Guided metacognitive* training, described in the next section.

Metacognitive Strategy Training

It may be that neither explicit strategy instruction nor practice alone is most optimal. Instead, guided use of effective strategies that integrates metacognitive components may extend benefits of strategy training by helping a person recognize when and why a particular strategy is applicable and thus increase the probability of generalization (see Schaeffner et al., this volume). Metacognitive strategy training focuses on the general process of how to go about a task, including analyzing task demands, strategy generation and selection, and self-monitoring and self-evaluation of performance (Toglia 2018). A learner-centered approach that actively engages the participant in a collaborative process of planning or choosing strategies and evaluating effectiveness can be integrated with metacognitive strategy training by using systematic questions and guided prompts to facilitate self-generation of strategies (McEwen et al. 2018; Toglia 2018). Learner-centered approaches, such as guided discovery, are rooted in constructivism theories of learning that suggest that learning is enhanced when the learner is actively engaged in the process of discovering solutions themselves (e.g., McDaniel and Schlager 1990).

Preliminary evidence supporting the use of guided metacognitive strategy techniques in enhancing transfer of learning or generalization has been reported for older adults (Bottiroli et al. 2013; Dawson et al. 2014) as well as for cognitive rehabilitation of executive functions in individuals with stroke or brain injury (Skidmore et al. 2014; Toglia et al. 2010). For example, Bottiroli et al. (2013) found that transfer of learning was facilitated in older adults by encouraging active involvement in analyzing memory tasks involving lists, stories, locations, or paired-associates and adapting strategies to meet task demands. Guided metacognitive strategy training, however, has not been applied to PM training. Another important question, therefore, is whether PM strategies are best learned through explicit instruction or through guided metacognitive methods.

In the following experiment, we adapted a guided metacognitive strategy framework described by Toglia (2018) to the training of PM. The framework is outlined in Table 1 and consists of three components: (1) preactivity discussion on analyzing task demands, identifying similarities with meaningful activities, and self-generation of strategies; (2) mediation during the task to facilitate self-moni-

Table 1 Guided metacognitive strategy framework for prospective memory training

Treatment session components		Metacognitive focus
Preactivity discussion	Identify the type of PM	Analysis of task demands
	Identify everyday activities that involve similar PM requirements	Connect PM task with everyday activities. Identify similarities of task characteristics
	Generate strategies for PM	Plan and choose strategies that match task demands
During task	Stop and mediate after errors are observed. Guide generation of alternate strategies if needed	Self-monitoring skills. Strategy adjustment based on performance
After task	Participant summarizes methods used and comments on strategy effectiveness	Self-evaluation of performance

toring and the use of alternative strategies when needed; and (3) after-task questioning aimed at promoting self-evaluation of performance and strategy use.

Healthy older adults were assigned to one of four separate experimental groups: metacognitive strategy training, explicit strategy training, practice only, and a no-training control. Approximately 20 participants were assigned to each group (tested at both Washington University in St. Louis and Mercy College). To give some indication of the sample characteristics, participants' ages ranged from 60 to 90 with a mean age of 69.65, and all participants were living independently in the community. Montreal Cognitive Assessment (MoCA) scores ranged from 18 to 30, with a mean of 26.85 (for 61 out of 81 participants); 20 participants came from a subject pool with preexisting archival data (Knight Alzheimer's Disease Research Center at Washington University in St. Louis) and did not have MoCA scores, but were all screened as cognitively normal.

To assess PM, the Virtual Week task (VW task, previously described) was administered approximately 1 week before and after a single strategy training session. After completing the pretraining VW task, participants returned 1 week later for the training session (the retest control did not return to the lab at this point). This session included three different computerized PM games, previously described by Waldum et al. (2016), with increasing difficulty across the tasks (focal + time-based, nonfocal + time-based, a combination of focal + nonfocal + time-based).

For the metacognitive strategy group, after a general introduction to types of PM (i.e., time-based, focal, and nonfocal tasks), participants were then presented with PM tasks and asked to identify the type of PM required by the task. Next, guided questioning was used to help the person identify how the PM training task was similar to everyday activities or situations, and the person was given the opportunity to try the PM games using their own methods. During the activity, the examiner stopped and mediated performance as errors occurred and guided the person to reassess the effectiveness of their method. If needed, the person was encouraged to adjust or generate alternative strategies.

For the explicit strategy group, participants were instructed on different strategies depending on task demands (i.e., focal + time-based, nonfocal + time-based, or focal + nonfocal + time-based). The strategy training for focal tasks was to use implementation intention encoding (e.g., "When the focal target X occurs, I will remember to perform Y") repeated aloud and visualized (see McDaniel and Scullin 2010). The strategy training for time-based tasks encouraged participants to ramp up clock monitoring behavior when approaching the appropriate time (see Einstein et al. 1995). Finally, the strategy trained for nonfocal tasks was active monitoring, which involved trying to maintain a state of active cue-searching (Einstein et al., 2005).

The practice-only condition, after receiving a general introduction to types of PM, received no strategy instructions, and simply practiced the PM tasks during the training session. The control condition received no training. One week after completing the training session, participants completed the VW assessment again. The control completed the pre- and posttest VW assessments separated by 2 weeks.

The proportions of correctly detected PM targets as a function of assessment time (pre and post) and training condition (control, explicit, practice-only, and metacognitive) are shown in Fig. 1. There was a significant increase in scores from pretest ($M = 0.49$, $SE = 0.03$) to posttest ($M = 0.63$, $SE = 0.03$). However, there was no effect of training approach, nor was there any interaction between the two variables. The explicit and practice-only conditions obtained modest gains from pre- to posttest (0.08 and 0.10, respectively) and the metacognitive group obtained the greatest increase (0.18).

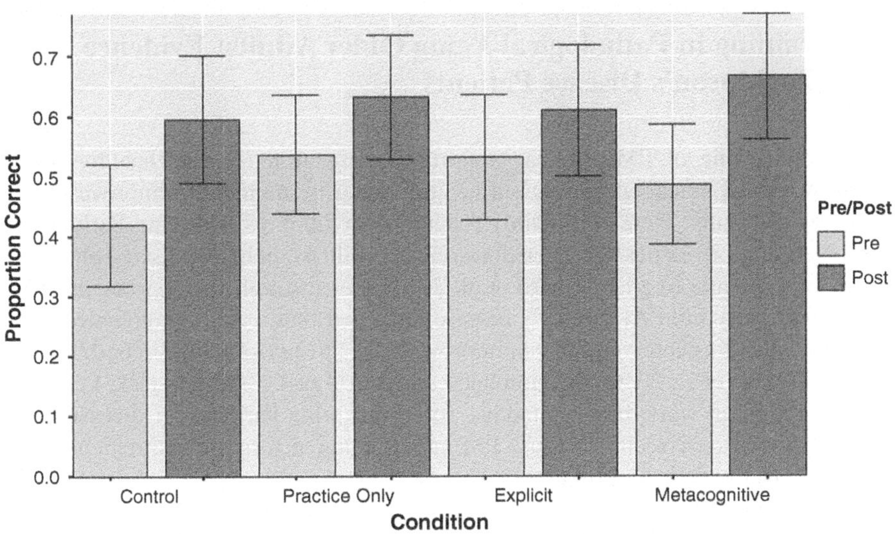

Fig. 1 Proportion of PM targets detected on Virtual Week from pre- to posttest as a function of training condition

This pattern is initially encouraging regarding the benefits of metacognitive training; however, the control group performed surprisingly well, too, also increasing by 0.18 from pre- to posttest. One interpretation is that, due to low sample size, random assignment did not adequately balance individual differences across groups, such that participants in the control group were by chance more able learners compared to those in the other groups. Another interpretation rests on the following feature of the experiment: The pre- and posttest VW versions were identical to one another. Accordingly, it is possible that the increases in performance on VW, for at least the control group, reflected practice of the specific PM tasks encountered during both pre- and posttesting, rather than acquisition of more general PM skills and strategies. We had not expected this improvement on VW in a no-trained control given previous research with repeated administration of VW (e.g., McDaniel et al. 2014); however, that research used intervals of 6 months between pre- and posttesting, not the 2 weeks used here. In retrospect, the experiment could have been more sensitive had we used different versions of VW at pre- and posttesting that incorporated different particular PM tasks.

Nevertheless, two speculative conclusions might be offered. First, the metacognitive strategy training seems more promising for training PM transfer than does practice alone or even explicit strategy training. The second conclusion follows from the observation that the improvement from pre- to posttest in the training groups was not more robust than that displayed in the control group. It may be that a brief one-session training is not sufficient to adequately train PM skills and strategies that significantly transfer. Clearly, these possibilities merit further research.

PM Training in Pathological Aging Older Adults: Evidence from Parkinson's Disease Patients

Effective training of PM also has important applications beyond healthy aging. Some work has extended findings in healthy aging to attempts to improve PM in pathological aging. Here, we mainly focus on our findings regarding Parkinson's disease, though work has also been done on very mild Alzheimer's disease and other forms of dementia (e.g., Burkhard et al. 2014). For example, prior work on older adults with very mild Alzheimer's disease indicated that a simple implementation intention encoding intervention can improve focal PM performance in both laboratory tasks (Lee et al. 2016) and simulated real-world tasks (the VW task; Lee et al. (2016)). Similar work has been done for those with Parkinson's disease (PD) because this disease seems to cause PM impairments in forming and remembering intentions (Kliegel et al. 2011).

Foster et al. (2017) studied individuals with mild to moderate PD without dementia on the VW PM task described above. First, participants completed the VW task without any special instructions. Then, a week later participants again performed the VW task. Prior to doing so, half were instructed to form implementation intentions. That is, they were told to create a "When X, I will do Y" statement, repeat it out loud three times, and then visualize performing the task at the correct time in the game. The other half simply repeated the PM tasks out loud three times. Regardless of the instructions, participants improved compared to their initial performance. This was especially true for event-based compared to the time-based tasks. More importantly, the implementation intention strategy training led to better performance than the verbal repetition task when participants completed nonrepeated tasks—tasks that were only presented once during the overall VM task—compared to the ones that were repeated.

These strategies were then extended to self-reports of naturalistic PM experiences. Goedeken et al. (2018) examined PD patients' experience of PM via the Prospective and Retrospective Memory Questionnaire Prospective Scale (PRMQ-Pro) 1 week before and 1 month after the same two training techniques: implementation intention strategy training and verbal repetition. The training occurred within the context of the VW task, but participants were then instructed to use the strategies as much as possible in their daily lives. Those in the verbal repetition actually showed a decline on the PRMQ-Pro, whereas those in the implementation intention group showed no change. Here, the effectiveness of the implementation intention training seemed to be in preventing decline rather than in improving PM. Of course, a limitation of this work is that it is based on patients' self-reports rather than actual performance on naturalistic PM tasks. Still, taken together, the findings are heartening in that training strategies can not only be taught and implemented by PD patients but can lead to maintenance of PM, if not even improvements. As progress is made in understanding the mechanisms and strategies for effective improvement of PM for healthy older adults, it appears fruitful to then test these techniques for those with clinical issues.

Conclusions

A unique aspect of our research is the appreciation of different types of PM tasks, with training oriented toward informing learners of these differences and highlighting particular strategies targeted at the different types of tasks. It seems that a parallel approach for retrospective memory training might be considered to improve outcomes for assisting older adults with their everyday retrospective memory challenges (cf. McDaniel and Bugg 2012). However, our new, though preliminary, results suggest that a relatively brief training session may not be enough to produce transfer of learned PM strategies to at least a simulation of real-world PM tasks. At this point, we remain optimistic that the present training approach, with training extended beyond one session, might benefit older (and younger) adults in improving their everyday prospective remembering. Clearly, however, a definitive conclusion awaits more complete experimental findings.

More generally, our research is attempting to examine and identify essential ingredients of cognitive training that enhance successful outcomes and generalization. There are many choices to be made in developing cognitive training, and as researchers, we need to be confident that those decisions will provide the greatest improvement (Cochrane and Green, this volume). Fundamentally of interest is what we are trying to train. Many programs have targeted cognitive capacities themselves (see Guye et al., Könen et al., Rueda et al., Wenger et al., this volume). Instead, our approach is to focus on teaching effective strategies that older adults can then use to tackle the PM situations they face.

One concern is how to implement this kind of strategy training, starting with how extensive the training ought to be. Though several sessions may be beneficial, the *right* kind of single training session may help older adults, which is a more practical proposition. In such a single session, the variability of the tasks that participants are exposed to in training is likely to be critical to later generalizability; experiencing a few different tasks may allow for more robust and flexible strategy development and application. In strategy training, it seems that guided metacognitive strategy training might be the best (see Schaeffner et al., this volume). Having such support in instruction has promise for older adults in comparison to allowing them to try and develop their own approach to PM tasks on their own.

Finally, the ecological validity of the training and the assessments of learning and transfer are critical. PM looks quite different inside and outside the laboratory. Thus, it is an important goal to foster the transfer of effective strategy use from training to the real world. As such, training programs must consider the balance and inclusion of laboratory training, homework, and simulated real-world activities during training such as the VW task. As these different considerations are explored, we are confident that an effective and efficient PM training for older adults will emerge, one that promotes transfer and generalizability to the real-world PM challenges.

Acknowledgments A special thanks is extended to Mercy College OT students Kim Havel, Mary Tremmel, Teresa Sanchez, and Hasena Punwasie for their assistance with the guided metacognitive strategy condition. The authors would also like to thank Deborah Koh and Amulya Joseph for their help with the prospective memory training work at Washington University. Some of the research discussed was supported by the National Institute on Aging Grant T32 AG000030 awarded to SU.

References

Bottiroli, S., Cavallini, E., Dunlosky, J., Vecchi, T., & Hertzog, C. (2013). The importance of training strategy adaptation: A learner-oriented approach for improving older adults' memory and transfer. *Journal of Experimental Psychology: Applied, 19*, 205–218.

Breneiser, J. (2007). *Implementation intentions and cost in prospective memory retrieval*. Unpublished doctoral dissertation. St. Louis: Washington University.

Burkhard, C., Rochat, L., Van der Linden, A. J., Gold, G., & Van der Linden, M. (2014). Is working memory necessary for implementation intentions to enhance prospective memory in older adults with cognitive problems? *Journal of Applied Research in Memory and Cognition, 3*, 37–43.

Craik, F. I. M., & Bialystok, E. (2006). Planning and task management in older adults: Cooking breakfast. *Memory & Cognition, 34*, 1236–2124.
Dawson, D., Richardson, J., Troyer, A., Binns, M., Clark, A., Polatajko, H., ... & Bar, Y. (2014). An occupation-based strategy training approach to managing age-related executive changes: a pilot randomized controlled trial. *Clinical Rehabilitation, 28*, 118–127.
Einstein, G. O., McDaniel, M. A., Richardson, S. L., Guynn, M. J., & Cunfer, A. R. (1995). Aging and prospective memory: Examining the influences of self-initiated retrieval processes. *Journal of Experimental Psychology: Learning, Memory, and Cognition, 21*, 996–1007.
Einstein, G. O., McDaniel, M. A., Thomas, R., Mayfield, S., Shank, H., Morrisette, N., & Breneiser, J. (2005). Multiple processes in prospective memory retrieval: Factors determining monitoring versus spontaneous retrieval. *Journal of Experimental Psychology: General, 134*, 327–342.
Foster, E. R., McDaniel, M. A., & Rendell, P. G. (2017). Improving prospective memory in persons with Parkinson disease: A randomized controlled trial. *Neurorehabilitation and Neural Repair, 31*, 451–461.
Goedeken, S., Potempa, C., Prager, E. M., & Foster, E. R. (2018). Encoding strategy training and self-reported everyday prospective memory in people with Parkinson disease: A randomized-controlled trial. *The Clinical Neuropsychologist, 32*, 1282–1302.
Goode, M. K., Geraci, L., & Roediger, H. L. (2008). Superiority of variable to repeated practice in transfer on anagram solution. *Psychonomic Bulletin & Review, 15*, 662–666.
Harrison, T. L., Shipstead, Z., Hicks, K. L., Hambrick, D. Z., Redick, T. S., & Engle, R. W. (2013). Working memory training may increase working memory capacity but not fluid intelligence. *Psychological Science, 24*, 2409–2419.
Healy, A. F., Wohldmann, E. L., Parker, J. T., & Bourne, L. E. (2005). Skill training, retention, and transfer: The effects of a concurrent secondary task. *Memory & Cognition, 33*, 1457–1471.
Henry, J. D., MacLeod, M. S., Phillips, L. H., & Crawford, J. R. (2004). A meta-analytic review of prospective memory and aging. *Psychology and Aging, 19*, 27–39.
Hertzog, C., Kramer, A. F., Wilson, R. S., & Lindenberger, U. (2009). Enrichment effects on adult cognitive development: Can the functional capacity of older adults be preserved and enhanced? *Psychological Science in the Public Interest, 9*, 1–65.
Hintzman, D. L., & Stern, L. D. (1978). Contextual variability and memory for frequency. *Journal of Experimental Psychology: Human Learning and Memory, 4*, 539–549.
Jennings, J. M., & Jacoby, L. L. (2003). Improving memory in older adults: Training recollection. *Neuropsychological Rehabilitation, 13*, 417–440.
Karbach, J., & Kray, J. (2009). How useful is executive control training? Age differences in near and far transfer of task-switching training. *Developmental Science, 12*, 978–990.
Kerr, R., & Booth, B. (1978). Specific and varied practice of a motor skill. *Perceptual and Motor Skills, 46*, 395–401.
Kliegel, M., Altgassen, M., Hering, A., & Rose, N. S. (2011). A process-model based approach to prospective memory impairment in Parkinson's disease. *Neuropsychologia, 49*, 2166–2177.
Kramer, A. F., Larish, J. F., & Strayer, D. L. (1995). Training for attentional control in dual task settings: A comparison of young and old adults. *Journal of Experimental Psychology: Applied, 1*, 50–76.
Lee, J. H., Shelton, J. T., Scullin, M. K., & McDaniel, M. A. (2016). An implementation intention strategy can improve prospective memory in older adults with very mild Alzheimer's disease. *British Journal of Clinical Psychology, 55*, 154–166.
Liu, L. L., & Park, D. C. (2004). Aging and medical adherence: The use of automatic processes to achieve effortful things. *Psychology and Aging, 19*, 318–325.
Lustig, C., Shah, P., Seidler, R., & Reuter-Lorenz, P. A. (2009). Aging, training, and the brain: A review and future directions. *Neuropsychology Review, 19*, 504–522.
MacKay-Brandt, A. (2011). Training attentional control in older adults. *Aging, Neuropsychology, and Cognition, 18*, 432–451.
McDaniel, M. A., & Bugg, J. M. (2012). Memory training interventions: What has been forgotten? *Journal of Applied Research in Memory and Cognition, 1*, 45–50.

McDaniel, M. A., & Einstein, G. O. (2007). *Prospective memory: An overview and synthesis of an emerging field.* Thousand Oaks: Sage Publications, Inc.
McDaniel, M. A., & Schlager, M. S. (1990). Discovery learning and transfer of problem solving skills. *Cognition and Instruction, 7*, 129–159.
McDaniel, M. A., & Scullin, M. K. (2010). Implementation intention encoding does not automatize prospective memory responding. *Memory & Cognition, 38*, 221–232.
McDaniel, M. A., Binder, E. F., Bugg, J. M., Waldum, E. R., Dufault, C., Meyer, A., Johanning, J., Zheng, J., Schechtman, K. B., & Kudelka, C. (2014). Effects of cognitive training with and without aerobic exercise in cognitively-demanding everyday activities. *Psychology and Aging, 29*, 717–730.
McEwen, S. E., Mandich, A., & Polatajko, H. J. (2018). CO-OP approach™: A cognitive-based intervention for children and adults. In N. Katz & J. Toglia (Eds.), *Cognition, occupation, and participation across the lifespan* (4th. ed., pp. 315–334). Bethesda: American Occupational Therapy Association. https://doi.org/10.7139/2017.978-1-56900-479-1.
Park, D., Herzog, C., Kiddor, D., Morrell, R., & Mayhorn, C. (1997). Effect of age on eventbased and time-based prospective memory. *Psychology and Aging, 12*, 314–327.
Redick, T. S., Shipstead, Z., Harrison, T. L., Hicks, K. L., Fried, D. E., Hambrick, D. Z., Kane, M. J., & Engle, R. W. (2013). No evidence of intelligence improvement after working memory training: A randomized, placebo-controlled study. *Journal of Experimental Psychology: General, 142*, 359–379.
Rose, N., Rendell, P. G., McDaniel, M. A., Aberle, I., & Kliegel, M. (2010). Age and individual differences in prospective memory during a "Virtual Week": The roles of working memory, vigilance, task regularity, and cue focality. *Psychology and Aging, 25*, 595–605.
Rose, N. S., Rendell, P. G., Hering, A., Kliegel, M., Bidelman, G. M., & Craik, F. I. M. (2015). Cognitive and neural plasticity in older adults' prospective memory following training with the virtual week computer game. *Frontiers in Human Neuroscience, 9*, 592.
Schmidt, W., Berg, I. J., & Deelman, B. G. (2001). Prospective memory training in older adults. *Educational Gerontology, 27*, 455–478.
Shelton, J. T., Lee, J. H., Scullin, M. K., Rose, N. S., Rendell, P. G., & McDaniel, M. A. (2016). Improving prospective memory in healthy older adults and very mild Alzheimer's disease Patients. *Journal of the American Geriatrics Society, 64*, 1307–1312.
Skidmore, E. R., Dawson, D. R., Butters, M. A., Grattan, E. S., Juengst, S. B., Whyte, E. M., Begley, A., Holm, M. B., & Becker, J. T. (2014). Strategy training shows promise for addressing disability in the first 6 months after stroke. *Neurorehabilitation and Neural Repair*, 1–9.
Toglia, J. P. (2011). The dynamic interactional model of cognition in cognitive rehabilitation. In N. Katz (Ed.), Cognition, occupation and participation across the life span: *Neuroscience, neurorehabilitation and models of intervention in occupational therapy* (3rd ed., pp. 161–201). Bethesda, MD: AOTA Press.
Toglia, J. (2018). The dynamic interactional model and the multicontext approach. In N. Katz & J. Toglia (Eds.), *Cognition, occupation, and participation across the lifespan* (4th ed., pp. 355–385). Bethesda: American Occupational Therapy Association. https://doi.org/10.7139/2017.978-1-56900-479-1.
Toglia, J., Johnston, M. V., Goverover, Y., & Dain, B. (2010). A multicontext approach to promoting transfer of strategy use and self-regulation after brain injury: An exploratory study. *Brain Injury, 24*, 664–677.
Waldum, E. R., Dufault, C. L., & McDaniel, M. A. (2016). Prospective memory training: Outlining a new approach. *Journal of Applied Gerontology*, 35, 1211–1234.

Executive Function Training

Julia Karbach and Jutta Kray

Contents

Abstract Executive function (EF) refers to the ability to control one's processing along external and internal goals, including working memory, inhibition, cognitive flexibility, and multitasking. Numerous studies showed that EF can be improved by training across a wide range of ages. Some of them also reported performance improvements on untrained tasks measuring the same construct (near transfer) and even on tasks measuring other cognitive abilities (far transfer). However, especially results regarding far transfer have been very inconsistent and seem to vary as a function of intervention type, training intensity, and target population. In this chapter, we first introduce definitions and models of EF and present their implications for EF training. Afterwards, we review findings from studies focusing on the training of multitasking, inhibition, and cognitive flexibility (for working memory training see Könen et al., this volume) and describe individual differences in the effects of these training interventions. We close by discussing the current state of research and proposing important issues for future research.

J. Karbach (✉)
University of Koblenz-Landau, Landau, Germany
e-mail: karbach@uni-landau.de

J. Kray
Saarland University, Saarbrücken, Germany

T. Strobach, J. Karbach (eds.), *Cognitive Training*,
https://doi.org/10.1007/978-3-030-39292-5_14

The ability to control behavioral activity and to adapt to changing environmental demands is described as executive functions and includes a set of higher-level cognitive control processes. Many of these control processes gradually develop over childhood and well into adolescence while declining in older age (for a review, see Wiebe and Karbach 2017; see also Guye et al., Rueda et al., this volume). In this chapter, we will review evidence indicating that executive control functions can be improved by training across a wide range of ages. In the first section, we introduce definitions and models of executive functions and present their implications for the training of executive control. Secondly, we review findings from studies focusing on the training of multitasking, inhibition, and cognitive flexibility and discuss individual differences in the effects of these training interventions. The last section summarizes the current state of research and suggests important issues for future research.

Definition and Models

The term executive function (EF) refers to a set of cognitive processes serving to guide thoughts and actions along internal or external goals, tasks, and intentions in order to optimally adapt to changes in the environment. There is relative consensus in the literature what kind of tasks and situations require EF, such as multitasking, task switching and scheduling, and performance monitoring. However, there exist different views on the structure of EF and the question of whether they are better described as a unitary or modular construct. Within traditional models from working-memory research, EF is often conceptualized as a unitary system that is involved in coordinating and controlling of processing and storing of information in working memory (see also Könen et al., this volume). For instance, in the well-known working-memory model by Baddeley (1996) a central executive system is considered to be responsible for coordinating the information flow between the phonological loop, visual-spatial sketchpad, and the episodic buffer. Hence, different functions need to be coordinated such as the active maintenance of task-relevant information, retrieval from long-term memory, attending to relevant information and inhibiting irrelevant ones, and scheduling of multiple tasks. Coming from a neuropsychological research tradition, Norman and Shallice (1986) suggested the Supervisory Attentional Model that distinguishes between automatic and controlled information processing modes by assuming two separate systems, the contention scheduling and the supervisory attentional system (SAS). The SAS is especially required, for instance, in situations in which tasks and demands are novel, difficult, and dangerous, or in situations in which the suppressing of a dominant but currently inappropriate behavior and action tendency is needed. Hence, its main function is to bias the contention scheduling system in a way that task-relevant information is activated while task-irrelevant information is inhibited.

In contrast to the view of a unitary system of EF, other researchers have stressed a modular view assuming a set of several distinct functions. These functions include

(a) initiation of complex behavior, (b) active maintenance of task-relevant information, (c) shifting between tasks and mind sets, (d) planning and scheduling multiple steps of complex tasks and events, (e) inhibition of inappropriate behavior and response tendencies, (f) performance monitoring and adjustment, (g) coordination multiple task requirements, and so on (cf. Miller and Cohen 2001; Smith and Jonides 1999). Such a modular view was further supported by empirical evidence of discrete neuronal systems involved in these different functions that were reported in clinical studies including patients with particular lesions as well as in neuroimaging studies. For instance, it has been found that maintaining and biasing task-relevant task sets during task preparation was specifically associated with activity changes in the dorsolateral prefrontal cortex while processing of response conflict during task execution was specifically associated with activity in the anterior cingulated cortex (ACC) (Botvinick et al. 2001). Such double dissociations of neuronal activations in different executive functions contributed to a modular view.

Using a structural equation modeling approach, Miyake et al. (2000) integrated these opponent theoretical views on EF. They found evidence supporting both the unity and diversity of executive control by investigating individual differences in indicators of three core components of EF, working memory, shifting, and inhibition. Results of this first modeling study revealed that all three constructs were separable but moderately correlated, indicating that they were not fully independent of each other. In their later work they found only a shifting-specific and a working-memory updating specific factor as well as a common EF factor, reflecting the ability to maintain task-relevant information (see also Miyake and Friedman 2012).

Interestingly, this theoretical view is now also supported by meta-analytic evidence on the basis of neuronal data. In this study, Niendam et al. (2012) included about 200 neuroimaging studies on EF examining shifting, planning, working memory, initiation, and vigilance and determined common as well as domain-specific brain activations across these functions. The results indicated a common pattern of activation including the prefrontal, dorsal anterior cingulate, and parietal cortices in line with the idea of the higher-order control system. At the same time they also found evidence for domain-specific activation depending on the involved executive task in anterior prefrontal, anterior, and midcingulate as well as subcortical brain regions. Note that individual variation in fluid intelligence has been linked to EF and the efficiency of recruiting the dorsolateral prefrontal cortex (e.g., Verhaeghen and Cerella 2002).

Implications for the Expected Effects of Executive Function Training

Considering the different theoretical views on EF described above, which positive effects are to be expected after a training of EF? As discussed in detail in this volume (e.g., Könen and Auerswald, Taatgen, this volume), a cognitive training intervention should not only result in performance gains on the trained components of

EF; it should also generalize to other cognitive functions or even to everyday activities and school and academic performance. Finally, these gains should also persist over longer periods of time. However, if and to what degree transfer of training in cognitive functioning is possible has fueled a heated debate in recent years, especially regarding working-memory training (e.g., Melby-Lervåg and Hulme 2013; Sala and Gobet 2017; Shipstead et al. 2012; see also Guye et al., Katz et al., Könen et al., this volume).

From the perspective of a unitary view on EF we would expect to see relatively broad transfer of EF training to a large variety of other aspects of EF because training the higher-order control system should enhance a broad range of different EFs. Taking a more modular perspective, the generalization of training gains in one EF should not necessarily result in benefits in another EF and the scope of transfer should depend on the degree of domain-specific overlap (processing and neural) between the trained and the transfer task (i.e., the more shared resources, the more likely the transfer should be). Assuming both common and domain-specific aspects of EF we would expect to see larger transfer if the prefrontal-dorsal-parietal control network is trained.

Aside from the transfer scope, Lövdén et al. (2010) proposed a fundamental precondition for the success of a training intervention in their theoretical framework of cognitive plasticity: They assumed that the environmental demands during training should cause an imbalance with the actual brain supply. In case of a substantial and prolonged mismatch between environmental demands and brain supply, the brain will react to this mismatch with functional or structural brain changes. Taking a lifespan view, age-related deficits in EF that are typically present in childhood and older age suggest that this mismatch will occur more often in these age groups, indicating that especially children and older adults may gain from a moderate mismatch induced by training interventions targeting EF. As a consequence, they will benefit more than younger ones from brain alteration in their less efficient prefrontal lobe system (see below).

Another recent account focusing on transfer of working memory training is based on a similar idea. The cognitive routine framework (Gathercole et al. 2019; see also Könen et al., this volume) assumes that during training, participants are faced with task features that induce unfamiliar and challenging cognitive demands, resulting in the need to develop new cognitive routines because existing mechanisms are not sufficient to meet these demands. These newly acquired cognitive routines are thought to be automated cognitive procedures rather than task-specific strategies. They can then be applied to novel tasks sharing the same requirements, which is considered the basis for any transfer of training. In addition, the framework includes specific assumptions regarding common task features generating (or impairing) transfer: In terms of working-memory training, for instance, transfer to tasks requiring interference control will only occur after working-memory training tasks including the requirement to suppress distracting information, such as complex span tasks (in contrast to simple span tasks).

To sum up the theoretical considerations, EF training interventions that (1) enhance a superordinate fronto–cingulo–parietal network in groups showing alterations in these brain regions, (2) induce an individually adapted, prolonged mismatch between the current ability level and task demands, and (3) generate new automated cognitive routines rather than task-specific strategies should have a good chance to generalize to other EF and even to fluid abilities. As most of the training studies on EF in the last decade aimed at enhancing working memory/updating and various reviews and meta-analyses exist (see also Guye et al., Könen et al., Rueda et al., this volume), we will focus our summary of research findings on recent evidence regarding the effectiveness of training in multitasking, flexibility, and inhibition, and will discuss them in the light of the theoretical considerations presented above.

Multitasking Training

The majority of multitasking training studies has applied dual-task trainings and reported robust practice-related improvements in the ability to coordinate two simultaneously presented and overlapping component tasks (for a review, see Strobach et al. 2014). These training-induced gains generalized to new, untrained dual-task situations. While no such gains were found after single-task practice (i.e., practice on the component tasks one at a time), hybrid training including both single-task and dual-task situations increased transfer to a visual-auditory dual task (e.g., Liepelt et al. 2011; see also Bherer et al. 2008). The second auditory response selection started earlier and more efficiently after hybrid training than after single-task practice, resulting in reduced dual-task costs. This suggests that the training indeed resulted in an optimized and efficient processing of two simultaneously presented tasks (see Strobach 2020).

While these and other findings provided evidence for the acquisition of transferable task-coordination skills, the specific nature of such mechanisms is still under debate. One recent model, the Efficient Task Instantiation (ETI) model (Schubert and Strobach 2018; Strobach et al. 2014) assumes that dual-task performance improves after dual-task practice because relevant task information (such as stimulus-response mapping rules) is efficiently and conjointly instantiated in working memory at the onset of a dual-task trial. Several findings from the dual-task literature support the ETI model. First, De Jong (1995) showed that incomplete instantiation of relevant task information resulted in impaired dual-task performance. Second, investigating older adults with age-related working-memory impairments and younger adults without these deficits showed significant age differences: In contrast to younger adults, older participants did not learn the instantiation of a complex second task unless its complexity, and therefore the associated working-memory load was reduced (Maquestiaux et al. 2004). Third, the efficient instantiation of task information during training was impaired in young adults when a working-memory overload was induced by highly complex training tasks (Ruthruff et al. 2006). These findings are also in line with the theoretical framework outlined above as they underline that training demands have to induce a moderate mismatch that differs across age in order to induce practice-related improvements in EF.

Aside from working-memory processes, the ability to resolve task interference plays an important role for the effectiveness of dual-task training. Evidence for this view comes from a study by Anguera et al. (2013). They examined the effectiveness of dual-task training in older adults by embedding the training in an adaptively designed game simulation called NeuroRacer. The authors compared training and transfer gains in a group of older adults in three training conditions: (1) participants were to perform both tasks simultaneously (dual-task training condition); (2) to perform each task separately (single-task condition; active control group), or (3) to perform none of the tasks (passive control group). Importantly, the dual-task condition included stimuli that were relevant for both tasks and participants were to constantly control interference from one task in order to correctly perform the other. Results revealed a larger reduction of dual-task costs under the dual-task training condition than under the active or passive control condition. After 12 h of training, older adults even performed better in this task than untrained younger adults. Moreover, training gains not only generalized to a new, untrained dual task but also to measures of sustained attention and working memory. Importantly, the study also revealed the first robust correlations between behavioral improvement and changes on neural signatures of cognitive control (enhanced midline frontal theta power and frontal–posterior theta coherence).

Flexibility Training

Most of the studies that aimed at enhancing cognitive flexibility have applied task-switching training in pretest-training-posttest designs with one or more treatment groups that practiced to switch between tasks in random or predictable task orders (for a review, Kray and Dörrenbächer in press). Active control groups often performed the same tasks but practiced them in separate blocks (i.e., single task blocks) (see Minear and Shah 2008, who introduced this type of design).

Meanwhile, a number of training studies demonstrated robust and substantial improvements in task-switching performance across various age groups such as children and adolescents (for a review, see Karbach and Unger 2014; Rueda et al., this volume) to younger and older adults (for meta-analyses, see Karbach and Verhaeghen 2014; Nguyen et al. 2019; see also Guye et al., this volume) and also in clinical groups such as children with ADHD (e.g., Dörrenbächer and Kray, 2019; Kray et al. 2012). For instance, Karbach and Kray (2009) showed that costs of switching between tasks were substantially reduced after four sessions of practice. Training net gains varied across training conditions from 0.85 SD to 1.88 SD and a variable training (on a new set of stimuli and task rules in each training session) showed the smallest training gains (cf. Sabah et al. 2018). In a meta-analysis, Karbach and Verhaeghen (2014) reported training improvements in task switching in old age, with raw gains of about 0.90 SD and net gains (after subtracting the effects of active controls) of about 0.50 SD. Although nearly all studies report a reduction of latency switch costs, it should be noted that the findings on the reduction of error switch costs were rather mixed, probably because these costs are usually already relatively low at the beginning of the training.

Similar to the practice effects of training in task switching, most of the studies also reported near transfer, that is, training gains in an untrained switching task for various age ranges (e.g., Minear and Shah 2008). In line with the theoretical view that a considerable supply-demand mismatch is a precondition for inducing plasticity in task-switching performance, the amount of near transfer varies with age when training conditions are constant across age groups. For instance, in a task-switching training study of Karbach and Kray (2009), near-transfer gains on task switching costs were much larger in healthy children and older adults than in younger adults. Given that the training was not adaptive, it may have induced a larger mismatch in children and older adults who also show age-related alterations in brain regions associated with task switching. Moreover, in the meta-analysis of Karbach and Verhaeghen (2014) clear near-transfer gains of EF training were shown for older adults. However, there is also evidence that for younger adults and adolescents near-transfer gains are sometimes restricted to the general level of task switching (mixing costs) or to an uncued switching training (Pereg et al. 2013; Zinke et al. 2012), which again may induce a larger mismatch between task demands and brain supply. In line with this view are also results of Kray and Fehér (2017). In their study, demands on switching (single task vs. mixed task blocks), interference control (unambiguous or ambiguous stimuli), and working-memory demands (with or without task cues) varied between different training conditions. Interestingly, for younger adults transfer gains did not vary across these conditions. In contrast, older participants that were trained in resolving task interference (ambiguous groups) showed larger transfer gains than the participants that were trained on unambiguous stimuli inducing low task interference.

Results on far-transfer effects of training in task switching are rather mixed. While some studies found a relatively broad transfer to other EF and even to measures of fluid intelligence (e.g., Karbach and Kray 2009), others found no far transfer effects at all (Kray and Fehér 2017; Pereg et al. 2013). In one of our first task-switching training studies (Karbach and Kray 2009), we found far transfer to measures of inhibition, working memory, and even to fluid intelligence for children and younger and older adults. Children with ADHD also showed far transfer to measures of inhibition and verbal working memory (Kray et al. 2012). In contrast, adolescents and young adults did not show broad far transfer to other task domains (Pereg et al. 2013; Zinke et al. 2012) in other studies. A recent study investigating normally developing children (8–11 years of age) tested the effects of flexibility training, including task-switching, focus-switching, and dual-task training (Johann and Karbach 2019; see also Johann and Karbach, this volume). Aside from near transfer to untrained flexibility tasks, the training also benefitted reading ability (sentence comprehension), but only when it was embedded in a game-like training environment, and not in a standard version without game elements (cf. Dörrenbächer et al. 2014; Prins et al. 2011).

The mixed pattern of findings regarding far transfer is also reflected in the results of the recent meta-analyses. They suggest that far transfer of flexibility training to other EF and fluid intelligence is small but significant especially in older adults (Karbach and Verhaeghen 2014; Nguyen et al. 2019), supporting the pattern of compensation effects reported in previous studies (see below).

Inhibition Training

Training studies investigating inhibition training are relatively scarce and only very few of them examined the effects of inhibition training in childhood (cf. Kray and Ferdinand 2013). Results of these studies showed improved inhibitory control after training in early childhood. Given that intensive and adaptive working-memory training resulted in enhanced performance on measures of inhibitory control and reasoning in children with and without cognitive control impairments (for reviews, see Karbach and Unger 2014; Könen et al., Rueda et al., this volume), Thorell et al. (2009) tested whether inhibition and interference control training produced similar transfer gains. Preschoolers practiced three different tasks: the go/no-go, stop-signal, and flanker task. The results indicated that this inhibition training, in contrast to working-memory training, did not lead to improvements on other cognitive tasks. One possible explanation for the lack of transfer might be that the variability of training was much larger in the inhibition and interference control training group than in the working-memory training group and previous work indicated that variability of training may hinder transfer in childhood (Karbach and Kray 2009). In contrast, a study from Rueda et al. (2005) applied a training battery including interference resolution and inhibition tasks to 4- and 6-year old children. They reported transfer of training to reasoning tasks, a finding supporting the view that high demands on interference resolution during training may increase transfer (cf. Anguera et al. 2013). Recently, Johann and Karbach (2019) trained children (8–11 years of age) on three different inhibition tasks, a Stroop-like task, a Flanker task, and a go/no-go task. Similar to the results regarding flexibility training reported above, the training resulted in significant performance gains on the training tasks as well as in significant near transfer. Interestingly, the inhibition training also improved reading abilities (sentence comprehension and reading speed), but these benefits were again restricted to the game-based version of the training and not present in a standard version (see Johann and Karbach, this volume).

Training studies including younger and older adults often relied on training on the Stroop task and showed better inhibitory control after practice (e.g., Davidson et al. 2003; Dotson et al. 2014; Wilkinson and Yang, 2012). However, results regarding transfer of inhibitory control training to untrained tasks and abilities are ambiguous: Some studies reported transfer to new, untrained tasks, such as processing speed tasks and dual tasks (e.g., Mozolic et al. 2011) while others found no transfer (e.g., Reisberg et al. 1980; for a review, see Strobach et al. 2014).

Individual Differences

The studies reviewed so far showed that cognitive training can have positive effects on cognitive functions on the group level. However, individual differences in training-induced gains are often very large. This is particularly critical in children and older adults, because they are likely to differ more from each other than young adults and between-group comparisons do little justice to individuals' strengths and weaknesses. Therefore, the question who benefits most from cognitive interventions has been more and more acknowledged (see Katz et al., Schmiedek, this volume). Its importance is obvious from an applied point of view, especially when it comes to the adaptation of training interventions to populations with specific needs, such as children with neurocognitive disorders or older adults with specific cognitive impairments. Moreover, it is also of interest on the theoretical level, because individual differences in training-related benefits may help us understand the underpinnings of cognitive and neural plasticity. Also, the mismatch between environmental demands and brain supply that is induced may strongly vary between age groups and therefore needs to be considered for creating optimal training interventions.

Two prominent accounts have been put forward to describe and explain individual differences in training-related performance gains: First, the magnification account (also Matthew effect or scissor effect) assumes that individuals that are already performing very well will also benefit the most from cognitive interventions. It is assumed that high-performing and well-educated participants have more efficient cognitive resources to acquire and implement new strategies and abilities. Thus, baseline cognitive performance at pretest should be positively correlated with the training-related gains and the training should result in a magnification of age differences and individual differences (see Fig. 1, left panel). In fact, there are a number of earlier studies supporting this account, most of them from the field of memory strategy training (see Rebok et al. 2007, for a meta-analysis).

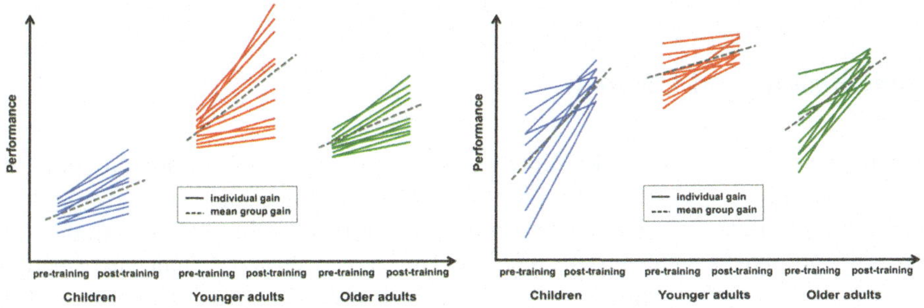

Fig. 1 Illustration of magnification (left panel) and compensation (right panel) effects after EF training: (1) changes in interindividual differences in performance from pretraining to posttraining, (2) changes of age group from pretraining to posttraining, and (3) correlation between baseline cognitive performance at pretest and training gain

Second, the compensation account assumes that high-performing individuals will benefit less from cognitive interventions, because they are already functioning at the optimal level and have less room for improvement. Thus, baseline cognitive performance should be negatively correlated with training gains and age differences and individual differences should be reduced after the intervention (see Fig. 1, right panel). Evidence supporting this account comes from numerous studies focusing on EF training, revealing that training-related benefits were larger in children and older adults than in younger adults (e.g., Kray and Lindenberger 2000; for a review, see Karbach and Unger 2014). While these studies were based on comparisons at the group level, recent studies also have analyzed correlations between baseline cognitive ability and training-related benefits, indicating that working-memory training yielded larger training and transfer effects in older adults with low cognitive performance at pretest (e.g., Zinke et al. 2014). Moreover, recent work has applied latent variable approaches to analyze individual differences in performance changes as well as correlations between baseline cognitive ability and training-related benefits (see also Könen and Auerwald, this volume). These studies also provided evidence for the magnification effects after memory-strategy training and compensation effects after task-switching training (cf. Könen and Karbach 2015).

A recent study including children, younger adults, and older adults directly tested the magnification account against the compensation account based on a task-switching training (Karbach et al. 2017). The authors applied latent-change modeling (cf. Könen and Auerwald, this volume; Lövdén et al., 2012; Schmiedek et al. 2010; Schmiedek, this volume) and tested changes in individual differences and age differences from pretraining to posttraining as well as the correlation between baseline cognitive abilities at pretest and training gains/transfer gains. Results showed that both individual differences and age differences were reduced after the training and that lower baseline abilities were associated with larger training-induced gains. Importantly, this correlation was higher in the training group than in the active control group, indicating that they were more likely based on the effects of executive control training than on nonfocal effects (e.g., regression to the mean or retest effects).

Conclusions and Outlook

In sum, research focusing on EF training consistently showed that intensive training resulted in robust performance improvements on the training tasks, indicating that cognitive plasticity is considerable up to old age (see in this volume: Belleville et al., Gajewski and Falkenstein, Guye et al., Könen et al., Pothier and Bherer, Umanath et al., Verhaeghen, Wenger et al.). Most studies also reported near transfer of training to tasks measuring the same construct as the training task and some studies even reported far transfer to different cognitive abilities in older age. However, especially when it comes to far transfer, the existing evidence is mixed and has recently inspired heated debates in the field. Yet, many inconsistent findings can be explained by large differences in the type and intensity of the training as well as in the research design and the analytical methods that have been applied (see Könen and Auwerwald, Schmiedek, Schmiedek, Taatgen, this volume). For instance, transfer seems to occur more consistently if (1) the training is process-based and engages higher-order control processes, such as EF, instead of task-specific strategies, (2) the training and transfer tasks engage overlapping cognitive and neural resources, (3) the training is adaptive or variable (e.g., by including changing tasks and processes), and (4) the training puts a high load on the ability to resolve task interference (e.g., with high stimulus ambiguity or changing task modalities) (e.g., Anguera et al. 2013; Au et al. 2014; Gathercole et al. 2019; Karbach and Kray 2009; Karbach and Unger 2014; Kray and Fehér 2017; Kray and Ferdinand 2013; Nguyen et al. 2019; Strobach et al. 2014; Karbach and Verhaeghen 2014).

Current meta-analytic evidence suggests that there is small but significant far transfer of EF training across the adult lifespan, including transfer to other components of EF, attention, and fluid intelligence (e.g., Au et al. 2014; Karbach and Verhaeghen 2014; Nguyen et al. 2019; but see Melby-Lervåg and Hulme 2013; Sala and Gobet 2017). Thus, given that the effect sizes for far transfer seem to be relatively small, the question really is whether these effects should be considered relevant? In accordance with many other researchers (e.g. Green et al. 2019; Oberauer 2015), we think that these effects indeed are extremely relevant. From a scientific point of view, they are very informative for our theoretical conception of EF. The fact that EF training transfers to other components of EF but that these transfer effects are smaller than the gains on the training tasks is more in line with the idea that EF is a set of separable but highly correlated control functions. From a more applied point of view, even small improvements in cognitive performance can be extremely relevant for individuals with cognitive deficits (see Johann and Karbach, Belleville et al., this volume). Moreover, these small effects could be much increased if we understood exactly which features of training moderate the effectiveness of the intervention and how this effectiveness can be maximized (cf. Oberauer 2015). Future studies will have to focus on these issues, for instance by considering individual differences in motivational (e.g., training motivation and self-efficacy; see Johann and Karbach, Katz et al., this volume) and social aspects (e.g., socio-emotional processes, educational background, or socioeconomic status; see Johann and Karbach, Thompson and Steinbeis et al., this volume) as well as genetic predispositions of the participants. Current evidence suggests those and other variables may significantly moderate the amount of training-induced gains and the scope of transfer, but clearly more research is needed to understand how they contribute to the effectiveness of EF training (see Colzato and Hommel, this volume).

References

Anguera, J. A., Boccanfuso, J., Rintoul, J. L., Al-Hashimi, O., Faraji, F., Janowich, J., ... & Gazzaley, A. (2013). Video game training enhances cognitive control in older adults. *Nature, 501*, 97–101.

Au, J., Sheehan, E., Tsai, N., Duncan, G. J., Buschkuehl, M., & Jaeggi, S. M. (2014). Improving fluid intelligence with training on working memory: A meta-analysis. *Psychonomic Bulletin & Review, 22*, 366–377.

Baddeley, A. (1996). Exploring the central executive. *The Quarterly Journal of Experimental Psychology, 49*, 5–28.

Bherer, L., Kramer, A. F., Peterson, M. S., Colcombe, S., Erickson, K., & Becic, E. (2008). Transfer effects in task-set cost and dual-task cost after dual-task training in older and younger adults: Further evidence for cognitive plasticity in attentional control in late adulthood. *Experimental Aging Research, 34*, 188–219.

Botvinick, M. M., Braver, T. S., Barch, D. M., Carter, C. S., & Cohen, J. D. (2001). Conflict monitoring and cognitive control. *Psychological Review, 108*, 624–652.

Davidson, D. J., Zacks, R. T., & Williams, C. C. (2003). Stroop interference, practice, and aging. *Aging, Neuropsychology, and Cognition, 10*, 85–98.

De Jong, R. (1995). The role of preparation in overlapping-task performance. *Quarterly Journal of Experimental Psychology: Human Experimental Psychology, 48A*, 2–25.

Dörrenbächer, S., & Kray, J. (2019). The impact of game-based task-shifting training on motivation and executive control in children with ADHD. *Journal of Cognitive Enhancement, 3*, 64–84.

Dörrenbächer, S., Müller, P. M., Tröger, J., & Kray, J. (2014). Dissociable effects of game elements on motivation and cognition in a task –switching training in middle childhood. *Frontiers in Psychology, 5*, 1275.

Dotson, V. M., Sozda, C. N., Marsiske, M., & Perlstein, W. M. (2014). Within-session practice eliminates age differences in cognitive control. *Aging, Neuropsychology, and Cognition, 20*, 522–531.

Gathercole, S. E., Dunning, D. L., Holmes, J., & Norris, D. (2019). Working memory training involves learning new skills. *Journal of Memory and Language, 105*, 19–42.

Green, C. S., Bavelier, D., Kramer, A. F., Vinogradov, S., Ansorge, U., Ball, K. K., ... & Witt, C. M. (2019). Improving methodological standards in behavioral interventions for cognitive enhancement. *Journal of Cognitive Enhancement, 3*, 2–19.

Johann, V. E., & Karbach, J. (2019). Effects of game-based and standard executive control training on cognitive and academic abilities in elementary school children. *Developmental Science*, e12866.

Karbach, J., & Kray, J. (2009). How useful is executive control training? Age differences in near and far transfer of task –switching training. *Developmental Science, 12*, 978–990.

Karbach, J., & Unger, K. (2014). Executive control training from middle childhood to adolescence. *Frontiers in Psychology, 5*, 390.

Karbach, J., & Verhaeghen, P. (2014). Making working memory work a meta-Analysis of executive-control and working memory training in older adults. *Psychological Science, 25*, 2027–2037.

Karbach, J., Könen, T., & Spengler, M. (2017). Who benefits the most? Individual differences in the transfer of executive control training across the lifespan. *Journal of Cognitive Enhancement, 1*, 394–405.

Könen, T., & Karbach, J. (2015). The benefits of looking at intraindividual dynamics in cognitive training data. *Frontiers in Psychology, 6*, 615.

Kray, J., & Dörrenbächer, S. (in press). The effectiveness of training in task switching: New insights and open issues from a lifespan view. To appear. In M. Bunting & J. Novick (Eds.), *Cognitive training*. Oxford: Oxford University Press.

Kray, J., & Fehér, B. (2017). Age differences in the transfer and maintenance of practice-induced improvements in task switching: The impact of working-memory and inhibition demands. *Frontiers in Psychology, 8*, 410.

Kray, J., & Ferdinand, N. K. (2013). How to improve cognitive control in development during childhood: Potentials and limits of cognitive interventions. *Child Development Perspectives, 7*, 121–125.

Kray, J., & Lindenberger, U. (2000). Adult age differences in task switching. *Psychology and Aging, 15*, 126–147.

Kray, J., Karbach, J., Haenig, S., & Freitag, C. (2012). Can task–switching training enhance executive control functioning in children with attention deficit/–hyperactivity disorder? *Frontiers in Human Neuroscience, 5*, 180.

Liepelt, R., Strobach, T., Frensch, P., & Schubert, T. (2011). Improved intertask coordination after extensive dual-task practice. *The Quarterly Journal of Experimental Psychology, 64*, 1251–1272.

Lövdén, M., Bäckman, L., Lindenberger, U., Schaefer, S., & Schmiedek, F. (2010). A theoretical framework for the study of adult cognitive plasticity. *Psychological Bulletin, 136*, 659–676.

Lövdén, M., Brehmer, Y., Li, S. C., & Lindenberger, U. (2012). Training-induced compensation versus magnification of individual differences in memory performance. *Frontiers in Human Neuroscience, 6*, 141.

Maquestiaux, F., Hartley, A. A., & Bertsch, J. (2004). Can practice overcome age-related differences in the psychological refractory period effect? *Psychology and Aging, 19*, 649–667.

Melby-Lervåg, M., & Hulme, C. (2013). Is working memory training effective? A meta-analytic review. *Developmental Psychology, 49*, 270–291.

Miller, E. K., & Cohen, J. D. (2001). An integrative theory of prefrontal cortex function. *Annual Review of Neuroscience, 24*, 167–202.

Minear, M., & Shah, P. (2008). Training and transfer effects in task switching. *Memory & Cognition, 36*, 1470–1483.

Miyake, A., & Friedman, N. P. (2012). The nature and organization of individual differences in executive functions: Four general conclusions. *Current Directions in Psychological Science, 21*, 8–14.

Miyake, A., Friedman, N. P., Emerson, M. J., Witzki, A. H., Howerter, A., & Wagner, T. (2000). The unity and diversity of executive functions and their contributions to complex "frontal lobe" tasks: A latent variable analysis. *Cognitive Psychology, 41*, 49–100.

Mozolic, J. L., Long, A. B., Morgan, A. R., Rawley-Payne, M., & Laurienti, P. J. (2011). A cognitive training intervention improves modality-specific attention in a randomized controlled trial of healthy older adults. *Neurobiology of Aging, 32*, 655–668.

Nguyen, L., Murphy, K., & Andrews, G. (2019). Immediate and long-term efficacy of executive functions cognitive training in older adults: A systematic review and meta-analysis. *Psychological Bulletin, 145*, 698–733.

Niendam, T. A., Laird, A. R., Ray, K. L., Dean, Y. M., Glahn, D. C., & Carter, C. S. (2012). Meta-analytic evidence for a superordinate cognitive control network subserving diverse executive functions. *Cognitive, Affective, & Behavioral Neuroscience, 12*, 241–268.

Norman, D. A., & Shallice, T. (1986). Attention to action: Willed and automatic control of behavior. In R. J. Davidson (Ed.), *Consciousness and self-regulation* (pp. 1–18). New York: Plenum Press.

Oberauer, K. (2015). Training of intelligence: A question of intelligent training. http://www.psychonomic.org/featured-content-detail/training-of-intelligence-question-of-intelligent-t.

Pereg, M., Shahar, N., & Meiran, N. (2013). Task switching training effects are mediated by working-memory management. *Intelligence, 41*, 467–478.

Prins, P. J., Dovis, S., Ponsioen, A., ten Brink, E., & Van Der Oord, S. (2011). Does computerized working memory training with game elements enhance motivation and training efficacy in children with ADHD? *Cyberpsychology, Behavior and Social Networking, 14*, 115–122.

Rebok, G. W., Carlson, M. C., & Langbaum, J. B. (2007). Training and maintaining memory abilities in healthy older adults: Traditional and novel approaches. *The Journals of Gerontology Series B: Psychological Sciences and Social Sciences, 62*, 53–61.

Reisberg, D., Baron, J., & Kemler, D. G. (1980). Overcoming stroop interference: The effects of practice on distractor potency. *Journal of Experimental Psychology: Human Perception and Performance, 6*, 140–150.

Rueda, M. R., Rothbart, M. K., McCandliss, M. D., Saccomanno, L., & Posner, M. L. (2005). Training, maturation, and genetic influences on the development of executive attention. *Proceedings of the National Academy of Science, 102*, 14931–14936.

Ruthruff, E., Van Selst, M., Johnston, J. C., & Remington, R. W. (2006). How does practice reduce dual-task interference: Integration, automatization, or simply stage-shortening? *Psychological Research, 70*, 125–142.

Sabah, K., Dolk, T., Meiran, N., & Dreisbach, G. (2018). When less is more: Costs and benefits of varied vs. fixed content and structure in short-term task switching training. *Psychological Research, 83*, 1–12.

Sala, G., & Gobet, F. (2017). Does far transfer exist? Negative evidence from chess, music, and working memory training. *Current Directions in Psychological Science, 26*, 515–520.

Schmiedek, F., Lövdén, M., & Lindenberger, U. (2010). Hundred days of cognitive training enhance broad cognitive abilities in adulthood: Findings from the COGITO study. *Frontiers in Aging Neuroscience, 2*, 27.

Schubert, T., & Strobach, T. (2018). Practice-related optimization of dual-task performance: Efficient task instantiation during overlapping task processing. *Journal of Experimental Psychology: Human Perception and Performance, 44*, 1884–1904.

Shipstead, Z., Redick, T. S., & Engle, R. W. (2012). Is working memory training effective? *Psychological Bulletin, 138*, 628–654.

Smith, E. E., & Jonides, J. (1999). Storage and executive processes in the frontal lobes. *Science, 283*, 1657–1661.

Strobach, T. (2020). The dual-task practice advantage: Empirical evidence and cognitive mechanisms. *Psychonomic Bulletin & Review, 27*, 3–14.

Strobach, T., Salminen, T., Karbach, J., & Schubert, T. (2014). Practice-related optimization and transfer of executive functions: A general review and a specific realization of their mechanisms in dual tasks. *Psychological Research, 78*, 836–851.

Thorell, L. B., Lindqvist, S., Bergman Nutley, S., Bohlin, G., & Klingberg, T. (2009). Training and transfer effects of executive functions in preschool children. *Developmental Science, 12*, 106–113.

Verhaeghen, P., & Cerella, J. (2002). Aging, executive control, and attention: A review of meta-analyses. *Neuroscience & Biobehavioral Reviews, 26*, 849–857.

Wiebe, S. A., & Karbach, J. (Eds.). (2017). Executive function: Development across the life span. New York: Routledge.

Wilkinson, A. J., & Yang, L. (2012). Plasticity of inhibition in older adults: Retest practice and transfer effects. *Psychology and Aging, 27*, 606–615.

Zinke, K., Einert, M., Pfennig, L., & Kliegel, M. (2012). Plasticity of executive control through task switching training in adolescents. *Frontiers in Human Neuroscience, 6*, 41.

Zinke, K., Zeintl, M., Rose, N. S., Putzmann, J., Pydde, A., & Kliegel, M. (2014). Working memory training and transfer in older adults: Effects of age, baseline performance, and training gains. *Developmental Psychology, 50*, 304–315.

Part IV
Multidomain Trainings

Action Video Game Training and Its Effects on Perception and Attentional Control

Benoit Bediou, Daphne Bavelier, and C. Shawn Green

Contents

Abstract Over the past two decades, a large body of research has examined the impact of playing one particular type of video game, known as action video games, on cognitive function. This work has demonstrated, in both correlational and intervention work, positive relationships between action video game experience and performance on a host of perceptual, attentional, and cognitive tasks. However, like all areas of research in the social sciences, such positive relationships have not always been observed. Furthermore, the massive expansion in terms of the number of studies and heterogeneity in terms of methods has made the literature difficult to summarize qualitatively. As such, the past few years have seen a number of published meta-analyses of the field, which have sought to examine not only the main effect-type questions (e.g., is action video game experience associated with increases in cognitive skill?), but also to utilize the heterogeneity in the research to address more subtle questions (e.g., which subdomains of cognition are most positively associated with action video game experience?). Overall the various published meta-analyses have offered largely convergent evidence – generally finding that action video game experience (in both cross-sectional and intervention work) is associated

B. Bediou · D. Bavelier
Université de Genève, Geneva, Switzerland

C. S. Green (✉)
University of Wisconsin–Madison, Madison, WI, USA
e-mail: cshawn.green@wisc.edu

T. Strobach, J. Karbach (eds.), *Cognitive Training*,
https://doi.org/10.1007/978-3-030-39292-5_15

with increases in cognitive skill, but that certain subdomains of cognition (e.g., perception, top-down attention, etc.) are associated with disproportionately large effects. Yet, some differences in inferences do exist across these meta-analyses, which we explore in the context of how the field may need to adapt going forward.

Introduction

Essentially as soon as video games began to grow in mass popularity in the early 1980s, their interactivity, rewarding properties, and clear perceptual, attentional, spatial, and motor demands attracted the attention of psychologists interested in their possible impact on cognitive functions. This early research in the field made use of the two main methodological approaches that, at least in terms of broad strokes, continue to dominate the field today. The first such approach is cross-sectional in nature. Here, individuals who play a great deal of video games as part of their normal daily life are contrasted in terms of cognitive skill against individuals who play few to no video games. For example, in one of the first works outlining the association between video game play and cognitive function, heavy video game players were found to possess enhanced sensorimotor skills as compared to non-players (Griffith et al. 1983). While cross-sectional studies are relatively easy to run and can indicate whether video game play is associated with increased cognitive abilities, like all correlational approaches, it cannot establish whether any observed relationships are causal. The cross-sectional method has thus traditionally been supplemented by a second approach, the true experiment or intervention design. Here cognitive skills are measured before and after purposeful training on a given video game. For instance, in one of the first such examples of this methodological approach in the field, nonvideo game players who were trained to play video games (*Targ* and *Battlezone*) showed improvements in spatial abilities from pretest to post-test relative to a control group who did not play, thus establishing a causal relationship between the video game play and the observed enhancements in cognitive function (Gagnon 1985).

While early work in the field either lumped all video games together or else considered single exemplar games, as the video game industry developed through the early 1990s and into the early 2000s, it became increasingly possible to categorize games into discrete genres. Critically, these genres not only served to separate games in terms of their narrative structure or viewpoint, they also served to differentiate games according to their cognitive demands (Spence and Feng 2010). This was a key development with respect to cognitive psychology, as essentially all theories outlining how experience could improve cognitive functions have emphasized the need to place sustained heavy load upon the given functions of interest. Specifically, it was during this time period that games now known as the "action video game" genre, which primarily encompassed first-person shooter (FPS) and third-person shooter games, became of primary interest to cognitive psychologists, as these games placed heavy load upon speed of processing, perception, divided and focused attention, multitasking, and spatial cognitive abilities in a manner not seen in most other game types at the time (Dale and Green 2017; see also Strobach and Schubert this volume).

Given these changes in the game industry, the typical methodological approaches utilized in the field therefore shifted slightly, with cross-sectional work coming to mainly contrast avid action video game players (AVGPs) against nonaction video game players (NVGPs), and intervention work examining the impact of training on action video games (with other nonaction commercial games being used as a control training experience). Over the past 15 years, a substantial number of reports have been published demonstrating a positive relationship between action video game experience and a host of perceptual, attentional, and cognitive functions utilizing both cross-sectional and intervention methods (Bavelier et al. 2012; Green and Bavelier 2012). However, as is true in essentially all areas in the social sciences, such effects have not been universally observed in all studies (Bisoglio et al. 2014; Boot et al. 2011, 2013). And overall, the proliferation of studies utilizing an increasingly diverse set of methods and addressing a wide variety of specific questions has resulted in a rich, but relatively heterogeneous, literature that is difficult to summarize qualitatively. The literature though is now an excellent candidate for quantitative summarization via meta-analytic methods that can address not only the extent to which major main effects are observed (e.g., whether action video game experience is associated with an overall increase in cognitive function), but also utilize the heterogeneity in the field to ask more subtle questions related to, for instance, which subdomains of cognition are most/least impacted, or how certain methodological choices affect the strength of the observed results.

Meta-analyses of the Action Video Game Literature

Over just the past few years, a number of meta-analyses have been published examining the impact of video game play on cognitive function (Table 1). As we will review below, these have generally found substantial agreement. However, differences in meta-analytic methods, including in terms of which games have been considered and/or have been aggregated across, have led to slightly different outcomes, which may together suggest interesting avenues for future research to address. In particular, one major difference across meta-analyses has been the extent to which they have considered only the cognitive impact of action video games alone or have attempted to consider the impact of all video games together (and then from there, perhaps attempting to isolate whether certain types of games lead to differential impact). Our disposition has been that the approach of considering all games together tests a hypothesis that is simply not reasonable from first principles. The term "video game" encompasses an exceedingly wide array of experiences that differ markedly in terms of their cognitive demands. The term "video games" for

Table 1 Meta-analyses results: effect of action games on cognition

Reference	Design	Games/age group	E.S.	CI	*p*	*m*	*k*
Powers et al. (2013)	Quasi-experiments	Action/violent games	0.62	0.53, 0.72	<0.001		196
Sala et al. (2018)	Correlational	Action games/ frequency + skills	0.11	0.06, 0.16	<0.001		69
	Cross-sectional	Action games	0.4	0.33, 0.47	<0.001		199
Bediou et al. (2018)	Cross-sectional	FPS-TPS	0.55	0.42, 0.68	<0.001	89	194
Powers & Brooks (2014)	True experiments	FPS games	0.23	0.07, 0.39	0.005		61
Powers et al. (2013)	True experiments	Action/violent games	0.22	0.13, 0.3	<0.001		135
Sala et al. (2018)	Intervention	Action vs. nonaction	0.1	−0.01, 0.2	0.068		96
	Intervention	Action vs. no-game	−0.12	−0.25, 0.01	0.072		88
Wang et al. 2017	Intervention	Action games/all ages	0.58	0.37, 0.78	<0.001	20	20
	Intervention	Action games/young adults	0.75	0.43, 1.07	<0.001	12	12
	Intervention	Action games/old adults	0.38	0.12, 0.64	<0.001	8	8
Bediou et al. (2018)	Intervention	FPS-TPS vs. Nonaction	0.29	0.08, 0.51	0.01	18	101
	Intervention	FPS-TPS vs. Nonaction/ young adults	0.34	0.07, 0.61	0.02	16	90
	Intervention	FPS-TPS vs. Nonaction/ older adults	−0.36	−1.16, 0.43	0.16	2	11

Notes. *ES* effect size, *CI* 95% confidence interval, *p p* value, *m* number of studies (random effects) or clusters (RVE analysis), *k* number of effect sizes

instance, would include both the game *Cookie Clicker* (where players just click the mouse to make a cookie, with the "goal" of clicking as many times as possible) and the game *Skyrim* (where the player navigates a rich fantasy environment, chooses to advance their character along a multitude of dimensions, buys and sells items, attacks enemies in various ways, converses with nonplayer characters, etc.). It is not remotely evident why there should be an "overall effect" of an experience category that encompasses both of these games (noting that these are just two exemplars at different extremes in the space of video games; if one considers the variety of experience across the entire space of video games, it, if anything, makes this point even more clearly).

Given this perspective, we therefore first consider just those meta-analyses that have examined the action video game literature. In particular, we will examine the extent to which these meta-analyses have converged on three main questions: (1) is action video game play, in cross-sectional/correlational work, associated with increased cognitive abilities? (2) Is action video game training, in intervention work, associated with increased cognitive abilities? (3) Which subdomains of cognitive skill appear to be most associated with action video game experience? After reviewing the meta-analyses of action video games, we then briefly turn to meta-analyses of all video games that have considered the impact of action games in their moderator or subgroup analyses. Given the existing differences in methods, especially regarding the analysis and interpretation of publication bias, we focus on uncorrected effects.

Meta-analyses of cross-sectional studies contrasting action video game players with nonaction players Cross-sectional studies are perhaps the most common type of study in this domain. As one example of such a study, Feng et al. compared a group of action video game players (who had played action games more than 4 hours per week in the previous 3 years) with a group of nonvideo game players (who had no video game experience in the previous 3 years) on a Useful Field of View task measuring visuo-spatial abilities (Feng et al. 2007). The group of action game players outperformed the nonplayers, thereby replicating previous work (Green and Bavelier 2003). The authors also reported superior performance in males compared to females, and in science majors compared to arts majors.

Across the literature, cross-sectional studies have differed widely in their methods, especially regarding participant recruitment (e.g., overt vs. covert methods) and selection criteria (minimum and maximum weekly hours allowed in the action and nonaction groups). Studies in this domain have also differed markedly in terms of the specific measures of cognition utilized, the critical dependent variables assessed (e.g., reaction time, accuracy, or both), and whether the predicted outcome involves a difference in overall performance (i.e., main effects) or just in specific conditions (e.g., an interaction term, such as the difference between congruent and incongruent in two groups). Meta-analysis is an appropriate way to summarize this methodological variability.

The cross-sectional meta-analysis by Bediou et al. is the one (to our knowledge) meta-analysis that has focused on studies contrasting action gamers –defined as those who have played 3 or more hours of FPS or third-person shooter games in the past 6 months – with nonplayers – who have played less than 1 hour of action video games or less than 3 hours of all video games in the past 6 months (Bediou et al. 2018). Overall, the meta-analysis captured data from 73 reports, encompassing 3833 participants, and 199 effect sizes from 91 independent samples of participants (note that in some cases, single reports produced multiple effect sizes). In this analysis, an overall difference in cognition of moderate size ($g = 0.55$) was in favor of the action game group.

Furthermore, and consistent with existing theory in the field suggesting that transferable gains in abilities require sustained load placed upon those abilities, the magnitude of the observed effect sizes differed depending on the subdomain of cognition that was considered (Cardoso-Leite et al. 2020; Dobrowolski et al. 2015; Powers and Brooks 2014; Spence and Feng 2010). Action video games place particularly heavy load upon perceptual, visual attentional, and spatial cognitive abilities. And these were indeed the subdomains where the largest effect sizes were seen.

Moderate to large action video game player advantages were observed in perception ($g = 0.78$), spatial abilities ($g = 0.75$), top-down attention ($g = 0.63$), and multitasking ($g = 0.55$; note that each subdomain encompasses various tasks) (see also Strobach and Schubert this volume). Small to medium effects were observed for inhibition ($g = 0.31$) and verbal tasks ($g = 0.30$). No reliable effect was observed for problem solving (though this could be due to a lack of studies).

Meta-analyses of true experiments examining the impact of action video game training Intervention studies are unfortunately much rarer as compared to cross-sectional work. Although this is perhaps not surprising given the extreme difference in cost and difficulty in running a long-term intervention study as compared to a cross-sectional study, the smaller number of intervention studies limits our ability to draw conclusions as firmly as in the case of cross-sectional work. One example of an intervention study was a follow-up to the example cross-sectional study discussed above by Feng, Spence and colleagues (Feng et al. 2007). Here, a group of nonvideo game players (with no video gaming experience in the past 4 years) were recruited and their pretraining performance was measured on a UFOV task and a mental rotation task. Pairs of same-gender participants were formed based on pretest performance, and one member of each pair was randomly allocated to the experimental (action game Medal of Honor) group, whereas the other member was allocated to the control (nonaction puzzle game 3D Ballance) group. Participants in both groups went through a total of 10-hours of individually supervised video game training that was conducted in the lab in sessions of 1–2 hours over a period of up to 4 weeks. Posttest performance was then measured after the training and at follow-up, with an average delay of 5 months. Performance on both tasks improved in the experimental group but not in the control group, and the improvement remained significant at follow-up.

As was true in the case of cross-sectional work, across the broader literature, intervention studies differ in myriad dimensions including not only participant selection criteria and cognitive ability measures, but also, in factors related to training, including the choice of experimental and control games, how training was implemented (e.g., training duration and distributed vs. massed practice), and how performance was measured. Again, meta-analyses offer a unique way to summarize this variability.

Two different meta-analyses have focused on true experiments examining whether action video game experience is causally related to enhancements in cognitive function. Bediou et al. (2018) focused on intervention studies contrasting changes in performance from pre to posttest between a group trained on a commercially available action video game and a group trained on a commercially available non-action video game. Furthermore, a minimum of 8 hours of training distributed over a minimum of eight sessions was required. Overall, this meta-analysis considered 23 reports, encompassing 713 participants, and 111 effect sizes from 20 independent samples of participants (data from some samples were reported in multiple papers), thus making a much smaller dataset compared to the cross-sectional work. The meta-analysis by Wang et al. (2017), meanwhile, was more lenient in its inclusion criteria (e.g., it included studies without an active control). Yet, despite these (and other) methodological differences, the top-line findings were quite similar. Both found a moderate impact of action video game training on cognition in healthy adults (Hedge's $g = 0.40$; Bediou et al. 2018; Cohen's $d = 0.58$; Wang et al. 2017; note that Hedge's g is conceptually similar to Cohen's d, but includes a correction for small sample sizes).

In terms of subdomains of cognition, a moderator analysis in Bediou et al. (2018) pointed toward a beneficial effect of action video game training on top-down attention ($g = 0.31$) and spatial cognition ($g = 0.45$), with (nonsignificant) trends pointing in the same direction for perception, multitasking and verbal cognition (though all should be taken with caution given the lack of statistical power) (see also Strobach and Schubert this volume). These effects fit squarely with those observed by Wang and colleagues, who divided the cognitive space slightly differently and found moderate benefits of action game training on visuospatial ability ($d = 0.54$) and processing speed/attention ($d = 0.50$), as well as executive functions ($d = 0.49$) and a small improvement in memory ($d = 0.33$) (see also Strobach and Schubert this volume).

Finally, in terms of which participants benefitted the most from action video game training, both analyses suggested that larger effects had been found in younger (largely college-age individuals) than older participants (i.e., individuals over the age of 60). In Bediou et al. (2018), preliminary analyses revealed a small to moderate effect of action game training in young adults ($g = 0.40$), but a trend toward negative effect in older adults (which could not be reliably estimated given the limited number of studies involving older adults). Wang et al. found that effects were stronger in younger adults ($d = 0.75$) relative to older adults ($d = 0.38$), though the age ranges were not specified. Altogether this finding is perfectly aligned with the principles of learning (Stafford and Dewar 2014) and may reflect the fact that action games and in particular, FPS games, are tailored toward young adults at the peak of their cognitive performance, and thus are too difficult for older adults because they exceed their zone of proximal development.

Meta-analyses of All Video Games: Converging Results

While a full discussion of the entirety of the video game research space is outside the scope of this chapter, several meta-analyses examining the impact of video games (all encompassing) have also examined the effect of action video games as part of their moderator or subgroup analyses and thus provide converging evidence to that discussed above. Powers and colleagues (Powers et al. 2013) found that action/violent games were associated with large cognitive effects in quasi-experiments ($g = 0.62$), but small effects in true experiments ($g = 0.22$). This latter value was consistent with a later reanalysis of true experiments where Powers and Brooks (2014) categorized games into a greater number of genres (Powers and Brooks 2014). Of these new genres, the closest match to the "action genre" was the FPS genre, which was associated with a small effect in true experiments ($d = 0.23$).

In terms of subdomains impacted, these authors found that FPS training significantly improved perceptual processing ($d = 0.45$) and spatial imagery ($d = 0.17$), but not motor skills ($d = 0.07$) or executive functions ($d = -0.17$) (see also Strobach and Schubert, this volume).

More recently, Sala and colleagues conducted a series of meta-analyses looking at correlational, cross-sectional and experimental studies (Sala et al. 2018). A positive effect of video game on cognition was found in cross-sectional meta-analysis. Players of action games showed an overall cognitive benefit ($g = 0.40$) with significant enhancements in all subdomains (ranging between $g = 0.31$ and $g = 0.45$) except intelligence/reasoning. An intervention meta-analysis focusing on true experiments found marginal improvements when action video game training was contrasted with nonaction game training ($g = 0.10$, $p < 0.068$). The effect was significant only for visual attention/processing, $g = 0.22$. However, we note that six of the eight effects in this analysis were from studies in older adults. For example, their analysis included a study by Boot et al. (2013), which compared a group trained for 22 hours on Mario Kart (which would not typically be considered an action video game), with a control group who played 60 hours of brain training. Moreover, the two effects in young adults included in these analyses were obtained with less than 3 hours of training (Valadez and Ferguson 2012).

Issues and Challenges: Methodological Choices in Video Game Research

How to Categorize Games for Research

Perhaps the single most important point from the data presented above is that different games have different cognitive impact. Action video games in particular have been consistently associated with positive effects on a variety of cognitive abilities, despite some disagreement regarding the magnitude of effects and their interpretation (especially regarding analyses of moderator effects and publication bias in intervention studies). Probably the largest issue identified when comparing and contrasting across meta-analyses is one that has been increasingly recognized throughout the space – the problem of categorizing games into genres and from genres into predicted cognitive outcomes (Table 1). For example, the category of action/violent games in Powers et al. (2013) was relabeled as FPS in Powers and Brooks (2014) and corresponds roughly (but not perfectly) with the action video game definition used in Bediou et al. (2018). Meanwhile, the action game category in Wang et al. included games that differ substantially from those that would have been labeled as "action" by Powers and Brooks (2014) or Bediou et al. (2018). These include various types of platform games (e.g., Donkey Kong or Pac Man; Clark et al. 1987; Seçer and Satyen 2014), strategy games (e.g., Rise of Nations; Basak et al. 2008), Wii games (Cherney 2008), puzzle games (Professor Layton and The Pandora's

Box; Colom et al. 2012), and even a computer program used primarily to test theories of skill acquisition, Space Fortress (Stern et al. 2011). Furthermore, the exact same games (Professor Layton and the Pandora's box; Pac Man and Donkey Kong) were considered action games in Wang et al. (2017) but as nonaction games in Sala et al. (2018). Obviously, the inclusion of games that lack significant load upon the cognitive functions of interest is likely to impact the overall conclusions that are reached.

Adding complexity to the issue is that the video game industry is not static. A game type that might have been a "nonaction game" in 2003 could be an "action game" in 2019. Indeed, emerging work over just the past few years has shown that shifts in genres, in particular wherein action characteristics are mixed into nonaction genres to create so called "hybrid-genres," must be taken into account (e.g., action-role-playing games, action-real-time-strategy games, etc.). As such, when coding for meta-analyses, it is not sufficient to know the genre alone (Dale and Green 2017). The year must also be considered (e.g., a role-playing game in the year 1995 would likely not be an action game; a role-playing game in the year 2019 very likely will be).

Cross-Sectional Studies: How to Define Gamers

Beyond the definition of game genres, the question of how video game experience (or skills) is measured (e.g., in correlational studies) and how this measure is then used to identify groups (e.g., action video game players vs. nonplayers in cross-sectional studies) has posed an enduring challenge to researchers trying to relate video game experience with cognitive skills. Identifying gamers with specific game-playing habits has been increasingly difficult given (i) the growing tendency of people to play multiple genres and (ii) the need to take into account a longer history of game play. The same is true for nonvideo game players, who make up a smaller proportion of the population each year.

The criteria used to define action video game players and nonvideo game players in cross-sectional studies have varied extensively. For example, the same participants who are classified as action gamers in some studies (e.g., Mack and Ilg 2014 defined players as those playing more than 1 hour/week of action games) may be considered nonplayers in other studies (e.g., Schenk et al. 2017 defined NVGPs as those playing less than 4 hours/week; Özçetin et al. 2019 and Matern 2018 used a cutoff of 5 hours/week). The fact that the gamers in one study could play less than the nongamers in another study certainly poses problems for meta-analyses trying to isolate the difference between groups.

Meanwhile, some studies have combined a low cutoff criterion for gamers (action gamers in studies from the Bavelier lab must play at least 5 hours per week in the past year) with a high cutoff for nongamers (nongamers must not exceed 3 hours of video game play per week in total and play no more than 1 hour of action games). This ensures that the gamers and nongamers differ substantially in their action video game experience and avoids any overlap in gaming hours between the two groups (i.e., extreme group approach). The selection criteria used by the Bavelier lab also emphasize (lack of) experience with other genres (action gamers must play at least 5 hours of action games per week, but not more than 3 hours per week of nonaction games), as well as past experience (participants can be excluded based on their prior video game experience). This, together with other methodological specificities, could account for the stronger effects (and smaller sample sizes) reported by this lab (see Bediou et al. 2018 and also Hilgard et al. 2019). Unfortunately, however, in many cases throughout the literature, specific recruitment details at this level have not been systematically reported, and thus cannot be properly evaluated with meta-analytic methods.

Future Directions and Concluding Remarks

The study of the cognitive effects of action video games poses a number of issues, which relate to the complex and continuously changing ecosystem of video games and video game players. The meta-analyses reviewed here have begun to speak of some of these issues. However, all of the most pressing questions in the field, for instance those concerning the cognitive enhancing properties of action video game training (and thus the causality of the effects, which cannot be assessed in correlational and cross-sectional work), require substantially more data to be properly evaluated. There are simply not enough studies to speak with strong confidence regarding the central tendency in the field.

Several directions (beyond continued exploration and refinement of the main topics discussed here) are of interest going forward. For instance, some authors (Boot et al. 2011, 2013; Boot and Simons 2012; Sobczyk et al. 2015) have recently argued that the observed effects in cross-sectional studies (i.e., action game players outperforming nonplayers) could be at least partially explained (or in the strongest form of the argument, fully explained) by differences in participant expectations (see also Cochrane and Green, Katz et al. this volume). Such a hypothesis requires that participants (1) understand why they were selected (e.g., because they are an avid action game player or a nonplayer), (2) can intuit the expected results and how to modify their behavior accordingly, and (3) are actually capable of making those modifications to their behavior (e.g., responding 10% faster without a reduction in accuracy). While there are reasons to think, at least in many cases, that this is unlikely to be possible, it remains a supposition of interest in the field.

Although explicit tests of this hypothesis are largely lacking in the domain, Bediou et al. made use of the heterogeneity in the literature to at least begin to broach the question (Bediou et al. 2018). For instance, their meta-analysis found that how performance was measured in studies (e.g., speed vs. accuracy and main effect vs. interaction) made no difference in the size of the observed effects (with the idea being that if expectations were driving effects in the field, it might be easier for participants to change certain types of behaviors or to intuit certain types of effects). Another way their meta-analyses tackled this issue is through the analysis of covert (where participants are not made aware that their gaming status is important until after the study ends) versus overt (where participants are aware that their gaming status is important) recruitment methods. Here Bediou et al. found that while overt recruitment might magnify the size of the observed effects, positive effects were observed regardless of the recruitment method. However, while the results above are suggestive, it is absolutely the case that without dedicated studies on the topic, it remains very difficult to make strong claims. Studies that directly manipulate expectations should thus be encouraged.

With regard to meta-analytic specific issues of interest, one major issue within all of social science is publication bias. Publication bias refers to the fact that significant results are more likely to be published than nonsignificant results, especially when sample sizes are small. Numerous methods are available to test for the presence of publication bias and to attempt to estimate unbiased or corrected effects. Most of these methods, however, attempt to test and correct for possible small study effects (Carter et al. 2019; Debray et al. 2018; McShane et al. 2016). The problem with this approach is that small study effect can arise from causes other than publication bias, including genuine differences in methods. For example, longer training durations are expected to produce larger effects. However, these studies are more expensive and are thus more likely to involve smaller samples, due to their cost, complexity, and possible drop-out rates that increase with study duration. Removing this type of "small study effect" via meta-analysis would be clearly inappropriate. The situation is further complicated because other factors may also covary with training duration, such as the consideration of the subdomain of cognition, how performance is measured, and which games are used as experimental and control games. Finally, most methods for correcting publication bias effects are developed using simulated data and imply assumptions regarding the number and distribution of effect sizes and their heterogeneity, which are rarely met or often difficult to verify. In all, this is one area that is expected to be greatly benefited by the growing emphasis on open-science and reporting practices. Indeed, if the results of all studies, not just "successful" studies were reported, this would eliminate the need to correct for publication bias.

As a conclusion, we note that the issues reviewed here are not specific to the literature on cognitive effects of video games. We hope that this chapter will give the keys to understanding some of the debates that exist in other fields. For example, the meta-analyses of the relationship between violent videogames and aggression have been characterized by intense debates around the choice of studies included/excluded and also the methods for analyzing publication bias (Anderson et al. 2010; Ferguson 2007; Ferguson and Kilburn 2010; Hilgard et al. 2017; Kepes et al. 2017). Other debates have concerned other methodological aspects, such as the choice of measures (e.g., taking covariates or partial correlations) (Boxer et al. 2015; Ferguson 2015a, b; Markey 2015; Prescott et al. 2018; Rothstein and Bushman 2015; Valkenburg 2015).

References

Anderson, C. A., Shibuya, A., Ihori, N., Swing, E. L., Bushman, B. J., Sakamoto, A., Rothstein, H. R., & Saleem, M. (2010). Violent video game effects on aggression, empathy, and prosocial behavior in eastern and western countries: A meta-analytic review. *Psychological Bulletin, 136*, 151–173.

Basak, C., Boot, W. R., Voss, M. W., & Kramer, A. F. (2008). Can training in a real-time strategy video game attenuate cognitive decline in older adults? *Psychological Aging, 23*, 765–777.

Bavelier, D., Green, C. S., Pouget, A., & Schrater, P. (2012). Brain plasticity through the life span: Learning to learn and action video games. *Annual Review of Neuroscience, 35*, 391–416.

Bediou, B., Adams, D. M., Mayer, R. E., Tipton, E., Green, C. S., & Bavelier, D. (2018). Meta-analysis of action video game impact on perceptual, attentional, and cognitive skills. *Psychological Bulletin, 144*, 77–110.

Bisoglio, J., Michaels, T. I., Mervis, J. E., & Ashinoff, B. K. (2014). Cognitive enhancement through action video game training: Great expectations require greater evidence. *Frontiers in Psychology, 5*, 136.

Boot, W. R., & Simons, D. J. (2012). Advances in video game methods and reporting practices (but still room for improvement): A commentary on Strobach, Frensch, and Schubert (2012). *Acta Psychologica, 141*, 276–277.

Boot, W. R., Blakely, D. P., & Simons, D. J. (2011). Do action video games improve perception and cognition? *Frontiers in Psychology, 2*, 226.

Boot, W. R., Simons, D. J., Stothart, C., & Stutts, C. (2013). The pervasive problem with placebos in psychology: Why active control groups are not sufficient to rule out placebo effects. *Perspectives on Psychological Science, 8*, 445–454.

Boxer, P., Groves, C. L., & Docherty, M. (2015). Video games do indeed influence children and adolescents' aggression, prosocial behavior, and academic performance: A clearer reading of Ferguson (2015). *Perspectives on Psychological Science, 10*, 671–673.

Cardoso-Leite, P., Joessel, A., & Bavelier, D. (2020). Games for enhancing cognitive abilities. In *MIT handbook of game-based learning*. Boston: MIT Press.

Carter, E. C., Schönbrodt, F. D., Gervais, W. M., & Hilgard, J. (2019). Correcting for bias in psychology: A comparison of meta-analytic methods. *Advances in Methods and Practices in Psychological Science, 2*, 115–144.

Cherney, I. D. (2008). Mom, let me play more computer games: They improve my mental rotation skills. *Sex Roles, 59*, 776–786.

Clark, J. E., Lanphear, A. K., & Riddick, C. C. (1987). The effects of videogame playing on the response selection processing of elderly adults. *Journal of Gerontology, 42*, 82–85.

Colom, R., Quiroga, M. Á., Solana, A. B., Burgaleta, M., Román, F. J., Privado, J., Escorial, S., Martínez, K., Álvarez-Linera, J., Alfayate, E., García, F., Lepage, C., Hernández-Tamames, J. A., & Karama, S. (2012). Structural changes after videogame practice related to a brain network associated with intelligence. *Intelligence, 40*, 479–489.

Dale, G., & Green, C. S. (2017). The changing face of video games and video gamers: Future directions in the scientific study of video game play and cognitive performance. *Journal of Cognitive Enhancement, 1*, 280–294.

Debray, T. P. A., Moons, K. G. M., & Riley, R. D. (2018). Detecting small-study effects and funnel plot asymmetry in meta-analysis of survival data: A comparison of new and existing tests: Detecting small-study effects. *Research Synthesis Methods, 9*, 41–50.

Dobrowolski, P., Hanusz, K., Sobczyk, B., Skorko, M., & Wiatrow, A. (2015). Cognitive enhancement in video game players: The role of video game genre. *Computers in Human Behavior, 44*, 59–63.

Feng, J., Spence, I., & Pratt, J. (2007). Playing an action video game reduces gender differences in spatial cognition. *Psychological Science, 18*, 850–855.

Ferguson, C. J. (2007). Evidence for publication bias in video game violence effects literature: A meta-analytic review. *Aggression and Violent Behavior, 12*, 470–482.

Ferguson, C. J. (2015a). Do angry birds make for angry children? A meta-analysis of video game influences on children's and adolescents' aggression, mental health, prosocial behavior, and academic performance. *Perspectives on Psychological Science, 10*, 646–666.

Ferguson, C. J. (2015b). Pay no attention to that data behind the curtain: On angry birds, happy children, scholarly squabbles, publication bias, and why betas rule metas. *Perspectives on Psychological Science, 10*, 683–691.

Ferguson, C. J., & Kilburn, J. (2010). Much ado about nothing: The misestimation and overinterpretation of violent video game effects in eastern and western nations: Comment on Anderson et al. (2010). *Psychological Bulletin, 136*, 174–178.

Gagnon, D. M. (1985). Videogames and spatial skills: An exploratory study. *Educational Communication and Technology, 33*, 263–275.

Green, C. S., & Bavelier, D. (2003). Action video game modifies visual selective attention. *Nature, 423*, 534–537.

Green, C. S. S., & Bavelier, D. (2012). Learning, attentional control, and action video games. *Current Biology, 22*, R197–R206.

Griffith, J. L., Voloschin, P., Gibb, G. D., & Bailey, J. R. (1983). Differences in Eye-hand motor coordination of video-game users and non-users. *Perceptual and Motor Skills, 57*, 155–158.

Hilgard, J., Engelhardt, C. R., & Rouder, J. N. (2017). Overstated evidence for short-term effects of violent games on affect and behavior: A reanalysis of Anderson et al. (2010). *Psychological Bulletin, 143*, 757–774.

Hilgard, J., Sala, G., Boot, W. R., & Simons, D. J. (2019). Overestimation of action-game training effects: Publication bias and Salami slicing. *Collabra: Psychology, 5*, 30.

Kepes, S., Bushman, B. J., & Anderson, C. A. (2017). Violent video game effects remain a societal concern: Reply to Hilgard, Engelhardt, and Rouder (2017). *Psychological Bulletin, 143*, 775–782.

Mack, D. J., & Ilg, U. J. (2014). The effects of video game play on the characteristics of saccadic eye movements. *Vision Research, 102*, 26–32.

Markey, P. M. (2015). Finding the middle ground in violent video game research: Lessons from Ferguson (2015). *Perspectives on Psychological Science, 10*, 667–670.

Matern, M. F. (2018). *The effects of video gaming on visual selective attention in early adulthood* (Doctoral dissertation, University of Pretoria).

McShane, B. B., Böckenholt, U., & Hansen, K. T. (2016). Adjusting for publication bias in meta-analysis: An evaluation of selection methods and some cautionary notes. *Perspectives on Psychological Science, 11*, 730–749.

Özçetin, M., Gümüştaş, F., Çağ, Y., Gökbay, İ. Z., & Özmel, A. (2019). The relationships between video game experience and cognitive abilities in adolescents. *Neuropsychiatric Disease and Treatment, 15*, 1171–1180.

Powers, K. L., & Brooks, P. J. (2014). Evaluating the specificity of effects of video game training. In F. Blumberg (Ed.), *Learning by playing: Frontiers of video gaming in education* (pp. 302–329). Oxford: Oxford University Press.

Powers, K. L., Brooks, P. J., Aldrich, N. J., Palladino, M. A., & Alfieri, L. (2013). Effects of video-game play on information processing: A meta-analytic investigation. *Psychonomic Bulletin & Review, 20*, 1055–1079.

Prescott, A. T., Sargent, J. D., & Hull, J. G. (2018). Metaanalysis of the relationship between violent video game play and physical aggression over time. *Proceedings of the National Academy of Sciences, 115*, 9882–9888.

Rothstein, H. R., & Bushman, B. J. (2015). Methodological and reporting errors in meta-analytic reviews make other meta-analysts Angry: A commentary on Ferguson (2015). *Perspectives on Psychological Science, 10*, 677–679.

Sala, G., Tatlidil, K. S., & Gobet, F. (2018). Video game training does not enhance cognitive ability: A comprehensive meta-analytic investigation. *Psychological Bulletin, 144*, 111–139.

Schenk, S., Lech, R. K., & Suchan, B. (2017). Games people play: How video games improve probabilistic learning. *Behavioural Brain Research, 335*, 208–214.

Seçer, I., & Satyen, L. (2014). Video game training and reaction time skills among older adults. *Activities, Adaptation, & Aging, 38*, 220–236.
Sobczyk, B., Dobrowolski, P., Skorko, M., Michalak, J., & Brzezicka, A. (2015). Issues and advances in research methods on video games and cognitive abilities. *Frontiers in Psychology, 6*, 1451.
Spence, I., & Feng, J. (2010). Video games and spatial cognition. *Review of General Psychology, 14*, 92–104.
Stafford, T., & Dewar, M. (2014). Tracing the trajectory of skill learning with a very large sample of online game players. *Psychological Science, 25*, 511–518.
Stern, Y., Blumen, H. M., Rich, L. W., Richards, A., Herzberg, G., & Gopher, D. (2011). Space fortress game training and executive control in older adults: A pilot intervention. *Aging Neuropsychology and Cognition, 18*, 653–677.
Valadez, J. J., & Ferguson, C. J. (2012). Just a game after all: Violent video game exposure and time spent playing effects on hostile feelings, depression, and visuospatial cognition. *Computers in Human Behavior, 28*, 608–616.
Valkenburg, P. M. (2015). The limited informativeness of meta-analyses of media effects. *Perspectives on Psychological Science, 10*, 680–682.
Wang P, Liu H-H, Zhu X-T, Meng T, Li H-J and Zuo X-N (2017) Action Video Game Training for Healthy Adults: A Meta-Analytic Study. *Frontiers in Psychology* 7:907.

Video Game Training and Effects on Executive Functions

Tilo Strobach and Torsten Schubert

Contents

Abstract In the present chapter, we reviewed studies investigating the effects of video game training (particularly action video games) on the executive functions shifting, dual tasking, updating, and inhibition. These studies provide evidence that video game training improves the performance in task-switching (i.e., shifting) and dual-task situations. Evidence for an effect of video game training on working memory updating is mixed, and this effect might not be a consequence of video game training. The literature on effects of action video gaming rather suggests no relation between training in action video games and improved inhibition. In sum, this set of findings is consistent with the assumption that transfer from action video game training to executive function measures is domain-specific and might depend on similarities between the trained video game and the laboratory task.

T. Strobach (✉)
Department Psychology, Medical School Hamburg, Hamburg, Germany
e-mail: tilo.strobach@medicalschool-hamburg.de

T. Schubert
Martin-Luther University Halle-Wittenberg, Halle, Germany

T. Strobach, J. Karbach (eds.), *Cognitive Training*,
https://doi.org/10.1007/978-3-030-39292-5_16

Introduction

The video game industry expands as its sales volume and the number of the industry's clients constantly increase. Surveys show that the annual video game sales exceeded 43.4 billion (in the USA exclusively) and more than 1.2 billion individuals worldwide are considered as video gamers (Spil Games 2013), including more than 164 million Americans (The Entertainment Software Association 2018); this frequent use of video games is largely independent of gender, education, and income (e.g., BIU 2012). Cognitive research provided evidence in recent years that experienced video gamers outperform non-experienced people in a number of basic cognitive functions (e.g., Bavelier et al. 2012, 2018; Bediou et al., this volume; however, see Simons et al. 2016, for a more skeptical view on training-related plasticity). These positive effects in video gamers led us to focus on the particular effects of video game experience on executive functions (see also Karbach and Kray, this volume).

Executive functions typically control our behavior when we perform in demanding and complex situations including situations in which the management of different tasks or task sequences is required. These functions define a set of general-purpose control mechanisms, often linked to the prefrontal cortex of the brain, that modulate the operation of various cognitive subprocesses and thereby regulate the dynamics of human cognition (Baddeley 1986; Miyake and Friedman 2012; Miyake et al. 2000). Different types of executive functions have been classified by different authors, for example, shifting, dual tasking, updating, and inhibition. While their processing can be time-consuming and inefficient under unpracticed conditions, recent studies suggest that executive functions can be improved as a result of extensive training and training-induced improvements can be transferred to non-practiced situations (Anguera et al. 2013; Strobach 2020; Strobach et al. 2014). This training-related plasticity is particularly relevant when aiming to compensate for the strong age-related declines in executive functions and frontal lobe tasks (Raz 2000; Strobach et al. 2015).

The present chapter includes a concise review of empirical studies and meta-analyses investigating the potential optimization and transfer of different types of executive functions as a result of video game experience. Here, we primarily focus on studies within the action video game genre. Since many studies have been concerned with assessing the impact of this game genre on executive function as action video game playing seems highly adequate for training executive control skills. In action video games, gamers have to control and conduct multiple simultaneous tasks at a high speed. Important information, such as interim targets and assignments, must be updated all the time (Spence and Feng 2010) and gamers need to adapt their actions and action goals under permanently changing task conditions (Bavelier et al. 2012). The most prominent action games are first-person shooters such as the *Counter-Strike, Unreal Tournament, Call of Duty, or Medal of Honor* series of games and third-person shooters like the *Grand Theft Auto* series. In these games, gamers play in an open virtual world with a first-person or third-person perspective on the main character. They usually have to fight against enemies, find objects, and navigate through this world.

The relation between action video games and executive functions (as well as other game genres and mental domains) is usually investigated from two methodological perspectives. First, there are cross-sectional comparisons between individuals self-reporting a high amount of experience with these games. The executive function performance in these habitual video gamers is typically contrasted with the one of individuals reporting no such experience; these individuals are either unexperienced in video games in general or are not experienced in action video games in particular but built up experience in other game genres (for the sake of simplicity, we refer to these latter individuals as non-gamers). However, if comparisons between gamers and non-gamers show performance differences, in particular performance advantages in gamers, do these advantages in gamers mean that there is a causal link between video game experience and optimized executive functions? The answer is no, not exactly (Green et al. 2014, 2019). Advantages in gamers do not necessarily have to be a result of video game experience (Schubert and Strobach 2012). The advantage could be, for instance, inherited or just given before they started playing video games (which would mean that the advantage would then be independent of the video game experience). As a consequence, research on video games has implemented more and more well-controlled training interventions with non-gamers in order to assess potential causal links between game experience and optimized executive functions. Usually, these training studies have a pretest–training–posttest design with tests on executive functions during pretest and posttest and one group of non-gamers with training in an action video game across several hours. To control for methodological impacts such as test-retest effects or general motivational issues, one or more control groups complete a similar general design of pretest–training–posttest. During training, these groups usually perform control procedures different from action video gaming but perform (again in the present case) tests assessing executive functions during pretest and posttest.

In the video game literature, two theoretical perspectives were introduced to generally explain mechanisms of transfer effects from video gaming to situations beyond the game context (e.g., laboratory-based transfers to measures of executive functioning). The first explanation to account for potential transfer effects is that these effects are all due to a single more general level of improvement, which then aids performance in all transfer tasks. One proposal of general training-related transfer is that video gamers improve in probabilistic inference, or "learning to learn." As a result of training, according to this "learning to learn" account, action video gamers generally become more effective in using evidence from repeated presentations of a task to guide their decision-making and allocation of cognitive resources (Bavelier et al. 2012). This "learning to learn" account predicts that, as a result of appropriate action video game training, there should be transfer effects to all types of executive functions, that is, shifting, dual tasking, updating, and inhibition.

In contrast, transfer effects may be due to video games having several separate demands in common with laboratory tasks that measure perception, attention, or cognition (Oei and Patterson 2015). According to this "common demands" account (Dahlin et al. 2008), transfer from action video games to executive function measures is specific and depends on similarities between the trained video game and the laboratory task. There may be some specific learned properties of the game, but there may also be higher level more abstract procedures that are developed during the game that may allow transfer from the game to behavioral measures. Taatgen (this volume) argues that skills required to perform a task can be broken down into "primitive information processing elements" of which some are task-general and some are specific. Only if two tasks share overlapping elements, those learned from training can be applied in test situations, producing transfer (see also Salminen et al. 2016, for the case of transfer in working memory updating). As a consequence, alternatively to the prediction of the "learning to learn" account, the "common demands" account predicts that transfer effects might not be general for all types of executive functions (i.e., shifting, dual tasking, updating, and inhibition) but might be specific for the functions where the game and task share common demands. In the final section of this chapter, we will evaluate the literature on action game experience and effects on executive functions regarding these accounts (i.e., "learning to learn" versus "common demands"), explaining general mechanisms of transfer effects.

Shifting

Also referred to as "attention switching" or "task switching," this type of executive functions concerns the ability to shift back and forth between multiple tasks, operations, or mental sets (Monsell 2003). Shifting involves the disengagement from irrelevant information (e.g., the task set of a previous task) and/or the active engagement in relevant information (e.g., the task set of an upcoming task). Evidence for optimized shifting derives from studies on task-switching practice (e.g., Berryhill and Hughes 2009; Karbach and Kray 2009; Strobach et al. 2012a, Wendt et al. 2017): These studies showed that performance costs associated with the shifting processes (e.g., task-switch costs reflected by larger reaction times [RTs] in trials with switches between different tasks in contrast to trials with task repetitions) are reduced with practice and, consequently, illustrate optimization of executive functioning of shifting.

Before we go into detailed studies and the theoretical explanations about training effects from these studies, we give a meta-analytic overview of the relation between action video gaming and shifting. Powers et al. (2013) showed a moderate benefit of experience in action video gaming in the shifting domain with Bediou et al. (2018) replicating this finding with rather upper-medium benefits of this experience type. The latter study could show that this effect was moderated by age with larger effects in young than in older adults. While the latter study did not test for shifting in an intervention training perspective, the former study could not show a benefit for this executive function domain.

Focusing on individual empirical studies, persons with experience in action video games showed less switch costs than non-gamers in a paradigm including predictable switches and repetitions (e.g., Colzato et al. 2010). This finding is the first – although cross-sectional – example of evidence for optimized executive functions in terms of improved shifting abilities. Strobach et al.'s (2012b) training intervention in young adults consisted of fifteen 1-hour sessions, in which two groups of non-gamers played different games. The first group worked on a puzzle game with only one main task and only low executive function demands. The second group played an action game with high executive function demands. In a test on task-switching performance before the training started, the switch costs do not differ between both groups of puzzle and action gamers. Afterwards, however, the results indicated lower switch costs in the action game group in comparison to the puzzle group. This training study shows that switch costs can be reduced with action game training specifically and that this reduction cannot be traced to inherited, given, or previously acquired attributes. These results provide evidence for a causal link between video game experience and optimized executive functions for shifting between different tasks. Further, this finding was generalized to numerous alternative task-switching situations, sharing varying numbers of input and output processors with typical action video games (Cain et al. 2012; Green et al. 2012).

However, the task-switching advantage of non-gamers after action video game training might be limited to situations with predictable task switches and the requirement to constantly update working memory: how many trials have been completed in the current task and to count down for the upcoming switch (Green et al. 2012; Strobach et al. 2012b). In a task-switching paradigm with the random and unpredictable occurrence of switch and repetition trials (e.g., the particular task is cued), updating of working memory is not required, and participants do not need to take into account the nature of previous trials. There is no evidence for superior shifting between tasks in action video gamers versus non-gamers as well as after training of an action video game, strategy game, or puzzle game in such an unpredictable task-switching situation (Boot et al. 2008; Oei and Patterson 2014). The observation of advanced task-switching performance of action video gamers predominantly in situations with predictable task switches might point to an impact of superior updating functions related to this group of participants. In our view, this assumption may represent an issue for fruitful future investigations (see also the updating section).

A further mechanism that may specifically explain action video gamers' improved performance in task-switching situations is a superior ability to control selective attention and thus active engagement in relevant information of an upcoming task (i.e., selective attention-dependent preparation, Karle et al. 2010). The effectiveness of engagement might be that relevant information of an upcoming task is only activated to a degree in working memory that is necessary to efficiently perform this task. In such a case, the following effort for an effective disengagement of this task information is reduced to a minimal degree. The reduced effort for task disengagement might free processing resources for alternative tasks, a potentially effective strategy for successful performance in complex gaming contexts.

Dual Tasking

Do action video gamers also have advantages when they perform different tasks simultaneously at the same time (instead of a sequential performance of different tasks as in the task-switching paradigm)? Are there any signs of optimized executive functions when the gamers are put in dual-task situations? Dual-task situations require the coordination of different tasks and task information due to executive functions (among others, dual tasks require the control of which task is performed first and which task second [Schubert 2008; Szameitat et al. 2006]). For instance, this coordination leads to longer RTs in dual-task situations compared to single-task situations, leading to dual-task performance costs.

Similar to the shifting domain, Powers et al. (2013) showed a moderate cross-sectional benefit of experience in action video gaming in the dual-task domain. Bediou et al. (2018) replicated this finding with rather upper-medium benefits of this experience, and this effect was moderated by age with larger effects in younger than in older adults. Focusing on individual empirical studies, Gaspar et al. (2014) were not able to find evidence for different dual-task costs between action video gamers vs. non-gamers however. In detail, a simulated street-crossing scenario was combined with a working memory task in their dual-task situation. The number of trials on which participants successfully crossed the street and the latency of initiating the crossing were impaired in the dual task compared to performance in the isolated crossing task under single-task conditions. However, there was no reduction of dual-task costs specific for action video gamers. These findings of a lacking dual-task advantage in this group were consistent with those of Donohue et al. (2012) that combined a multiple object tracking task, a paper and pencil search task, and a driving tracking task with answering trivia questions. Although these tasks are certainly relevant in daily life, they are no established measures of dual-task performance and differ considerably from reliable and valid laboratory paradigms.

This conclusion is supported by a number of studies, which showed positive effects of action video gaming on dual-task skills (Chiappe et al. 2013; Strobach et al. 2012b). For example, Strobach et al. (2012b) compared the performance of gamers and non-gamers in dual- and single-task situations including speeded and well-controllable choice RT tasks. There was no difference in single-task RTs between gamers and non-gamers. However, there was a difference in dual tasks: Gamers showed lower RTs and therefore a better performance particularly in dual-task situations compared to non-gamers. This result confirmed the assumption of an optimization of executive functions associated with the coordination of two simultaneous tasks. Also, with focus on dual-task performance, non-gamers increasingly benefitted from action video game training more than from puzzle training, which indicates a causal link between video game experience and optimized executive functions in dual-task situations (see also Schubert and Strobach 2012). These conclusions were supported from a dual-search situation combining an identification and comparison search task (Wu and Spence 2013); performance in this dual-task situation was specifically improved after non-gamers' action video game training vs. puzzle game training. The possible effect of video gaming on dual tasking is still a matter of debate, as a meta-analysis showed no robust effects of action video game training (in comparison to active control interventions) on dual tasking (Bediou et al. 2018). Nevertheless, the reported null effect might be explained by the small number of studies in this field, which requires further meta-analyses with larger samples of included studies.

Updating

Updating and monitoring of representations and information in working memory is another dimension of executive functions (Miyake et al. 2000). In detail, this dimension is related to the monitoring and coding of incoming information that is related to a task at hand. Further, updating processes serve to revise items held in working memory by replacing old information that is no longer relevant with newer, more relevant information. For instance, updating plays an important role in working memory tasks of the *n*-back type, in which a participant is presented with a sequence of stimuli and instructed to indicate when the currently presented stimulus matches the one from *n* steps earlier in the sequence (Jonides and Smith 1997).

Action video gamers show faster and more correct responses than non-gamers in the *n*-back paradigm, which indicates an optimized functionality of the updating function (Colzato et al. 2013). Further, even puzzle game training in non-gamers was effective in producing superior performance in a mental rotation task; transfer in this case is plausible, given that the mental rotation task was both visually and conceptually similar to this training game (Boot et al. 2008). However, in a spatial *n*-back task and a Corsi block-tapping task, no increase in accuracy could be registered, neither in action video gamers versus non-gamers nor after non-gamers' action game training, strategy game training, and puzzle game training (Boot et al. 2008). In sum, given the current state of the literature in the field, it remains unclear whether there really is a (causal) link between game experience and the executive function updating. This unclear conclusion is supported by meta-analyses in the field. While these analyses showed at least small effects of experience in action video games in cross-sectional studies, there is no meta-analytic evidence for an effect of video game training in longitudinal studies on updating (Powers et al. 2013).

Inhibition

A further executive function is inhibition, which is related to the ability to deliberately inhibit or stop dominant, automatic, or prepotent responses when necessary. A prototypical inhibition task is the color Stroop task (MacLeod 1991). In this task, participants are instructed to respond to the ink of color words; these color words are congruent (e.g., GREEN in green ink) or incongruent (e.g., GREEN in red ink). Typically, RTs in incongruent trials are larger than in congruent trials (i.e., the Stroop effect), indicating the requirement to inhibit or to override the tendency to produce a more dominant or automatic response on naming the color word. However, practice of a Stroop tasks results in a reduction of the Stroop effect within this task, indicating a task-specific training effect by an increased RT reduction in congruent versus incongruent trials (e.g., Davidson et al. 2003; Wilkinson and Yang 2012).

Given the current state of the literature, we are however skeptical about a positive effect of action video gaming on inhibition. In individual empirical cross-sectional studies, the Stroop effect was not reduced in participants that played a difficult version of an action video game versus a non-difficult version of such a game in the study of Engelhardt et al. (2015). This finding demonstrates no evidence for an impact of action video gaming on inhibition, which is also supported by the results of studies with alternative paradigms testing varying facets of inhibition. That is, action video gamers in contrast to non-gamer controls showed no superior performance in a Go/No-Go task (in this task, participants have to press a button [Go] given certain stimuli and inhibit that action under a different set of stimuli [No-Go], Oei and Patterson 2014) and in a stop-signal task (in this task, participants are presented with a stimulus prompting them to execute a particular manual response, and this stimulus may or may not be followed by a stop signal calling for the immediate abortion of that response, Colzato et al. 2013). Consistently, from a meta-analytic perspective, findings from longitudinal studies showed that action video game trainings had no impact on inhibition performance (Powers and Brook 2014). In sum, at the current state there is no convincing evidence that experience in action video games can improve executive functioning associated with the inhibition of responses when necessary.

Meta-Analyses on General Executive Functioning

Due to the increasing number of empirical studies, recently several meta-analyses have been conducted investigating the relationship between action video games and general executive functioning; in this regard, the term general executive functioning means that these studies performed analyses on executive functioning without disentangling the relation between video gaming and specific executive function domains. In habitual gamers versus non-gamers, Powers et al.'s (2013) combination of executive functions comprised executive function batteries, dual/multitasking, inhibition tasks (e.g., Stroop task, Simon task, Flanker task), intelligence tests, task switching, and working/short-term memory measures. Their meta-analysis showed a small but robust effect of experience in action video gamers versus non-gamers. Realizing more strict inclusion criteria on empirical studies and investigating the impact of publication biases, Sala et al. (2018) found only very small effects of experience in action video gamers on executive functions. However, this effect could be only very small since Sala et al. applied a categorization of executive functions in different domains. While their cognitive control domain included tests such as task switching, Go/No-Go, Simon, and Stroop tasks (thus rather exclusively shifting and inhibition), updating was categorized as memory in combination with tests such as span, *n*-back, and recall tasks (i.e., a combination of rather short-term and long-term memory aspects as well as working memory updating). This study thus divided executive functions in the system of Miyake and Friedman (2012) and applied in this chapter (see also Karbach and Kray, this volume) across different categories of analysis and combined it with long-term memory processes.

The meta-analysis of Wang et al. (2016) showed moderate effects of training of action video games on executive functioning; in this analysis executive functioning combined planning, working memory, reasoning, inhibition, mental flexibility, as well as monitoring of action as was primarily assessed by working memory tasks, stopping tasks, the Trail Making Test – Part B, Stroop tasks, the flanker task, and Raven's Advanced Progressive test. The effect of training on executive functions was moderated by age (younger adults showed an increased benefit than older adults), education, session duration, number of sessions, total training duration, and the type of the control group. Powers et al. (2013) found rather small to even only negligible effects of action video game interventions which was replicated in a later meta-analysis of the same group (Powers and Brook 2014). However, as we have seen above when discussing the individual executive function domains, follow-up analyses identified clear effects in specific domains.

Conclusions

To wrap up the previous sections, we reviewed empirical studies and meta-analyses investigating the effect of experience in video games (in particular action video games) on the executive functions shifting, dual tasking, updating, and inhibition. There is evidence that, at least under particular task conditions, massive video game experience may improve the performance in task-switching (i.e., shifting) and dual-task situations. Further, preliminary evidence for experience-based improvement in working memory updating exists. In contrast, the literature on effects of action video gaming rather suggests no relation between experience in action video games and improved inhibition.

Let's consider the general mechanism that may explain transfer effects from video gaming to test situations on executive functions. While the introduced version of the "learning to learn" account predicts a transfer from action video game experience to all types of executive functions (Bavelier et al. 2012), the "common demands" account rather predicts a specific transfer, depending on similarities between the trained video game and the laboratory task (Oei and Patterson 2015). First, from a more general perspective, there is evidence for transfer effects on shifting, dual tasking, and updating, while there is no such evidence for the case of inhibition. The observation of different validities of transfer effects across the executive function domains is consistent with the "common demands" account and indicates that switching between different sequential tasks, performing simultaneous tasks, as well as the updating of task information represent relevant demands in (action) video games. In contrast, the inhibition of responses seems to be no essential component in playing these games when applying the logic of the "common demands" account. This might be surprising given the usual characteristics of action video games. A closer look at these games suggests that withholding of motor responses and their interruption represent indeed important demands of action video games. Therefore, the fact that currently no valid evidence for effects of action video games on inhibition demands has been reported may be suggestive for two conclusions: It may suggest that the fast interruption and withholding of motor responses is not trainable and transferable at all (Strobach et al. 2014). Alternatively, it may suggest that the current experimental paradigms, which had been used in action video game studies, do not reflect the type of particular inhibition demands inherent to action video games.

Second, from a more detailed perspective, the observation of differential effects of video games on different types of executive functions is also consistent with this theory. For example, there is evidence that puzzle game training, but not action video game training, is able to improve performance in mental rotation (Boot et al. 2008); while the first training type shares common elements with the mental rotation task, the latter ones do not. Further, performance in dual-task situations with speeded, well-controllable component tasks is affected by action video game experience (e.g., Strobach et al. 2012b), while such experience does not seem to affect dual-task situations that are less similar to the gaming environment (e.g., paper and pencil search; Donohue et al. 2012). We are sure that these observations can be complemented with other type of training games and other different functions as well, if a careful analysis is conducted on the type of overlap between training and transfer function.

In sum, we evaluated the existing literature on action video games and executive functions as demonstrating evidence for transfers on the executive functions shifting, dual tasking, and updating, while this literature shows no evidence for transfer to the inhibition function. However, it is also obvious that each type of executive function requires attempts to replicate existing findings as well as additional analyses in future studies (Colzato and Hommel, this volume). These analyses should specify the effects of action video games and other game genres on different executive function types using different experimental paradigms. Preferably, this specification should be realized in the context of training experiments in order to make conclusions about the causal links between game experience and potential changes in executive functioning.

References

Anguera, J. A., Boccanfuso, J., Rintoul, J. L., Al-Hashimi, O., Faraji, F., Janowich, J., ... & Gazzaley, A. (2013). Video game training enhances cognitive control in older adults. *Nature, 501*, 97–101.

Baddeley, A. (1986). *Working memory*. New York: Clarendon Press/Oxford University Press.

Bavelier, D., Green, C. S., Pouget, A., & Schrater, P. (2012). Brain plasticity through the life span: Learning to learn and action video games. *Annual Review of Neuroscience, 35*, 391–416.

Bavelier, D., Bediou, B., & Green, C. S. (2018). Expertise and generalization: Lessons from action video games. *Current Opinion in Behavioral Sciences, 20*, 169–173.
Bediou, B., Adams, D. M., Mayer, R. E., Tipton, E., Green, C. S., & Bavelier, D. (2018). Meta-analysis of action video game impact on perceptual, attentional, and cognitive skills. *Psychological Bulletin, 144*, 77–110.
Berryhill, M. E., & Hughes, H. C. (2009). On the minimization of task switch costs following long-term training. *Attention, Perception & Psychophysics, 71*, 503–514.
Boot, W. R., Kramer, A. F., Simons, D. J., Fabiani, M., & Gratton, G. (2008). The effects of video game playing on attention, memory, and executive control. *Acta Psychologica, 129*, 387–398.
Bundesverband Interaktive Unterhaltungssoftware e. V. (BIU) & Gesellschaft für Konsumforschung (GfK). (2012). Gamer in Deutschland. Retrieved from http://www.biu-online.de/de/fakten/marktzahlen/der-deutsche-markt-fuer-computer-und-videospiele-in-der-uebersicht.html (17.08.2015).
Cain, M. S., Landau, A. N., & Shimamura, A. P. (2012). Action video game experience reduces the cost of switching tasks. *Attention, Perception, & Psychophysics, 74*, 641–647.
Chiappe, D., Conger, M., Liao, J., Caldwell, J., & Vu, K. L. (2013). Improving multi-tasking ability through action videogames. *Applied Ergonomics, 44*, 278–284.
Colzato, L. S., van Leeuwen, P. J. A., van den Wildenberg, W., & Hommel, B. (2010). DOOM'd to switch: Superior cognitive flexibility in players of first person shooter games. *Frontiers in Psychology, 1*, 8.
Colzato, L. S., van den Wildenberg, W. M., Zmigrod, S., & Hommel, B. (2013). Action video gaming and cognitive control: Playing first person shooter games is associated with improvement in working memory but not action inhibition. *Psychological Research, 77*, 234–239.
Dahlin, E., Neely, A. S., Larsson, A., Bäckman, L., & Nyberg, L. (2008). Transfer of learning after updating training mediated by the striatum. *Science, 320*, 1510–1512.
Davidson, D. J., Zacks, R. T., & Williams, C. C. (2003). Stroop interference, practice, and aging. *Aging, Neuropsychology, and Cognition, 10*, 85–98.
Donohue, S. E., James, B., Eslick, A. N., & Mitroff, S. R. (2012). Cognitive pitfall! Videogame players are not immune to dual-task costs. *Attention, Perception, & Psychophysics, 74*, 803–809.
Engelhardt, C. R., Hilgard, J., & Bartholow, B. D. (2015). Acute exposure to difficult (but not violent) video games dysregulates cognitive control. *Computers in Human Behavior, 45*, 85–92.
Gaspar, J. G., Neider, M. B., Crowell, J. A., Lutz, A., Kaczmarski, H., & Kramer, A. F. (2014). Are gamers better crossers? An examination of action video game experience and dual task effects in a simulated street crossing task. *Human Factors, 56*, 443–452.
Green, C. S., Sugarman, M. A., Medford, K., Klobusicky, E., & Bavelier, D. (2012). The effect of action video game experience on task-switching. *Computers in Human Behavior, 28*, 984–994.
Green, C. S., Strobach, T., & Schubert, T. (2014). On methodological standards in training and transfer experiments. *Psychological Research, 78*, 756–772.
Green, C. S., Bavelier, D., Kramer, A. F., Vinogradov, S., Ansorge, U., Ball, K. K., & Facoetti, A. (2019). Improving methodological standards in behavioral interventions for cognitive enhancement. *Journal of Cognitive Enhancement, 3*, 2–29.
Jonides, J., & Smith, E. E. (1997). The architecture of working memory. In M. D. Rugg (Ed.), *Cognitive neuroscience* (pp. 243–276). Cambridge, MA: The MIT Press.
Karbach, J., & Kray, J. (2009). How useful is executive control training? Age differences in near and far transfer of task-switching training. *Developmental Science, 12*, 978–990.
Karle, J. W., Watter, S., & Shedden, J. M. (2010). Task switching in video game players: Benefits of selective attention but not resistance to proactive interference. *Acta Psychologica, 134*, 70–78.
MacLeod, C. M. (1991). Half a century of research on the Stroop effect: An integrative review. *Psychological Bulletin, 109*, 163–203.
Miyake, A., & Friedman, N. P. (2012). The nature and organization of individual differences in executive functions: Four general conclusions. *Current Directions in Psychological Science, 21*, 8–14.

Miyake, A., Friedman, N. P., Emerson, M. J., Witzki, A. H., & Howerter, A. (2000). The unity and diversity of executive functions and their contributions to complex 'frontal lobe' tasks: A latent variable analysis. *Cognitive Psychology, 41*, 49–100.

Monsell, S. (2003). Task switching. *Trends in Cognitive Sciences, 7*, 134–140.

Oei, A. C., & Patterson, M. D. (2014). Playing a puzzle video game with changing requirements improves executive functions. *Computers in Human Behavior, 37*, 216–228.

Oei, A. C., & Patterson, M. D. (2015). Enhancing perceptual and attentional skills requires common demands between the action video games and transfer tasks. *Frontiers in Psychology, 6*, 133.

Powers, K. L., & Brooks, P. J. (2014). Evaluating the specificity of effects of video game training. In F. Blumberg (Ed.), *Learning by playing: Frontiers of video gaming in education*. Oxford, UK: Oxford University Press.

Powers, K. L., Brooks, P. J., Aldrich, N. J., Palladino, M. A., & Alfieri, L. (2013). Effects of video-game play on information processing: A meta-analytic investigation. *Psychonomic Bulletin & Review, 20*, 1055–1079.

Raz, N. (2000). Aging of the brain and its impact on cognitive performance: Integration of structural and functional findings. In F. M. Craik & T. A. Salthouse (Eds.), *The handbook of aging and cognition* (2nd ed., pp. 1–90). Mahwah: Lawrence Erlbaum Associates Publishers.

Sala, G., Tatlidil, K. S., & Gobet, F. (2018). Video game training does not enhance cognitive ability: A comprehensive meta-analytic investigation. *Psychological Bulletin, 144*, 111–139.

Salminen, T., Kühn, S., Frensch, P. A., & Schubert, T. (2016). Transfer after dual n-back training depends on striatal activation change. *Journal of Neuroscience, 36*, 10198–10213.

Schubert, T. (2008). The central attentional limitation and executive control. *Frontiers of Bisocience, 13*, 3569–3580.

Schubert, T., & Strobach, T. (2012). Video game experience and optimized executive control skills—On false positives and false negatives: Reply to Boot and Simons (2012). *Acta Psychologica, 141*, 278–280.

Simons, D. J., Boot, W. R., Charness, N., Gathercole, S. E., Chabris, C. F., Hambrick, D. Z., & Stine-Morrow, E. A. (2016). Do "brain-training" programs work? *Psychological Science in the Public Interest, 17*, 103–186.

Spence, I., & Feng, J. (2010). Video games and spatial cognition. *Review of General Psychology, 14*, 92–104.

Spil Games. (2013). State of online gaming report. Retrieved from http://auth-83051f68-ec6c-44e0-afe5-bd8902acff57.cdn.spilcloud.com/v1/archives/1384952861.25_State_of_Gaming_2013_US_FINAL.pdf

Strobach, T. (2020). The dual-task practice advantage: Empirical evidence and cognitive mechanisms. *Psychonomic Bulletin & Review, 27*, 3–14.

Strobach, T., Liepelt, R., Schubert, T., & Kiesel, A. (2012a). Task switching: Effects of practice on switch and mixing costs. *Psychological Research, 76*, 74–83.

Strobach, T., Frensch, P. A., & Schubert, T. (2012b). Video game practice optimizes executive control skills in dual-task and task switching situations. *Acta Psychologica, 140*, 13–24.

Strobach, T., Salminen, T., Karbach, J., & Schubert, T. (2014). Practice-related optimization and transfer of executive functions: A general review and a specific realization of their mechanisms in dual tasks. *Psychological Research, 78*, 836–851.

Strobach, T., Frensch, P., Müller, H., & Schubert, T. (2015). Evidence for the acquisition of dual-task coordination skills in older adults. *Acta Psychologica, 160*, 104–116.

Szameitat, A. J., Lepsien, J., von Cramon, D., Sterr, A., & Schubert, T. (2006). Task-order coordination in dual-task performance and the lateral prefrontal cortex: An event-related fMRI study. *Psychological Research, 70*, 541–552.

The Entertainment Software Association. (2018). Essential facts about the computer and video game industry. Retrieved from https://www.theesa.com/wp-content/uploads/2019/05/2019-Essential-Facts-About-the-Computer-and-Video-Game-Industry.pdf

Wang, P., Liu, H. H., Zhu, X. T., Meng, T., Li, H. J., & Zuo, X. N. (2016). Action video game training for healthy adults: A meta-analytic study. *Frontiers in Psychology, 7*, 907.
Wendt, M., Klein, S., & Strobach, T. (2017). More than attentional tuning–Investigating the mechanisms underlying practice gains and preparation in task switching. *Frontiers in Psychology, 8*, 682.
Wilkinson, A. J., & Yang, L. (2012). Plasticity of inhibition in older adults: Retest practice and transfer effects. *Psychology and Aging, 27*, 606–615.
Wu, S., & Spence, I. (2013). Playing shooter and driving videogames improves top-down guidance in visual search. *Attention, Perception, & Psychophysics, 75*, 673–686.

Mindfulness and Meditation Training

Paul Verhaeghen

Contents

Abstract This chapter reviews evidence that practicing meditation positively impacts attention. Functional and structural enhancements in parts of the salience and executive networks are described. At the behavioral level, the effect of meditation on tasks of controlled attention (such as Stroop and go/no-go tasks) is found to be about 0.4 *SD*; a smaller effect of about 0.25 *SD* is noted on sustained attention; no effect is found on the alerting component of the ANT, although there are consistent effects on different aspects of nonjudgmental alerting (such as attentional blink and error processing), with an effect size of about 0.5 *SD* for attentional blink studies. Meditation also lowers perceptual thresholds. Dose–response relationship studies underscore the importance of frequency or amount of recent meditation practice, rather than accumulated hours of practice.

P. Verhaeghen (✉)
Georgia Institute of Technology, Atlanta, GA, USA
e-mail: verhaeghen@gatech.edu

T. Strobach, J. Karbach (eds.), *Cognitive Training*,
https://doi.org/10.1007/978-3-030-39292-5_17

Minding the Mind: Meditation as a Skills Training for Attention

Acquiring a skill is often a straightforward process: Repeat a task many, many times, and over time its component skills will most likely be performed with increased efficiency. This process is often done deliberately (e.g., in music practice), but sometimes the skills training is much more hidden. This chapter is about one of such hidden trainings: How meditation (more specifically mindfulness meditation) works as a form of attention training.

Mindfulness Meditation as Attention Training

Typically, mindfulness meditation practices fall into two main categories or styles. In the first style, *focused-attention meditation*, the meditator focuses her mind on a single object – often the breath – unwaveringly and clearly. The goal is to calm the mind and to teach it to stay a particular course for a long period of time, thus practicing both controlled and sustained attention. Practicing focused-attention meditation also implies monitoring the mind, that is, to continuously check for distractions. When distractions arise, the meditator simply meets them with patience and kindness and then returns to the object of concentration. In the second style, *open-monitoring meditation*, awareness is applied to whatever is present in an experience – an emotion, a percept, a memory, or a thought – as it arises moment to moment, and simply observes this presence. Along the way, the meditator learns to cultivate "reflexive awareness," that is, awareness that refers back on itself.

The end goal of these practices is not to train attention per se (for an overview of such work, see Karbach & Kray this volume) but to learn to apply careful and open, nonjudgmental attention (e.g., Kabat-Zinn 1990) to one's perceptions in order to see their impermanence and to ultimately transcend the sense of self. However, one can argue that the process of practice by itself produces an exquisite form of skills training.

How focused-attention practice can hone attentional skill is nicely illustrated in a study by Hasenkamp et al. (2012). In this study, 14 seasoned meditators (with on average 1400 hours of lifetime practice) meditated inside the scanner for 20 minutes, focusing on the breath. They pressed a button as soon as they realized that their mind had wandered away. The researchers used an event-related design centered on 3-second slices defined in reference to the button press. In a 3-second time window before the button press – when meditators were presumably mind-wandering – many regions of the default-mode network were activated, as one would expect. In the 3 seconds around the button press – when meditators became aware of their minds wandering – the insula and the anterior cingulate cortex, regions associated with the salience attention network, were activated. The 3-second slice after that – when participants were switching their attention back to the breath – was associated with activation in the executive attention network (the lateral prefrontal cortex and the inferior parietal lobe). One intriguing finding was how often this cycle is repeated: n average, meditators pressed the button 15.5 times over the course of 20 minutes. Thus, over the course of their 1400 hours of lifetime experience, Hasenkamp's meditators must have gone through 63,000 cycles of activating first the silence and then the executive attention networks in response to default-mode activation.

Meditation and Attention in the Brain

Hasenkamp's study is not the only study finding brain activation in attention centers during meditation. The largest meta-analysis on the topic (Tomasino et al. 2013) collected the results from 26 studies, reporting on a total of 313 meditators, with on average 11,552 lifetime hours of meditation experience. The main result was that meditators, while meditating, activate both parts of the salience network (the insula) and the executive attention network (the superior and inferior parietal lobe). Note that the inferior parietal lobe acts as a switch between the executive control network and the default network (Spreng et al. 2013). Activation in this brain region might then mean that meditators are not just focusing their attention on the object of their meditation but are also actively suppressing activation of the default-mode network. This assumption is confirmed in the finding that parts of the default-mode network (the angular gyrus, the middle temporal gyrus, and the precuneus) quiet down during meditation.

Further corroboration of the hypothesis that meditation impacts attention comes from functional connectivity research. Three studies (Brewer et al. 2011; Froeliger et al. 2012; Josipovic et al. 2012) found heightened functional connectivity between at least one part of the default-mode networks and at least one part of the attention network during meditation, compared to the couplings when participants were resting in the scanner. This suggests that meditators stay on task while meditating: When the default-mode network is active – that is, when the mind strays from its focus – the attention system notices, clamps down, and corrects. This tighter coupling during meditation reinforces the main point of Hasenkamp's study, namely, that focused-attention meditation really is a predictable dynamic process, a series of cycles that occur in a consistent manner.

Finally, the largest meta-analysis on brain morphology and meditation (Fox et al. 2014; 21 studies comparing a total of 503 meditators with on average 4664 hours of expertise with 472 non-meditators) found changes in brain morphology consistent with changes in attentional proficiency. First, meditators had higher gray matter volume and/or density in parts of the salience network (the anterior cingulate cortex and the insula) as well as in parts of the executive network (the rostrolateral prefrontal cortex). Second, major parts of the default-mode network (the posterior parietal cortex and the precuneus) were lower in volume and/or density in meditators than non-meditators.

Fox et al. (2014) also identified differences in two white-matter pathways. First, meditators have increased information transfer between the anterior parts of the two hemispheres, as suggested by an enlarged forceps minor and genu of the corpus callosum, perhaps a by-product of activation in frontal areas during meditation (such as the insula, the anterior cingulate, the rostrolateral prefrontal cortex, and the orbitofrontal cortex). Second, meditators have a more efficient superior longitudinal fasciculus. The different subcomponents of this tract are responsible for the sense of the body in space, the moment-to-moment understanding of the state of the body itself, spatial attention, and control over the focus of attention (Makris et al. 2005). This enhanced connection might then be the fruit of repeatedly and persistently paying close attention to fleeting sensations in the body (i.e., the breath and/or bodily sensations).

There is some evidence that changes in gray matter volume in the salience attention network accrue over time. One study reports a correlation between meditation experience and volume/density in the anterior cingulate of .32 (averaged over two subregions; Grant et al. 2010). The average correlation between experience and volume/density in the insula is .25 (Hölzel et al. 2008; Luders et al. 2012). Data are more disappointing with regard to the executive control network: The single study that analyzed the correlation between meditation experience and volume/density of the rostrolateral prefrontal cortex failed to find one; nothing is known about the precuneus and the posterior cingulate cortex.

Note that all these data are correlational and hence cannot give a definitive answer about the causal direction.

Meditation and Its Effects on Psychological Aspects of Attention

The largest meta-analysis on the psychological effects of meditation (Sedlmeier et al. 2012) gathered 22 studies (total sample size of 1307) relevant to attention (operationalized as concentration/attention, sustained attention, orienting, alerting, conflict monitoring, executive processing, and behavioral inhibition). All these studies compared performance of a group of meditators (most of them fresh out of an 8-week mindfulness-based stress reduction (MBSR)-type program) with a control group of non-meditators. The average effect in these studies was a mean standardized difference (*MSD*) of 0.58, implying that the average meditator is more attentive than 72% of non-meditators.

It makes sense to delve a little deeper here and detail the results in terms of different aspects of attention. I will review three literatures, closely tied to the cycle of meditation as found by Hasenkamp et al. (2012): control over attention, alerting, and sustained attention.

Meditation and Control Over Attention

Most studies on the effects of mindfulness and meditation on attention have focused on changes in attentional control. There are essentially two types of such studies. One type compares participants who completed a mindfulness intervention with participants who did not; the latter group can either be a waitlist control group or an active control (e.g., participants in a relaxation treatment). The other type compares long-term meditators with meditation novices.

In a recent meta-analysis (Verhaeghen in preparation), I was able to locate 29 studies of the former type ($n = 1347$ participants) and 14 of the latter ($n = 687$). All of these studies used relatively quick, objective laboratory tasks – the Stroop task (Stroop 1935), various variations on the flanker task (including the executive component of the Attention Network Test [ANT]; Fan et al. 2002), the anti-saccade task, versions of the go/no-go paradigm, and the Hayling test (Burgess and Shallice 1997).

Intervention studies ranged in duration from 3 days to 4 months, with the total duration of mindfulness practice over the course of the program reaching from 20 minutes to 44 hours. Some of the studies use standard MBSR or MBCT protocols (MBCT, mindfulness-based cognitive therapy, is an MBSR-like curriculum intended to reduce the relapse of major depression), but most are ad hoc combinations of techniques, most often of the focused-attention variety. The average effect size for these 29 studies was $MSD = 0.29$ (95% confidence interval from 0.18 to 0.39). The long-term meditators in the second type of study had accumulated, on average, about 9 years of meditation practice. The average effect size for these 14 studies was $MSD = 0.32$ (95% confidence interval from 0.09 to 0.54). The effect sizes are close, and the confidence intervals of the two sets of studies overlap, suggesting that (at least quantitatively) the effects of short-term interventions and those of a standard personal meditation practice are quite similar.

This set of results suggests two things. First, even 40 or so hours of practice already result in measurable changes in control over attention. The effect size of 0.29 implies that the average person completing a mindfulness intervention will have better executive control over their attention than 61% of individuals who did not complete such a program. This effect size is a bit smaller than what we usually find in behavioral, educational, and therapeutic endeavors, where the median effect size is 0.44 (Lipsey and Wilson 1993).

Second, the finding that the effects of many years of meditation practice are not all that different from those of relatively short-term interventions suggests that the number of years of accumulated practice may be less important than the amount of recent daily practice one engages in. Consistent with this view, one study (Teper and Inzlicht 2013) did find that although the number of years of meditation experience correlated (rather modestly) with the Stroop effect ($r = -.27$), so did meditation frequency ($r = -.23$). (The first correlation can be explained by the fact that more seasoned meditators tend to spend more time on the cushion.) Another study (Chan and Woollacott 2007) found a dose–response relationship with the number of minutes practiced per day ($r = -.17$) but not with total hours of lifetime meditation experience. However, a third study (Joseffson and Broberg 2011) failed to find any significant correlations between Stroop and either meditation frequency or meditation experience.

Meditation and Alerting

As far as I know, the only experimental paradigm to tap into the alerting function of attention is the alerting subcomponent of the Attention Network Test. Effects of mindfulness training on this test are small and nonsignificant (*MSD* = 0.07, 95% CI from −0.32 to 0.47 for the six extant intervention studies; *MSD* = 0.15, 95% CI from −0.37 to 0.67 for the three extant studies on long-term meditators; Verhaeghen in preparation), suggesting, somewhat surprisingly perhaps, that this aspect of attention is not easily trained using meditation interventions.

It can be argued, however, that the ANT's alerting paradigm, which measures a participant's response time when they are alerted that the stimulus is about to appear versus when no such alert is provided, is very different from the type of alerting the salience network would be involved in during meditation. The ANT is explicitly built on cues that are both exogenous (i.e., occurring outside the individual) and clear. In contrast, the types of cues meditators work with are endogenous and rather subtle – a drifting away of attention, or a being captured by something else than what one is supposed to be captured by.

As such, studies that look at the *quality* of alerting might hit closer to the mark. These studies are inspired by Kabat-Zinn's (1990) concept of nonjudgmental alerting, that is, taking a nonreactive stance toward whatever presents itself to the attentional field.

One set of such studies considered the attentional blink and meditation (Fabio and Towey 2018; May et al. 2011; Slagter et al. 2007; van Leeuwen et al. 2009; van Vugt and Slagter 2014). In the attentional blink paradigm, participants see a stream of 20 or so letters, presented at about 10 items per second. One or two digits are intermingled with the letters; participants press a button whenever they spot a digit. When the stream contains two digits, subjects often miss the second digit when it is shown within 500 ms of the first one. The standard explanation is that detecting the second digit is only possible if enough resources are available; this requires the subject to relinquish attention as soon as the first digit has been detected. Three studies compared the attentional blink effect in long-term practitioners with that in novices; the average effect size was *MSD* = 0.54 (95% CI from 0.14 to 0.93). One of these three studies also compared attentional blink before and after a 3-month retreat; the effect size for this comparison was *MSD* = 0.38 in beginning meditators and *MSD* = 1.17 in advanced meditators. Another finding is that the attentional blink effect is smaller during open-monitoring meditation than during focused-attention meditation, at least in highly experienced meditators, as one would expect if the decrease in attentional blink signals an increase in openness and nonreactivity (van Vugt and Slagter 2014).

Other paradigms confirm that meditators have lower levels of reactivity.

One study (Hodgins and Adair 2010) used the infamous gorilla-basketball video and found that meditators were 50% more likely to spot the interloping gorilla than non-meditators. At the same time, meditators were also about twice as accurate in keeping count of the passes, suggesting that they were able to both focus and be open-minded at the same time.

Another study (Van den Hurk et al. 2010) used a bimodal startle-type task. Participants stood in front of a screen. At one point, a light turned on either to the left or the right; participants were asked to turn their head toward the light as fast as possible. Participants typically speed up when the light is accompanied by a centrally presented uninformative sound; this is likely an arousal effect. Meditators were less likely to speed up than non-meditators, suggesting that they process the sound without attaching a startling, arousing quality to it – it is just a loud noise.

Two other studies that demonstrate that meditators may have lower reactivity are an ERP study on Stroop (Teper and Inzlicht 2013) and one on the flanker effect (Andreu et al. 2017). In both studies, the researchers were interested in error-related negativity (ERN) and error-related positivity (Pe). The ERN occurs about 100 ms after making an incorrect response, and it likely originates from the anterior cingulate, a part of the salience attention network. The Pe occurs a little later, about 200 ms after making an error, and it originates likely in the posterior cingulate, which is part of the core of the default-mode network; the Pe signifies awareness of the error. Teper and Inzlicht found a larger ERN effect in meditators than in non-meditators; both years and frequency of meditation correlated with ERN ($r = .37$ and .35, resp.), showing that meditators' brains are more alert to the mistakes they make. Interestingly, meditators did not show larger Pe values; thus, their increased sensitivity to errors did not lead to stronger awareness of errors. Andreu and colleagues likewise found a higher ERN amplitude in meditators, and no experience effect on Pe. They also obtained a larger CRN effect. The CRN is a smaller component, like the ERN originating in the anterior cingulate, and thought to be involved in performance monitoring. One possible interpretation of this pattern of results is that even though a meditator's brain quickly realizes its mistakes, it is also very quick to let go of that reaction.

In a fourth study, van Leeuwen et al. (2012) showed students a series of local-global stimuli – larger digits formed out of smaller digits. Participants pressed a button whenever they saw the digit 1 or 2, regardless of whether it was the global (large) or local (small) digit in the figure. Typically, subjects respond faster to global digits than local digits – in this study the difference was 56 milliseconds. The study also included eight Buddhist monks and nuns; these showed less of a global bias – the difference was only 21 milliseconds. This suggests that meditators have more openness to what is really there (viz., two different digits). ERP analysis also demonstrated that the meditators had stronger P1 and N1 responses, suggesting a quicker uptake of information; they also showed larger engagement in the attention networks that are typically implicated in this task.

This quicker uptake of information is confirmed in studies that directly examined perceptual thresholds. Jensen et al. (2011) had participants perform attention tasks before and after a standard 8-week MBSR program. Before training, the perceptual threshold for identifying a single letter was 15 ms; after MBSR training, this was 9 ms. Likewise, MacLean et al. (2010) tested seasoned meditators, people before, during, and after a 3-month retreat on line-length discrimination, and compared their performance with that of a no-retreat control group of equally seasoned meditators. Retreatants and non-retreatants did not differ in discrimination thresholds before the retreat, but retreatants were able to detect smaller differences between the lines both at the halfway point of the retreat and at the end of the retreat, as well at a follow-up session 5 months after the end of the retreat. There was a dose–response relationship: Those who spent more time in daily meditation during the after-retreat period could detect smaller differences between the two lines ($r = .36$).

Meditation and Sustained Attention

Studies on sustained attention and meditation use mostly the Sustained Attention to Response Test (SART; Robertson et al. 1997) or versions of the Continuous Performance Task (CPT; Rosvold et al. 1956). The 14 studies that examined the effects of intervention produced an average effect size of 0.32 (95% CI from 0.07 to 0.57). The effects of long-term meditation were, however, nonsignificant (*MSD* = 0.33, 95% CI from −0.05 to 0.71; note, however, that there were only five studies in this sample (Verhaeghen in preparation)). Only one of the studies (Josefsson and Broberg 2011) looked at the dose–response relationship; it did not find one. A potential complicating factor may be the type of mediation practiced: There is emerging behavioral and neuroimaging evidence that focused-attention meditation leads to improvements on sustained attention, but loving-kindness mediation does not (Lee et al. 2012).

I want to single out one additional study, by Carter et al. (2005), that used two rather exceptional tasks to measure stability of attention. The research team traveled all the way to the Himalayan mountains in Ladakh to test Tibetan Buddhist monks living in exile there. The first task was a binocular rivalry task. Binocular rivalry refers to the curious sensation that happens when participants are presented with two different stimuli, one presented to each eye (e.g., the right eye sees a house, the left eye sees a face): The two images tend to alternate in awareness every few seconds. Carter found that focused-attention meditation led to slower alterations (i.e., more stability) in over half of the monks, both during and right after meditation. The second task was a motion-induced blindness task. Participants saw a video of a blinking green dot at the center of a computer screen, which also has three stationary yellow dots arranged in a triangle closer to the edge of the screen, and a lattice of rotating crosses. With sustained attention, awareness of the yellow dots disappears after about 10 seconds, but the dots reappear as soon as attention is relaxed or eye movements are made. The average student volunteer was able to keep the three dots from reappearing for 2.6 seconds. The average monk was able to do this for 4.1 seconds, or about 50% longer. More importantly, the duration record in the group of students was 6 seconds; 10 out of 76 monks equaled or beat that record – one monk was able to stabilize the image for 128 seconds, and one even for 723 seconds.

Meditation and Attention: Conclusions

Meditation has an effect on all three forms of attention reviewed here: Its effect on controlled attention is around 0.3 *SD*; a similar effect is noted on sustained attention; and there are also consistent effects on nonjudgmental alerting, with an effect size of 0.65 *SD* for attentional blink studies but not on the alerting component of the ANT.

The most intriguing result is the evidence for nonjudgmental alerting, arguably a key component in the concept of mindfulness, as seen in the attentional blink task, the gorilla video task, the (non)startle effect, the local–global effect, and brain parameters like error-related negativity and positivity. This is also interesting because this form of openness or receptiveness is an aspect of attention that is often undervalued, to say the least, in standard cognitive psychology, which is much more concerned with the amount or acuity of attention rather than its quality. The present studies show that open-mindedness is a skill that operates in a process as basic as paying attention. It also demonstrates that this skill can be trained. Two studies even suggest that meditation practice can help lower the threshold of perception, literally letting more of the outside world enter the realm of awareness.

Attention is often considered the gateway to other aspects of cognition. Particularly, attention is important for working memory, helps with knowledge retrieval, and is important for real-life aspects of cognition. All of these aspects of cognition indeed benefit from meditation and mindfulness training (Verhaeghen 2017), although the number of studies and the number of participants involved in each of these studies is still too small to comfortably allow for definitive conclusions. These findings suggest that meditation may lead to a cognitive cascade where meditation leads to changes in attention, which in turn positively influences other aspects of cognition.

Finally, attention-and-meditation studies underscore the importance of frequency or amount of meditation, rather than accumulated hours of practice. Frequent practice appears to sharpen the focus of attention, to make one more alter to mistakes, to broaden the limits of perception, and to increase sustained attention. The finding that 8-week MBSR programs can have a meaningful impact on attention, often on par with the effects seen in very seasoned meditators, further underscores this point.

References

Andreu, C. I., Moënne-Loccoz, C., López, V., Slagter, H. A., Franken, I. H., & Cosmelli, D. (2017). Behavioral and electrophysiological evidence of enhanced performance monitoring in meditators. *Mindfulness, 8*, 1603–1614.

Brewer, J. A., Worhunsky, P. D., Gray, J. R., Tang, Y. Y., Weber, J., & Kober, H. (2011). Meditation experience is associated with differences in default mode network activity and connectivity. *Proceedings of the National Academy of Sciences, 108*, 20254–20259.

Burgess, P. W., & Shallice, T. (1997). *Hayling sentence completion test*. Suffolk: Thames Valley Test Co. Ltd..

Carter, O. L., Presti, D. E., Callistemon, C., Ungerer, Y., Liu, G. B., & Pettrigrew, J. D. (2005). Meditation alters perceptual rivalry in Tibetan Buddhist monks. *Current Biology, 15*, R412–R413.

Chan, D., & Woollacott, M. (2007). Effects of level of meditation experience on attentional focus: Is the efficiency of executive or orientation networks improved? *Journal of Alternative and Complementary Medicine, 13*, 651–658.

Fabio, R. A., & Towey, G. E. (2018). Long-term meditation: The relationship between cognitive processes, thinking styles and mindfulness. *Cognitive Processing, 19*, 73–85.

Fan, J., McCandliss, B. D., Sommer, T., Raz, A., & Posner, M. I. (2002). Testing the efficiency and independence of attentional networks. *Journal of Cognitive Neuroscience, 14*, 340–347.

Fox, K. C., Nijeboer, S., Dixon, M. L., Floman, J. L., Ellamil, M., Rumak, S. P., … & Christoff, K. (2014). Is meditation associated with altered brain structure? A systematic review and meta-analysis of morphometric neuroimaging in meditation practitioners. *Neuroscience & Biobehavioral Reviews, 43*, 48–73.

Froeliger, B., Garland, E. L., Kozink, R. V., Modlin, L. A., Chen, N. K., McClernon, F. J., ... & Sobin, P. (2012). Meditation-state functional connectivity (msFC): Strengthening of the dorsal attention network and beyond. *Evidence-based Complementary and Alternative Medicine, 2012*.

Grant, J., Courtemanche, J., Duerden, E. G., Duncan, G. H., & Rainville, P. (2010). Cortical thickness and pain sensitivity in Zen meditators. *Emotion, 10*, 43–53.

Hasenkamp, W., Wilson-Mendenhall, C. D., Duncan, E., & Barsalou, L. W. (2012). Mind wandering and attention during focused meditation: A fine-grained temporal analysis of fluctuating cognitive states. *NeuroImage, 59*, 750–760.

Hodgins, H. S., & Adair, K. C. (2010). Attentional processes and meditation. *Consciousness and Cognition: An International Journal, 19*, 872–878.

Hölzel, B. K., Ott, U., Gard, T., Hempel, H., Weygandt, M., Morgen, K., & Vaitl, D. (2008). Investigation of mindfulness meditation practitioners with voxel-based morphometry. *Social Cognitive and Affective Neuroscience, 3*, 55–61.

Jensen, C., Vangkilde, S., Frokjaer, V., & Hasselbalch, S. G. (2011). Mindfulness training affects attention—or is it attentional effort? *Journal of Experimental Psychology: General, 141*, 106–123.

Josefsson, T., & Broberg, A. (2011). Meditators and non-meditators on sustained and executive attentional performance. *Mental Health, Religion & Culture, 14*, 291–309.

Josipovic, Z., Dinstein, I., Weber, J., & Heeger, D. J. (2012). Influence of meditation on anti-correlated networks in the brain. *Frontiers in Human Neuroscience, 5*, 183.

Kabat-Zinn, J. (1990). *Full catastrophe living: The program of the stress reduction clinic at the University of Massachusetts Medical Center*. New York: Delta.

Lee, T. M., Leung, M. K., Hou, W. K., Tang, J. C., Yin, J., So, K. F., ... & Chan, C. C. (2012). Distinct neural activity associated with focused-attention meditation and loving-kindness meditation. *PLoS One, 7*, e40054.

Lipsey, M. W., & Wilson, D. B. (1993). The efficacy of psychological, educational, and behavioral treatment: Confirmation from meta-analysis. *The American Psychologist, 48*, 1181–1209.

Luders, E., Kurth, F., Mayer, E. A., Toga, A. W., Narr, K. L., & Gaser, C. (2012). The unique brain anatomy of meditation practitioners: Alterations in cortical gyrification. *Frontiers in Human Neuroscience, 6*, 34.

MacLean, K. A., Ferrer, E., Aichele, S. R., Bridwell, D. A., Zanesco, A. P., Jacobs, T. L., … & Wallace, B. A. (2010). Intensive meditation training improves perceptual discrimination and sustained attention. *Psychological Science, 21*, 829–839.

Makris, N., Kennedy, D. N., McInerney, S., Sorensen, A. G., Wang, R., Caviness, V. S., ... & Pandya, D. N. (2005). Segmentation of subcomponents within the superior longitudinal fascicle in humans: A quantitative, in vivo, DT-MRI study. *Cerebral Cortex, 15*, 854–869.

May, C. J., Burgard, M., Mena, M., Abbasi, I., Bernhardt, N., Clemens, S., … & Janz, A. (2011). Short-term training in loving-kindness meditation produces a state, but not a trait, alteration of attention. *Mindfulness, 2*, 143–153.

Robertson, I. H., Manly, T., Andrade, J., Baddeley, B. T., & Yiend, J. (1997). Oops!': Performance correlates of everyday attentional failures in traumatic brain injured and normal subjects. *Neuropsychologia, 35*, 747–758.

Rosvold, H., Mirsky, A., Sarason, I., Bransome, E., & Beck, L. (1956). A continuous performance test of brain damage. *Journal of Consulting Psychology, 20*, 343–350.

Sedlmeier, P., Eberth, J., Schwarz, M., Zimmermann, D., Haarig, F., Jaeger, S., & Kunze, S. (2012). The psychological effects of meditation: A meta-analysis. *Psychological Bulletin, 138*, 1139–1171.

Slagter, H. A., Lutz, A., Greischar, L. L., Francis, A. D., & Nieuwenhuis, S. (2007). Mental training affects distribution of limited brain resources. *PLoS Biology, 5*, 138.

Spreng, R. N., Sepulcre, J., Turner, G. R., Stevens, W. D., & Schacter, D. L. (2013). Intrinsic architecture underlying the relations among the default, dorsal attention, and frontoparietal control networks of the human brain. *Journal of Cognitive Neuroscience, 25*, 74–86.

Stroop, J. R. (1935). Studies of interference in serial verbal reactions. *Journal of Experimental Psychology, 18*, 643–662.

Teper, R., & Inzlicht, M. (2013). Meditation, mindfulness and executive control: The importance of emotional acceptance and brain-based performance monitoring. *Social Cognitive and Affective Neuroscience, 8*, 85–92.

Tomasino, B., Fregona, S., Skrap, M., & Fabbro, F. (2013). Meditation-related activations are modulated by the practices needed to obtain it and by the expertise: An ALE meta-analysis study. *Frontiers in Human Neuroscience, 6*, 346.

Van den Hurk, P. M., Giommi, F., Gielen, S. C., Speckens, A. M., & Barendregt, H. P. (2010). Greater efficiency in attentional processing related to mindfulness meditation. *Quarterly Journal of Experimental Psychology, 63*, 1168–1180.

van Leeuwen, S., Müller, N. G., & Melloni, L. (2009). Age effects on attentional blink performance in meditation. *Consciousness and Cognition, 18*, 593–599.

van Leeuwen, S., Singer, W., & Melloni, L. (2012). Meditation increases the depth of information processing and improves the allocation of attention in space. *Frontiers in Human Neuroscience, 6*, 133.

van Vugt, M. K., & Slagter, H. A. (2014). Control over experience? Magnitude of the attentional blink depends on meditative state. *Consciousness and Cognition, 23*, 32–39.

Verhaeghen, P. (2017). *Presence: How meditation and mindfulness shape mind, brain, and life*. New York: Oxford University Press.

Verhaeghen, P. (in preparation). *Mindfulness and attention: A meta-analysis.*

Metacognitive Training

Simone Schaeffner, Nicolas Chevalier, Maki Kubota, and Julia Karbach

Contents

Abstract Metacognition is usually defined as "thinking about thinking," and it refers to knowledge about factors that influence task performance and knowledge about strategies. Moreover, it includes metacognitive regulation processes such as planning and monitoring task performance as well as evaluating the efficiency of these planning and monitoring processes. Good metacognitive abilities are essential for academic success, and good metacognitive skills support a number of other cognitive processes that are necessary to perform a specific task. Thus, training of metacognitive skills has become an important element of different training programs in various domains. In the present chapter, we will give an overview of recent advancements in the knowledge about metacognitive training in the context of mathematical skills, reading abilities, and regarding executive function training. Research from all three domains reveals promising results, indicating that the integration of metacognitive training into more conventional training programs leads to greater improvements than conventional training alone. Metacognitive training is effective for many different age groups, via different methods, and in different contexts. At the same time, however, there are still a number of open questions like the question of interindividual differences or the question of long-term effects, indicating that the field of metacognitive training research is likely to keep in the future.

S. Schaeffner (✉) · J. Karbach
University of Koblenz-Landau, Department of Psychology, Landau in der Pfalz, Germany

Center for Research on Individual Development and Adaptive Education of Children at Risk (IDeA), Frankfurt, Germany
e-mail: Schaeffner@uni-landau.de

N. Chevalier · M. Kubota
Department of Psychology, University of Edinburgh, Edinburgh, UK

T. Strobach, J. Karbach (eds.), *Cognitive Training*,
https://doi.org/10.1007/978-3-030-39292-5_18

Introduction

Metacognition is broadly defined as "thinking about thinking" (Flavell 1979). It is a multidimensional construct referring to any knowledge or cognitive process that monitors or controls cognition. Typically, metacognition is divided into two subcomponents, which are assumed to be correlated: *knowledge* of cognition and *regulation* of cognition (cf. Brown 1980; Flavell 1979; Veenman et al. 2006). Knowledge of cognition refers to the declarative knowledge about oneself as a learner or problem-solver, the knowledge about the task and possible strategies that can be used for solving the task, and the knowledge about how and when to use a given strategy. In contrast, regulation of cognition refers to a set of activities that help to control one's thinking and learning processes such as planning, monitoring, and evaluation processes. Figure 1 gives an overview of the most important components of knowledge and regulation of cognition.

Metacognition improves consistently as a function of age and schooling (e.g., Justice 1986; Schneider 2008). It develops gradually in early childhood and becomes more and more explicit and effective the older a child gets (see also Kuhn 2000). For example, 3-year-old children begin to become aware of their own knowledge states when they start using verbs like "to think" and "to know" (Flavell 1999). Six-year-old children can already reflect with accuracy on their cognition (Schraw and Moshman 1995). The older the children are, the more accurately they can predict their future performance. This early metacognitive development serves as a basis for higher-order thinking processes that mature later. That is, individuals become more and more aware of their own knowledge and increasingly proficient in selecting the most efficient strategies to solve a specific task and manage demanding situations (e.g., Chen and Siegler 2000; Lemaire and Brun 2014).

Good metacognitive abilities seem to be especially essential for academic success as metacognitive skills support the cognitive skills that are necessary to perform a task. For example, the use of metacognitive strategies is related to enhanced learning outcomes (e.g., Jacobs and Paris 1987), and metacognitive regulation is a reliable predictor for student's success in college (Everson and Tobias 1998). That is, people with good metacognitive awareness are able to think about their own thinking as they engage in academic tasks, and improved regulatory skills and an

Metacognitive Knowledge			Metacognitive Regulation		
Knowledge about oneself as a learner and factors affecting cognition	Knowledge about the task and potential strategies	Knowledge about why and when to use a given strategy	Identification and selection of appropriate strategies and allocation of resources	Attending to and being aware of comprehension and task performance	Assessing the processes and products of one's learning, and revisiting and revising learning goals

Fig. 1 Overview of the different components of metacognitive knowledge and metacognitive regulation

understanding of how to use these skills can evoke significant improvements in learning (Brown and Palincsar 1989; Cross and Paris 1988). A high level of metacognition can even compensate for IQ with regard to problem-solving (Swanson 1990). Thus, improving metacognitive skills has been the goal of numerous training studies. Even though metacognition is sometimes seen as a stable individual trait (e.g., Fleming et al. 2010; McCurdy et al. 2013), several studies demonstrated that metacognitive skills are malleable and trainable via different methods, in different contexts, and in different age groups. Moreover, intervention studies have shown not only that it is possible to train metacognition but also that these improvements benefit other cognitive skills drawing on metacognitive abilities.

In the following we will provide examples for training studies focusing on metacognitive training. Our ambition is not to provide an exhaustive review of the literature but instead an overview of recent advancements in the knowledge about metacognitive training based on intervention studies from three different fields. Specifically, we will start with evidence for the effects of metacognitive training on mathematical skills. Then, we will turn to recent studies about metacognitive training designed to improve reading skills. Finally, we will report recent research about metacognitive training aiming to improve executive functions. In all three domains we will especially focus on the training methods and its effectiveness.

Metacognitive Training and Mathematical Abilities

Basic mathematical abilities are usually acquired across preschool and elementary school age and comprise a wide set of abilities, among them are arithmetic, geometry, and mathematical problem-solving. Numerous studies showed that young children initially struggle performing math tasks that require multiple steps or the prediction of task outcomes, leading researchers many years ago to assume that metacognitive processes play a defining role in the development of mathematical skills. Lester (1982), for instance, suggested that metacognitive knowledge is of particular importance for mathematical problem-solving and that metacognitive knowledge, monitoring, and self-regulation are crucial before, during, and after solving mathematical problems. The metacognitive activities supporting task performance include mathematical skills and experience as well as the ability to separate relevant from irrelevant information and to use heuristics representing task-relevant components. Similarly, Verschaffel (1999) assumed that metacognition is important not only during initial stages of mathematical problem-solving, when an appropriate representation of the problem needs to be built, but also at the final stage, when outcomes have to be checked and evaluated.

Since these early studies linking metacognition and mathematical abilities, many studies have explored the relationship between both domains and have highlighted the predictive value of metacognitive abilities for mathematics performance (for reviews see Desoete and Veenman 2006; Schneider and Artelt 2010). For instance, Veenman (2006) examined the role of metacognitive skills (assessed by systematic observation) and cognitive ability for the development of mathematical learning performance (assessed by a math test). He found that both metacognitive and cognitive abilities were associated with mathematics performance but that metacognition was a more reliable predictor than cognitive ability. This and other findings demonstrate the importance of metacognitive skills for mathematical abilities (see, e.g., Desoete et al. 2001a, b; Garofalo and Lester 1985; Lucangeli et al. 1997). Thus, it is not surprising that many training programs focused on metacognition in order to improve mathematical skills (e.g., Lucangeli et al. 1998; Özsoy and Ataman 2017). In this section, we will detail a number of intervention studies designed to improve mathematical abilities by training metacognitive knowledge and skills.

Cornoldi et al. (1995) implemented a training focusing on metacognitive awareness and control processes. One of their studies focusing on healthy children indicated that improvements in metacognition were associated with improvements in problem-solving and logical reasoning but not in geometry. Another study on children with a learning disorder struggling in mathematics showed that these children benefitted even more from the program, even if their teachers perceived them as severely learning disabled.

More recently, a number of training studies have focused on the MASTER program (Mathematics Strategy Training for Educational Remediation; Van Luit and Kroesbergen 2006). This program was specifically designed for children with mathematical learning disabilities and targets self-instruction during mathematical problem-solving. Van Luit and Kroesbergen (2006) trained small groups of children with mathematical disabilities across 16 weeks and compared their performance to children participating in mathematics training based on the standard curriculum. Children in the training group received lessons in multiplication and division with a focus on problem orientation (planning), understanding of the number system, control activities (e.g., checking answers and solution strategies), and the memorization of multiplication and division facts <100. Children in both groups were tested on a standardized mathematics test before and after the intervention as well as at a follow-up session. Results showed larger gains from pretest to posttest in the training group as compared to the control group, and this effect was stable at follow-up.

Similarly, Desoete et al. (2003) investigated the effects of metacognitive strategy instruction (five sessions) on mathematical problem-solving in third graders. They assessed prediction and evaluation assessments before and after instruction and showed that participants in the training group significantly improved their metacognitive skills and their problem-solving knowledge at follow-up. Moreover, individual differences in metacognitive abilities were predictive of mathematics performance, allowing a differentiation between good and moderately performing students and those with learning disabilities (Desoete et al. 2001a, b).

Other training studies were based on the IMPROVE program (Introducing new concepts, Metacognitive questioning, Practicing, Reviewing, Obtaining mastery on higher and lower cognitive processes, Verification, and Enrichment and remedial; Kramarski and Mevarech 2003; Mevarech and Kramarski 1997). The metacognitive instructions included several metacognitive strategies: (1) comprehension questions ("What am I supposed to do in this task?"), (2) connecting questions ("What are the differences and similarities between . . . and . . .?"), (3) strategic questions ("What strategy, tactic, or principle can be used to solve the problem or to complete the task? Why is this strategy, tactic, or principle most appropriate for this problem or task?"), and (4) reflection questions ("Does the result make sense? Can the problem be solved differently?"). In one of the intervention studies, Mevarech et al. (2006) trained students (8th grade) in learning settings either with or without cooperative learning environments. They found that the IMPROVE training on top of the cooperative environment resulted in better mathematical problem-solving than the cooperative environments alone. Moreover, students participating in IMPROVE showed increased planning and comprehension processes as well as better reflection skills.

Some intervention programs have also adopted computer-based approaches. Teong (2003), for instance, investigated the effects of metacognitive training on mathematical word problems. Results from low-achieving students (11–12 years of age) showed superior performance compared to controls on mathematics tests as well as more appropriate metacognitive decision-making. Focusing on younger children, Pennequin et al. (2010) tested whether metacognitive training improved metacognitive knowledge and skills as well as mathematical problem-solving in third graders. The interactive training program included five training sessions. Results showed higher metacognitive knowledge, metacognitive skills, and mathematical problem-solving scores in the training group compared to the control group. Interestingly, low-achieving children benefitted most and improved up to the level of normal achievers.

In sum, evidence for the effects of metacognitive training on mathematical abilities is limited, but existing findings suggest that metacognitive knowledge and regulation are associated with mathematics performance over and above cognitive ability. Results from intervention studies indicated that metacognitive training can effectively enhance different aspects of metacognition and mathematical abilities, especially mathematical problem-solving. However, more intervention-based research is needed in order to disentangle the effects of different types of metacognitive trainings as well as individual differences in training-induced gains. As much more research has focused on the effects of metacognitive training on language and reading comprehension, we illustrate important findings in this domain in the next section.

Metacognitive Training and Reading Comprehension

Reading comprehension is a complex task that requires execution of several mental processes. One of the most influential models to date depicting these processes is the Construction-Integration (CI) model (Kintsch 1988). According to this model the reader first recognizes words and understands the syntactic links between them, then generates meaning through the integration of propositions, and finally integrates textual information with additional information from the reader's prior knowledge (i.e., situational model). Although these processes are generally assumed to be automatic, studies have shown that efficient readers are able to consciously coordinate and strategically solve problems when comprehension breaks down (Baker and Brown 1984; Coté et al. 1998). This is often referred to as metacognitive control or comprehension monitoring, which involves evaluating one's understanding and taking appropriate steps to correct errors that are detected (Baker et al. 2015).

Thus, extensive research has been conducted on the relationship between metacognition and reading, using various measures of metacognition such as self-reports (Roeschl-Heils et al. 2003), interviews (Eme et al. 2006), and questionnaires (Kolić-Vehovec et al. 2014; Memiş and Bozkurt 2013; Van Kraayenoord et al. 2012). One of the first studies was the study by Myers and Paris (1978). They showed differences in metacognitive strategies between younger and older children. Older children tended to have greater understanding and awareness of strategies that they use when encountering unknown words or sentences or of effective ways to skim through a text for rapid comprehension. However, children who lag behind their peers in metacognitive knowledge and cognition in primary grades continue to do so in middle school (Roeschl-Heils et al. 2003).

Given the abundance of evidence suggesting a link between metacognition and reading comprehension, researchers increasingly examined whether metacognitively oriented interventions promote reading comprehension, especially among younger to older children. Torgesen (1977), for example, found that receiving strategy instruction on picture recall increased children's score on reading comprehension. Since then, several metacognitive methods and strategy trainings have been introduced in the literature in the hope of developing reading comprehension. These interventions involve practices of knowing what factors are influential, knowing how strategies and functions are applied, and knowing when, where, and why to apply strategies in reading (Paris et al. 1984), so that such training, if successful, enables children to better evaluate purposes and strategies in reading, plan relevant strategies to be applied, and constantly monitor their performance during problem-solving within reading tasks (Wright and Jacobs 2003).

Different methods and approaches such as self-questioning (Chan 1991; Palinscar and Brown 1984), creating a cognitive map (Boyle 1996), and comprehension monitoring (Lubliner and Smetana 2005) were investigated. Incorporated reciprocal teaching (Palinscar and Brown 1984) is an influential instructional approach that is still used to date, involving training of four different strategies: (1) predicting upcoming text, (2) clarifying unknown words and concepts, (3) summarizing the text, and (4) generating questions about the material. Intervention studies which employed reciprocal teaching have largely yielded positive results, stimulating many successful multiple-strategies interventions in the 1990s and early 2000s (Gajria and Salvia 1992; Klingner and Vaughn 1996; Moore and Scevak 1995; Souvignier and Mokhlesgerami 2006). For instance, Souvignier and Mokhlesgerami (2006) conducted metacognitive trainings with German fifth graders, which comprised of 20 lessons (45 minutes each). In these lessons, the pupils were taught to actively use metacognitive strategies such as summarizing a text and elaborating on its content. Upon pre-, post-, and delayed-posttest assessments, the findings revealed that pupils in the strategy-oriented instructional programs better improved their reading comprehension than the control group. In another large-scale study by Van Keer and Verhaeghe (2005), word-level and passage-level comprehension monitoring was taught to second and fifth graders in three conditions: (a) teacher-led instructions, (b) same-age peer tutoring, and (c) cross-age tutoring between second and fifth graders. The findings showed that second graders benefitted from teacher-led instructions and cross-age tutoring but not same-age peer tutoring, and the effects did not last 6 months after program instruction. The fifth graders, however, all improved their reading comprehension in the posttest, and the effect prolonged for 6 months (except for the cross-age tutoring group).

Given the sheer number of metacognitive intervention studies over the past 35–40 years, several comprehensive meta-analyses have been undertaken (Dignath and Büttner 2008; Haller et al. 1988; National Reading Panel 2000). The findings of these meta-analyses highlight three main points in regard to metacognition and reading (Baker et al. 2015). First, older children (in secondary levels) benefit more from metacognitive training than younger, primary level children, perhaps due to the fact that older children have already acquired basic reading skills, and therefore can better build on prior experiences. Second, metacognitive trainings are more effective with longer periods of training sessions, increasing the likelihood of transfer of strategies to new contexts. Third, children benefit from metacognitive training when it is provided by researchers rather than classroom teachers, emphasizing the need to also educate and instruct the teachers on how to implement metacognitive trainings in their classroom teaching.

In summary, past work on metacognitive training in the context of improving reading skills has produced promising results, especially when readers are introduced to multiple metacognitive strategies. However, the strength of effects was modulated by several different factors such as the age, length of intervention, and the context in which the strategies were practiced (e.g., who implements trainings, classroom or individual training, etc.). Although the effects of metacognitive training have been overwhelmingly successful, as noted earlier, metacognition is not the sole solution to enhancing one's reading comprehension, rather there seems to be a complex interplay of several factors such as working memory, vocabulary, and motivation that play a crucial role in addressing how effective comprehension and learning of a text could be supported among children and adolescents. Moreover, reading comprehension is also influenced by executive functions, which are in turn closely related to metacognition, as will be shown in the following section.

Metacognition and Executive Functions

Executive functions (EF) refer to the set of neurocognitive processes that ensure the goal-directed, effortful regulation of attention, thoughts, actions, and emotions. They are supported by a wide neural network including the prefrontal cortex, and they enable flexible and adaptive behaviors. Although the unity and diversity of EF is still debated, EF are generally thought to reflect a set of partially separable functions including inhibition of task-irrelevant information or actions, information maintenance and updating in working memory, and shifting between task sets or representations (Miyake et al. 2000; Miyake and Friedman 2012; see also Karbach and Kray, this volume).

Recently, attempts have been made to train executive functions in combination with metacognitive training due to the close conceptual ties between EF and metacognition. Not only is metacognitive control strikingly similar, if not identical, to the type of cognitive control supported by EF, but also is monitoring, especially of conflict and performance, considered as a central aspect of executive functioning (e.g., Botvinick et al. 2001). To flexibly tailor EF engagement to the specific demands of the to-be-performed task, individuals need to represent and use information about (a) cognitive demands, (b) available control strategies (and how much effort they require), and (c) likelihood of success of each strategy. Unlike adults who strategically avoid unnecessary cognitive effort when given the choice between higher and lower task demands (e.g., Kool et al. 2010; McGuire and Botvinick 2010), younger children seem oblivious to variations in task demands (Niebaum et al. 2019) but can strategically avoid cognitive effort when made aware of task-demand differences and provided feedback (O'Leary and Sloutsky 2017, 2019). Similarly, they engage EF in a more mature manner when prompted to reflect on their own performance (Hadley et al. 2019a).

Therefore, facilitating metacognitive reflection on EF engagement can successfully improve EF performance, at least in children, which has important implications for EF training. First, incorporating metacognitive reflection in EF training programs should promote near transfer by enhancing flexible EF engagement across task demands. Second, and perhaps most importantly, metacognitive reflection training may support generalization of training-elicited gains to novel situations and facilitate far transfer through metacognitive awareness of one's own skills as well as reflection on task demands and how to best respond to them (Zelazo et al. 2018). If so, it may help the field move beyond the limits of extent EF training programs, which show no consistent far-transfer effects (e.g., Kassai et al. 2019; see also Guye et al. this volume; Karbach and Kray, this volume; Könen et al., this volume). Consistently, greater EF performance is observed after preschoolers briefly practice for 15–30-minutes reflection on task rules by either decomposing the elements of these rules (Espinet et al. 2013) or teaching them to a puppet (Moriguchi et al. 2015). Importantly, behavioral improvement is accompanied by more mature neural activity. Specifically, one study showed reduced N2 amplitude in the EEG data, an event-related component associated with activity in the anterior cingulate cortex (ACC; Espinet et al. 2013). Reduced N2 may indicate greater conflict detection by the ACC, which would facilitate signaling to lateral prefrontal cortex (lPFC) the need for greater EF engagement. Consistently, metacognitive reflection was associated with greater lPFC activation in the other study (Moriguchi et al. 2015).

Metacognitive reflection and awareness have also been trained through contemplation and mindfulness, a practice consisting in attending to and reflecting on one's moment-to-moment experiences in a nonjudgmental manner. A growing body of research has shown EF improvement after such practice in both children and adults (see Shapiro et al. 2014, for a review; Verhaeghen, this volume). For instance, 2-month mindfulness training at school enhanced inhibition performance in preschoolers, and these benefits, relative to literacy training or business-as-usual classes, strengthened between the immediate posttest and follow-up session a month later (Zelazo et al. 2018). In another study, students with the initially lowest EF performance benefitted the most from mindfulness training at school in terms of both EF and metacognition (Flook et al. 2010), a notable finding given that these students are at greater risk for academic failure. Mindfulness may benefit EF through repeated practice of turning of attention inwards, sustaining attention, and increased awareness of attention lapses (Shapiro et al. 2014; Zelazo et al. 2018).

Metacognition training may be especially powerful when focused on reflection on how to best engage EF and combined with training of EF processes per se. Three recent studies, which have been conducted independently, have adopted this innovative approach with children ranging in age from 5 to 14 years (Hadley et al. 2019b; Jones et al. 2019; Pozuelos et al. 2019). In Pozuelos et al.'s study, 5-year-olds were trained on a broad range of tasks tapping multiple aspects of EF in 10 sessions over a month in Spain. In an ongoing study, we trained 7- to 11-year-olds on multiple tasks tapping working memory, inhibition, and set-shifting in 16 sessions over 2 months in the UK and Germany (see Hadley et al. 2019b, for the preliminary findings). Finally, in Jones et al., 9- to 14-year-olds were trained on working memory tasks in 20–25 sessions over 6–7 weeks in the UK. Importantly, all three studies compared EF-and-metacognitive-reflection training (MetaEF) to EF training alone (BasicEF) and included an active control group. Although metacognitive reflection activities (see Fig. 2 for an example) differed across the three studies, they all fostered reflection on task demands, generation and use of control strategies, and performance monitoring.

Together, the findings from the three studies largely converged toward a coherent set of conclusions. First of all, none of the studies showed any specific behavioral advantage of metacognitive reflection training at immediate posttest. Specifically, although both MetaEF and BasicEF groups showed greater behavioral gains than the active control group, there were no differences between MetaEF and BasicEF. Therefore, metacognitive reflection training did not elicit greater near transfer at the behavioral level. However, 5-year-olds in Pozuelos et al.'s study showed neural changes at immediate posttest, with more adultlike EEG markers in the MetaEF than the BasicEF group. Thus, metacognitive reflection training already yielded important changes in the way children approached the task even though these changes did not yet translate into behavioral benefits. Indeed, MetaEF training was associated with greater working memory performance than BasicEF training in a 3-month follow-up posttest in Jones et al.'s study (the only one to include a follow-up session), which is consistent with the previously reported sustained effect of mindfulness on EF over time (Zelazo et al. 2018). Therefore, metacognitive reflection training may set children on a virtuous trajectory, installing the habit of reflecting on task demands and how to respond to them. This may not necessarily facilitate performance on tasks relatively close to the trained tasks immediately after training

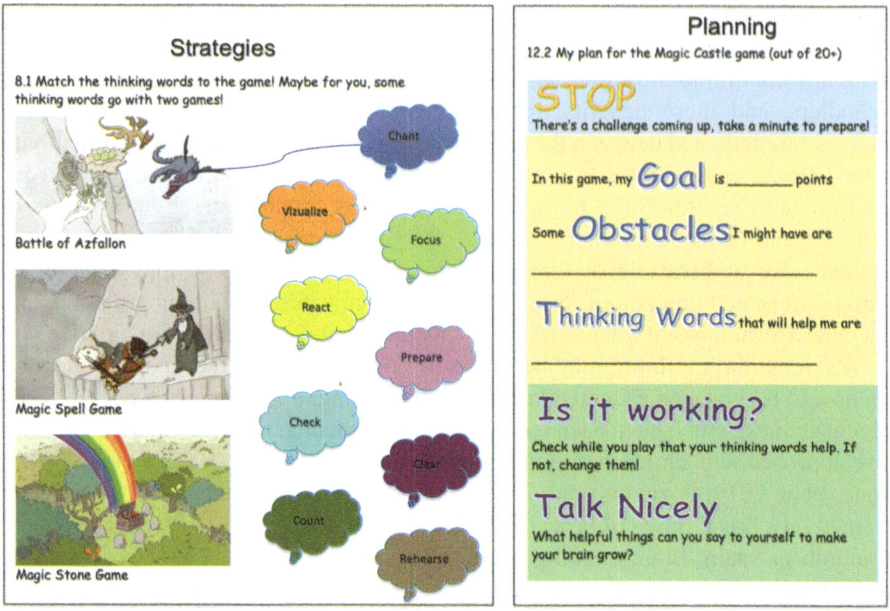

Fig. 2 Examples of an activity in which children had to identify what they found tricky in a specific game and discuss it with their partner. The activity on the left helped children think about strategy formats that could be applied to different games. The activity on the right helped children to prepare for an upcoming executive control task (Hadley et al. 2019b)

(relative to BasicEF), but the effect may build and strengthen over time and experiences, hence becoming more easily detectable after several months. Critically, although MetaEF did not yield immediately greater near transfer than BasicEF, it did elicit greater far transfer to nonverbal reasoning (progressive matrices; Pozuelos et al. 2019; Hadley et al. 2019b) as well as reading comprehension (Hadley et al. 2019b) at immediate posttest. The advantage of MetaEF over BasicEF may be immediately detectable for far-transfer tasks because these tasks are much less similar to the trained tasks than near-transfer tasks are, and thus, performance may better reflect the greater generalization of newly acquired skills that metacognitive reflection training instilled.

Therefore, training metacognitive reflection in conjunction with EF seems to be especially powerful to enhance EF in children. The clear far transfer to both nonverbal reasoning and academic skills (reading comprehension) and the sustained and even strengthening effects of metacognitive over time are very promising for the viability of this type of intervention. Indeed, in Pozuelos et al.'s study, 5-year-olds with lower EF skills at pretest showed the greatest gains from metacognitive reflection training, hence suggesting that children at risk may benefit the most from this type of intervention. That said, despite these promises, metacognitive reflection training is still in its early days, and much more research is needed to probe its efficacy in other populations, including young and older adults as well as children with developmental disorders such as autism and ADHD.

Conclusion and Outlook

To summarize, the present chapter demonstrates that integrating metacognitive training into common training methods leads to promising results across three different fields (training of mathematics, reading comprehension, and EF). Findings from all three domains correspond in several aspects that allow first conclusions regarding the efficiency of metacognitive training and also highlight implications for further research. Findings from all three domains indicate that metacognitive training is applicable across a wide range of children. Specifically, evidence for positive effects of metacognitive training has been found for children of many different age groups, ranging from preschoolers to adolescents, for healthy children, children with reduced skills in one specific domain (e.g., Pozuelos et al. 2019), or even for children with specific learning disorders (e.g., Cornoldi et al. 1995). However, it has to be noted that results across domains also provide first evidence for interindividual differences. For example, research on mathematical skills shows that children with learning disorders profit more from metacognitive training than healthy children, research on reading comprehension indicates age-related differences (cf., Baker et al. 2015), and research on math (Pennequin et al. 2010) as well as on EF training (Pozuelos et al. 2019) demonstrates that low-achieving children profit more from metacognitive training than high-achieving children. Thus, one important issue for further research might be the evaluation of different training programs for different groups of children in order to maximize gains after metacognitive training for each group.

Moreover, it has to be mentioned that existing findings are sometimes hard to compare due to considerable differences across studies – within and between the different domains. For example, the number of training sessions varies broadly, ranging from 5 sessions (e.g., Desoete et al. 2003) to 20–25 sessions (e.g., Jones et al. 2019), and also the length of the training period differs considerably, resulting in differing training intensity. Furthermore, also the training settings show a large variety of different possibilities. There are interventions taking place individually for each participant in a quiet room (e.g., Pozuelos et al. 2019), training sessions together with a training partner (Hadley et al. 2019b), training in small groups (e.g., Van Luit and Kroesbergen 2006), or even training within the classroom (e.g., Cornoldi et al. 2015). Thus, inconsistent findings might be due to these methodological differences. Hence, a systematic comparison might be an important subject of further research in order to gain further insights into the specific effects resulting from different methodological approaches.

Another subject of further research should be the investigation of long-term effects of metacognitive training. As reviewed above, there are indices that positive effects remain or even strengthen over time in all three domains. For example, in the context of mathematical problem-solving, it has been shown that improved performance from pre- to posttest can still be found at follow-up tests (e.g., Van Luit and Kroesbergen 2006). Regarding reading comprehension, Van Keer and Verhaeghe (2005) found improved reading comprehension 6 months after training, and Jones et al. (2019) demonstrated that MetaEF training was associated with greater working memory performance than BasicEF training in a 3-month follow-up posttest. So far, however, long-term effects of metacognitive training are only poorly explored, and findings are partly inconsistent. Long-term effects in the study by Van Keer and Verhaeghe (2005), for example, were only found for fifth graders but not for second graders, and in the context of executive function training, the only study including follow-up tests so far is the study by Jones et al. (2019). Thus, even though there are first promising results regarding long-term effects of metacognitive training from all three domains, further research is required in order to further clarify this issue.

Finally, we can say that there is compelling evidence indicating that training of metacognitive abilities is effective in different contexts, for different age groups, and via different methods. Moreover, improving metacognition has positive effects on other cognitive skills, so that the integration of metacognitive training into common training methods represents a promising approach. At the same time, however, it has to be said that research on metacognitive training is still scarce, and it will have to keep growing in order to further understand the complex interplay of the key influencing factors.

Acknowledgments This work was supported by a grant from the German Research Foundation (DFG) awarded to the last author (KA 3216/2-1) and a grant from the UK Economic and Social Research Council (ESRC, ES/N018877/1) awarded to the second author.

References

Baker, L., & Brown, A. L. (1984). Metacognitive skills and reading. In P. D. Pearson, R. Barr, M. L. Kamil, & P. Mosenthal (Eds.), *Handbook of reading research* (Vol. 1, pp. 353–394). New York: Longman.

Baker, L., DeWyngaert, L. U., & Zeliger-Kandasamy, A. (2015). Metacognition in comprehension instruction: New directions. In *Comprehension instruction: Research based practices* (3rd ed., pp. 72–87). New York: The Guilford Press.

Botvinick, M. M., Braver, T. S., Barch, D. M., Carter, C. S., & Cohen, J. D. (2001). Conflict monitoring and cognitive control. *Psychological Review, 108*, 624–652.

Boyle, J. R. (1996). The effects of a cognitive mapping strategy on the literal and inferential comprehension of students with mild disabilities. *Learning Disability Quarterly, 19*, 86–98.

Brown, A. L. (1980). Metacognitive development and reading. In R. J. Spiro, B. C. Bruce, & W. F. Brewer (Eds.), *Theoretical hues in reading comprehension*. Hillsdale: Erlbaum.

Brown, A. L., & Palincsar, A. S. (1989). Guided, cooperative learning and individual knowledge acquisition. In L. B. Resnick (Ed.), *Knowing, learning, and instruction: Essays in honor of Robert Glaser* (pp. 393–451). Hillsdale: Lawrence Erlbaum Associates.

Chan, L. K. (1991). Promoting strategy generalization through self-instructional training in students with reading disabilities. *Journal of Learning Disabilities, 24*, 427–433.

Chen, Z., & Siegler, R. S. (2000). VII. Components of strategic change. *Monographs of the Society for Research in Child Development, 65*, 43–58.

Cornoldi, C., Lucangeli, D., Caponi, B., Falco, G., Focchiatti, R., & Todeschini, M. (1995). *Matematica e Metacognizione*. Trento: Erickson.

Cornoldi, C., Carretti, B., Drusi, S., & Tencati, C. (2015). Improving problem solving in primary school students: The effect of a training programme focusing on metacognition and working memory. *British Journal of Educational Psychology, 85*, 424–439.

Coté, N., Goldman, S. R., & Saul, E. U. (1998). Students making sense of informational text: Relations between processing and representation. *Discourse Processes, 25*, 1–53.

Cross, D. R., & Paris, S. G. (1988). Developmental and instructional analyses of children's metacognition and reading comprehension. *Journal of Educational Psychology, 80*, 131–142.

Desoete, A., & Veenman, M. (Eds.). (2006). *Metacognition in mathematics education*. Haupauge: Nova Science.

Desoete, A., Roeyers, H., & Buysse, A. (2001a). Metacognition and mathematical problem solving in grade 3. *Journal of Learning Disabilities, 34*, 435–447.

Desoete, A., Roeyers, H., & De Clercq, A. (2001b). EPA2000: Een instrument om metacognitieve en rekenvaardigheden te meten. [EPA2000: An instrument for measuring metacognitive and arithmetic skills]. *Kind en Adolescent, 22*, 85–94.

Desoete, A., Roeyers, H., & De Clercq, A. (2003). Can offline metacognition enhance mathematical problem solving? *Journal of Educational Psychology, 95*, 188–200.

Dignath, C., & Büttner, G. (2008). Components of fostering self-regulated learning among students. A meta-analysis on intervention studies at primary and secondary school level. *Metacognition and Learning, 3*, 231–264.

Eme, E., Puustinen, M., & Coutelet, B. (2006). Individual and developmental differences in reading monitoring: When and how do children evaluate their comprehension? *European Journal of Psychology of Education, 21*, 91–115.

Espinet, S. D., Anderson, J. E., & Zelazo, P. D. (2013). Reflection training improves executive function in preschool-age children: Behavioral and neural effects. *Developmental Cognitive Neuroscience, 4*, 3–15.

Everson, H. T., & Tobias, S. (1998). The ability to estimate knowledge and performance in college: A metacognitive analysis. *Instructional Science, 26*, 65–79.

Flavell, J. H. (1979). Metacognition and cognitive monitoring: A new area of cognitive developmental inquiry. *American Psychology, 34*, 906–911.

Flavell, J. H. (1999). Cognitive development: Children's knowledge about the mind. *Annual Review of Psychology, 50*, 21–45.

Fleming, S. M., Weil, R. S., Nagy, Z., Dolan, R. J., & Rees, G. (2010). Relating introspective accuracy to individual differences in brain structure. *Science, 329*, 1541–1543.

Flook, L., Smalley, S. L., Kitil, M. J., Galla, B. M., Kaiser-Greenland, S., Locke, J., ... & Kasari, C. (2010). Effects of mindful awareness practices on executive functions in elementary school children. *Journal of Applied School Psychology, 26*, 70–95.

Gajria, M., & Salvia, J. (1992). The effects of summarization instruction on text comprehension of students with learning disabilities. *Exceptional Children, 58*, 508–516.

Garofalo, J., & Lester, F. K. (1985). Metacognition, cognitive monitoring, and mathematical performance. *Journal for Research in Mathematics Education, 16*, 163–176.

Hadley, L. V., Acluche, F., & Chevalier, N. (2019a). Encouraging performance monitoring promotes proactive control in children. *Developmental Science*, e12861.

Hadley, L. V., Schaeffner, S., Meaney, J. A., Könen, T., Morey, C. C., Auyeung, B., ... & Chevalier, N. (2019b). *Does training metacognition skills increase the benefit of executive function training?* Paper at the biennial meeting of the Society for Research in Child Development, Baltimore, United-States.

Haller, E. P., Child, D. A., & Walberg, H. J. (1988). Can comprehension be taught? A quantitative synthesis of "metacognitive" studies. *Educational Researcher, 17*, 5–8.

Jacobs, J. E., & Paris, S. G. (1987). Children's metacognition about reading: Issues in definition, measurement, and instruction. *Educational Psychologist, 22*, 255–278.

Jones, J., Milton, F., Mostazir, M., & Adlam, A. (2019). The academic outcomes of working memory and metacognitive strategy training in children: A double-blind randomised controlled trial. *Developmental Science*, e12870.

Justice, E. M. (1986). Developmental changes in judgements of relative strategy effectiveness. *British Journal of Developmental Psychology, 4*, 75–81.
Kassai, R., Futo, J., Demetrovics, Z., & Takacs, Z. K. (2019). A meta-analysis of the experimental evidence on the near- and far-transfer effects among children's executive function skills. *Psychological Bulletin, 145*, 165–188.
Kintsch, W. (1988). The role of knowledge in discourse comprehension: A construction-integration model. *Psychological Review, 95*, 163–182.
Klingner, J. K., & Vaughn, S. (1996). Reciprocal teaching of reading comprehension strategies for students with learning disabilities who use English as a second language. *The Elementary School Journal, 96*, 275–293.
Kolić-Vehovec, S., Rončević Zubković, B., & Pahljina-Reinić, R. (2014). Development of metacognitive knowledge of reading strategies and attitudes toward reading in early adolescence: The effect on reading comprehension. *Psihologijske Teme, 23*, 77–98.
Kool, W., McGuire, J. T., Rosen, Z. B., & Botvinick, M. M. (2010). Decision making and the avoidance of cognitive demand. *Journal of Experimental Psychology. General, 139*, 665–682.
Kramarski, B., & Mevarech, Z. R. (2003). Enhancing mathematical reasoning in the classroom: The effects of cooperative learning and metacognitive training. *American Educational Research Journal, 40*, 281–310.
Kuhn, D. (2000). Metacognitive development. *Current Directions in Psychological Science, 9*, 178–181.
Lemaire, P., & Brun, F. (2014). Effects of strategy sequences and response–stimulus intervals on children's strategy selection and strategy execution: A study in computational estimation. *Psychological Research, 78*, 506–519.
Lester, F. K. (1982). Building bridges between psychological and mathematics education research on problem solving. In F. K. Lester & J. Garofalo (Eds.), *Mathematical problem solving* (pp. 55–85). Philadelphia: The Franklin Institute Press.
Lubliner, S., & Smetana, L. (2005). The effects of comprehensive vocabulary instruction on Title I students' metacognitive word-learning skills and reading comprehension. *Journal of Literacy Research, 37*, 163–200.
Lucangeli, D., Coi, G., & Bosco, P. (1997). Metacognitive awareness in good and poor math problem solvers. *Learning Disabilities Research & Practice, 12*, 209–212.
Lucangeli, D., Cornoldi, C., & Tellarini, M. (1998). Metacognition and learning disabilities in mathematics. In T. E. Scruggs & M. A. Mastropieri (Eds.), *Advances in learning and behavioral disabilities* (pp. 219–285). Greenwich: JA.
McCurdy, L. Y., Maniscalco, B., Metcalfe, J., Liu, K. Y., De Lange, F. P., & Lau, H. (2013). Anatomical coupling between distinct metacognitive systems for memory and visual perception. *Journal of Neuroscience, 33*, 1897–1906.
McGuire, J. T., & Botvinick, M. M. (2010). Prefrontal cortex, cognitive control, and the registration of decision costs. *Proceedings of the National Academy of Sciences, 107*, 7922–7926.
Memiş, A., & Bozkurt, M. (2013). The relationship of reading comprehension success with metacognitive awareness, motivation, and reading levels of fifth grade students. *Educational Research and Reviews, 8*, 1242–1246.
Mevarech, Z. R., & Kramarski, B. (1997). IMPROVE: A multidimensional method for teaching mathematics in heterogeneous classrooms. *American Educational Research Journal, 34*, 365–394.
Mevarech, Z. R., Tabuk, A., & Sinai, O. (2006). Meta-cognitive instruction in mathematics classrooms: Effects on the solution of different kinds of problems. In A. Desoete & M. Veenman (Eds.), *Metacognition in mathematics education* (pp. 73–81). Haupauge: Nova Science.
Miyake, A., & Friedman, N. P. (2012). The nature and organization of individual differences in executive functions: Four general conclusions. *Current Directions in Psychological Science, 21*, 8–14.
Miyake, A., Friedman, N. P., Emerson, M. J., Witzki, A. H., Howerter, A., & Wager, T. D. (2000). The unity and diversity of executive functions and their contributions to complex "Frontal Lobe" tasks: A latent variable analysis. *Cognitive Psychology, 41*, 49–100.

Moore, P. J., & Scevak, J. J. (1995). The effects of strategy training on high school students' learning from science texts. *European Journal of Psychology of Education, 10*, 401–409.
Moriguchi, Y., Sakata, Y., Ishibashi, M., & Ishikawa, Y. (2015). Teaching others rule-use improves executive function and prefrontal activations in young children. *Frontiers in Psychology, 6*, 894.
Myers, M., & Paris, S. G. (1978). Children's metacognitive knowledge about reading. *Journal of Educational Psychology, 70*, 680–690.
National Reading Panel. (2000). *Report of the National Reading Panel: Teaching children to read: An evidence-based assessment of the scientific research literature on reading and its implications for reading instruction: Reports of the subgroups*. Rockville, MD: NICHD Clearinghouse.
Niebaum, J. C., Chevalier, N., Guild, R. M., & Munakata, Y. (2019). Adaptive control and the avoidance of cognitive control demands across development. *Neuropsychologia, 123*, 152–158.
O'Leary, A. P., & Sloutsky, V. M. (2017). Carving metacognition at its joints: Protracted development of component processes. *Child Development, 88*, 1015–1032.
O'Leary, A. P., & Sloutsky, V. M. (2019). Components of metacognition can function independently across development. *Developmental Psychology, 55*, 315–328.
Özsoy, G., & Ataman, A. (2017). The effect of metacognitive strategy training on mathematical problem solving achievement. *International Electronic Journal of Elementary Education, 1*, 67–82.
Palinscar, A. S., & Brown, A. L. (1984). Reciprocal teaching of comprehension-fostering and comprehension-monitoring activities. *Cognition and Instruction, 1*, 117–175.
Paris, S. G., Cross, D. R., & Lipson, M. Y. (1984). Informed strategies for learning: A program to improve children's reading awareness and comprehension. *Journal of Educational Psychology, 76*, 1239–1252.
Pennequin, V., Sorel, O., Nanty, I., & Fontaine, R. (2010). Metacognition and low achievement in mathematics: The effect of training in the use of metacognitive skills to solve mathematical word problems. *Thinking & Reasoning, 16*, 198–220.
Pozuelos, J. P., Combita, L. M., Abundis, A., Paz-Alonso, P. M., Conejero, Á., Guerra, S., & Rueda, M. R. (2019). Metacognitive scaffolding boosts cognitive and neural benefits following executive attention training in children. *Developmental Science, 22*, 1–15.
Roeschl-Heils, A., Schneider, W., & van Kraayenoord, C. E. (2003). Reading, metacognition and motivation: A follow-up study of German students in grades 7 and 8. *European Journal of Psychology of Education, 18*, 75–86.
Schneider, W. (2008). The development of metacognitive knowledge in children and adolescents: Major trends and implications for education. *Mind, Brain, and Education, 2*, 114–121.
Schneider, W., & Artelt, C. (2010). Metacognition and mathematics education. *ZDM Mathematics Education, 42*, 149–161.
Schraw, G., & Moshman, D. (1995). Metacognitive theories. *Educational Psychology Review, 7*, 351–371.
Shapiro, S. L., Lyons, K. E., Miller, R. C., Butler, B., Vieten, C., & Zelazo, P. D. (2014). Contemplation in the classroom: A new direction for improving childhood education. *Educational Psychology Review, 27*, 1–30.
Souvignier, E., & Mokhlesgerami, J. (2006). Using self-regulation as a framework for implementing strategy instruction to foster reading comprehension. *Learning and Instruction, 16*, 57–71.
Swanson, H. L. (1990). Influence of metacognitive knowledge and aptitude on problem solving. *Journal of Educational Psychology, 82*, 306–314.
Teong, S. K. (2003). The effect of metacognitive training on mathematical word-problem solving. *Journal of Computer Assisted Learning, 19*, 46–55.
Torgesen, J. K. (1977). Memorization processes in reading-disabled children. *Journal of Educational Psychology, 69*, 571–578.
Van Keer, H., & Verhaeghe, J. P. (2005). Effects of explicit reading strategies instruction and peer tutoring on second and fifth graders' reading comprehension and self-efficacy perceptions. *The Journal of Experimental Education, 73*, 291–329.

Van Kraayenoord, C. E., Beinicke, A., Schlagmuller, M., & Schneider, W. (2012). Word identification, metacognitive knowledge, motivation and reading comprehension: An Australian study of grade 3 and 4 pupils. *The Australian Journal of Language and Literacy, 35*, 51–68.

Van Luit, J. E. H., & Kroesbergen, E. H. (2006). Teaching metacognitive skills to students with mathematical disabilities. In A. Desoete & M. Veenman (Eds.), *Metacognition in mathematics education* (pp. 177–190). Haupauge: Nova Science.

Veenman, M. V. J. (2006). The role of intellectual and metacognitive skills in math problem solving. In A. Desoete & M. Veenman (Eds.), *Metacognition in mathematics education* (pp. 35–50). Haupauge: Nova Science.

Veenman, M. V., Van Hout-Wolters, B. H., & Afflerbach, P. (2006). Metacognition and learning: Conceptual and methodological considerations. *Metacognition and Learning, 1*, 3–14.

Verschaffel, L. (1999). Realistic mathematical modelling and problem solving in the upper elementary school: Analysis and improvement. In J. H. M. Hamers, J. E. H. Van Luit, & B. Csapo (Eds.), *Teaching and learning thinking skills. Contexts of learning* (pp. 215–240). Lisse: Swets & Zeitlinger.

Wright, J., & Jacobs, B. (2003). Teaching phonological awareness and metacognitive strategies to children with reading difficulties: A comparison of two instructional methods. *Educational Psychology, 23*, 17–47.

Zelazo, P. D., Forston, J. L., Masten, A. S., & Carlson, S. M. (2018). Mindfulness plus reflection training: Effects on executive function in early childhood. *Frontiers in Psychology, 9*, 208.

Socio-cognitive Processes Training

Abigail Thompson and Nikolaus Steinbeis

Contents

Abstract The potential to train social capacities could have wide-ranging positive effects for society and may be particularly relevant to clinical conditions in which social challenges impact on well-being and quality of life. Yet, the study of whether and how social skills can be trained has been neglected until recently. This chapter provides an overview of the most recent studies which have sought to train social abilities across different developmental populations. An overview is first provided of socio-cognitive (theory of mind) and socio-affective (empathy, compassion) processes, after which studies seeking to enhance these skills are reviewed. Studies are divided into those that directly target the particular skill or seek to enhance it by targeting an associated function. The neural mechanisms associated with training and impact on prosocial behaviours are highlighted, and methodological implications are discussed throughout. Overall, studies suggest training social capacities may be effective; however, further research will be needed to clarify the precise methodological features that lead to training success.

A. Thompson (✉) · N. Steinbeis
Division of Psychology and Language Sciences, University College London, London, UK
e-mail: abigail.thompson@ucl.ac.uk

T. Strobach, J. Karbach (eds.), *Cognitive Training*,
https://doi.org/10.1007/978-3-030-39292-5_19

Introduction

Understanding the emotions, beliefs and desires of others is a core human capacity that enables us to navigate the complex and interwoven social world in which humans are embedded. The potential to train social skills and increase human interpersonal understanding has an almost utopian appeal. A lack of understanding, or caring about the perspectives of others, has been blamed for such wide-ranging and devastating issues as political divide, war, ruthless capitalism and human suffering. Equally, good social abilities ensure increased social connectedness, which in turn is highly predictive of health and wellbeing (Cornwell and Waite 2009; Jose et al. 2012).

On a more practical level, the ability to train social capacities could improve the lives of individuals who experience day-to-day challenges due to social difficulties. Autism spectrum conditions have likely been most widely studied in this regard, but notably, challenges with social abilities are inherent to many other clinical conditions including schizophrenia, attention-deficit hyperactivity disorder, conduct disorder and depression, with social challenges affecting wellbeing and quality of life (see deVries and Geurts, this volume). It may also confer a number of advantages for professions in which understanding is a day-to-day aspect of the job (i.e. doctors). In fact, one exciting aspect of the current field is that in recent years researchers have made strides into defining the isolated components of socio-cognitive processes and in particular interpersonal understanding (Singer and Klimecki 2014). Much is known about training motor (Papale and Hooks 2018), perceptual (Broadbent et al. 2015; Bediou et al., this volume) and cognitive skills (Blair 2017; e.g. Guye et al., Rueda et al., this volume), yet whether and how socio-cognitive processes can be trained has been neglected until recently. The purpose of this chapter is to bring together the most recent advances in the field on training socio-cognitive abilities and behaviours.

This Chapter

Socio-affective and socio-cognitive processes include numerous components. For the purpose of the present review, we will cover those that are related to understanding others and prosocial behaviour. This chapter will first provide an overview of socio-cognitive and socio-affective routes to interpersonal understanding, including theory of mind (ToM), empathy and compassion, and will then go on to discuss the empirical studies that have been conducted to train these skills. Each of the routes to interpersonal understanding considered is supported by, or associated with, other mechanisms, such as executive functions and language in the case of ToM. For this reason, this chapter divides the reviewed into studies where the training directly targets the desired social skill or indirectly targets other abilities and reports an outcome on the target social skill.

Whilst training studies have often been conducted with the explicit aim of enhancing a particular social skill, such training can also be seen as a particularly useful tool to manipulate and tap into the underlying mechanisms. Thus, through training, many studies dissect and shed light on the underlying process itself (Hale and Tager-Flusberg 2003; Lohmann and Tomasello 2003; Santiesteban et al. 2012b). Following the progress of training studies targeting specific components of any given skill reveals an intriguing history of how researchers have thought about these concepts on the whole.

The plasticity of socio-affective and socio-cognitive processes is a new field of research. Although the majority of training studies have been conducted in adults, we will discuss developmental mechanisms and integrate developmental literature where possible. As with all cognitive training studies, the outcomes of training in this arena are closely related to methodological factors such as study design (including use of a control group), type and intensity of training, including number of training sessions (see Cochrane and Green, Schmiedek, Könen and Auerswald, this volume). Thus, where possible we will seek to discuss these factors in the context of the following questions: Is this interpersonal skill trainable? What key mechanisms are involved? What methodological factors appear to be most important in facilitating this?

Socio-cognitive and Socio-affective Routes to Interpersonal Understanding

The way in which we gain understanding of each other can be divided into two relatively distinct routes: cognitive and affective (Singer 2006)[1]. This distinction appears to be underpinned by relatively separate neural mechanisms (Shamay-Tsoory et al. 2009; Kanske et al. 2015), a primary distinction being that socio-affective may be considered ‘hot’ in the sense that they involve limbic regions of the brain whereas socio-cognitive does not and may be considered ‘cold’. These routes develop separately and are largely independent behaviourally although likely both play a role in complex (or real-life) social scenarios (Zaki and Ochsner 2009; Kanske et al. 2016).

[1] A note on terminology: Empathy and associated functions have been defined in various ways. For the purposes of the current chapter, we have chosen to use an adaptation of the classification system of Singer (2006), which takes advantage of a primary distinction between socio-cognitive and socio-affective abilities, because this distinction appears to be meaningful at a neural level. Some researchers classify experience sharing, mentalising and empathic/prosocial concern as subcomponents of ‘empathy’ (e.g. Zaki and Ochsner 2012). These terms relate to empathy, ToM and compassion, as used in the current study.

Socio-cognitive

We may rationalise about the desires, beliefs and intentions of others, in an effort to explain the causes behind their actions. This is the awareness that other people have different representations and beliefs about the world than we do and the ability to make assumptions about what these representations may be. Taking on the perspective of another person is a primarily cognitive process, referred to as possessing a ToM (Premack and Woodruff 1978) or engaging in 'mentalising' (Frith and Frith 2003). The false belief task is the most common task used to assess this (Wimmer and Perner 1983; Baron-Cohen et al. 1985).

Development of ToM

Although implicit understanding of false beliefs may be present in infancy (Oniski and Baillargeon 2005; Kovács et al. 2010) and continues to mature into adolescence (Grazzani and Ornaghi 2012; Devine and Hughes 2014), the developmental period between the ages of 3 and 5 appears to be critical for ToM development (Wellman et al. 2001). Numerous studies report that at the age of 3, children are not able to pass a customary measure of ToM skills, the false belief task, whilst at ages 4 and above, a majority of children are (Wimmer and Perner 1983; Astington and Gopnik 1991; Wellman and Lagattuta 2000), an effect that has been reported across cultures (Liu et al. 2008).

Understanding the possible reasons for this developmental shift in ability has been the focus of many studies over the past 20 years. It is now understood that ToM development is influenced by a wide range of factors including family size (Perner et al. 1994; McAlister and Peterson 2007), exposure to discussion of mental states and feelings within families (Cutting and Dunn 1999), engagement in pretend play (Lillard 2013) and family socioeconomic status (Cutting and Dunn 1999). Nonetheless, two primary factors have been most dominant: executive functions and language abilities.

Supporting Mechanisms: Executive Functions

A relationship between ToM and executive functions (EF; Karbach and Kray; Strobach and Schubert, this volume) was established by findings that show a correlational nature between their development (Perner and Lang 1999; Perner et al. 2002; Carlson et al. 2004). Two prominent theories have been proposed to explain this relationship. Firstly, that ToM tasks depend upon EF abilities; thus, the development of EF is a necessary prerequisite to be able to successfully complete ToM (and that failure on ToM tasks may actually reflect under-developed EF) (Russell 1997). This is known as the 'expression' account. Secondly, that EF support the development of ToM indirectly, by enabling children to maximise other experiences that support ToM development in day-to-day life, such as social interactions. This is the 'emergence' account.

Supporting Mechanisms: Language

Similar to the relationship between ToM and EF, early studies identifying a relationship between ToM and language skills reported correlations in their development (Jenkins and Astington 1996; Hughes and Dunn 1997; Cutting and Dunn 1999) and specifically that early language skills predict later ToM abilities, but not the other way around (Astington and Jenkins 1999). There are a number of different theories about the relationship, one of which is the conversational hypothesis – that ToM development is supported by discussing mental state terms during conversation (Siegal 1999; Hutto 2007). Both EF and language may specifically support explicit but not implicit ToM (Grosse Wiesmann et al. 2017).

Neural Mechanisms

At a neural level, ToM recruits the medial prefrontal cortex, superior temporal sulcus, precuneus and ventral temporo-parietal junction (TPJ) (Adolphs 2009; Schurz et al. 2014). The TPJ appears to play a critical role, potentially underpinning an ability that has been demonstrated to be necessary for cognitive perspective-taking – self-other distinction – the ability to inhibit and separate one's own response to a situation to that of another person (Saxe et al. 2004; Decety and Lamm 2007; Santiesteban et al. 2012a; Steinbeis 2016; Wiesmann et al. 2017).

Socio-affective: Empathy and Compassion

In contrast to cognitive routes to understanding, we may feel an instinctive resonance to the plight or suffering of others. This is a reflection of an emotional state and involves an affective processing route. Within this affective route, there are two key social capacities: empathy and compassion.

Empathy is the sharing of feelings of another, whether positive or negative, whilst maintaining awareness that those feelings are not your own (Eisenberg and Strayer 1987). Although there are widely recognised social benefits to empathy, in recent years researchers have discussed the potential downsides of empathy. For example, empathic responses are biased to in-group, out-group membership (Lamm et al. 2011) suggesting prejudice and are sensitive to moral blame (Decety et al. 2010). Furthermore, sharing the negative emotions of another can cause an aversive reaction known as empathic distress, which can lead to avoidance behaviours due to a desire to reduce the negative emotions felt by oneself (Batson et al. 1987).

In contrast, compassion (also known as empathic concern or sympathy) is characterised as concern for the welfare of others without necessarily feeling what they are feeling (Bloom 2017). This is associated with positive feelings, a focus on the other person rather than oneself, and motivation to engage in prosocial behaviours such as helping. A key determinant of whether an affective response is compassionate or empathetic may be self-regulation (Eisenberg et al. 1998; Eisenberg et al. 2006). Failure to regulate one's emotional response may lead to overarousal and empathic distress (Decety 2010).

Development

A developmental prerequisite of empathy is emotional contagion, the automatic ability to resonate with the emotions of others, which is present in infants (Heyes 2018). Empathic responding is present in children as young as 2 years old (Eisenberg et al. 2006). The emergence of self-other distinction is key for emotional contagion to develop into empathic responding. The role of self-other distinction was demonstrated in a study which reported that 16–24-month-old children who passed a self-recognition mirror test were more likely to help an adult who displayed distress in response to a teddy's arm breaking off (Bischof-Köhler 1994).

Neural Mechanisms

Empathising with another person activates regions of the brain that are also involved when experiencing something oneself, with the degree of activation associated with the degree to which participants also felt negative feelings when observing another in pain (Lamm et al. 2011) and anterior insula activation specifically correlates with empathic concern ratings (Singer et al. 2004). Empathy recruits limbic and paralimbic areas including the anterior insula and anterior cingulate cortex, as well as the dorso-lateral prefrontal cortex and supramarginal gyrus (Fan et al. 2011; Kanske et al. 2015). Activation in the anterior insula in particular is modulated by contextual factors, such as in-group, out-group membership, and predicts helping behaviour (Hein et al. 2010). In contrast, compassion has a distinct neural signature to that of empathy, specifically recruiting the ventral orbitofrontal cortex, ventral striatum and ventral tegmental area (Klimecki et al. 2014; Singer and Klimecki 2014). These are regions involved in positive emotions, including maternal and affiliative love, pointing towards the positive role of compassion.

Relation to Prosocial Behaviours

For training in any social domain to be impactful in a meaningful sense, it must relate to behaviour change in the realm of prosocial behaviours, which includes helping, comforting others in distress or sharing of resources. The distinction between the different components of social skills, and associated neural systems, is important in the context of cognitive training because they likely lead to distinct outcomes in terms of prosocial behaviour (as discussed above). Many studies report changes in the targeted skill rather than changes in objectively measured social behaviours. Throughout the chapter, we will highlight any studies that do report changes in terms of prosocial behaviour.

Training Studies

Training Socio-cognitive Skills

The first study to show a successful training effect on ToM was Slaughter and Gopnik (1996). Since then, a wealth of training studies have been conducted making use of a range of different training approaches including social cognition and perspective-taking, storytelling enriched with mental states, corrective feedback, role-play and language-based interventions such as narrative training and sentential complements. Due to the developmental studies that have shown a relatively robust shift in ToM understanding between the ages of 3 and 4, a majority of training studies targeting ToM have focused on this age group, both for typical development and groups of autistic or hearing-impaired children.

A recent meta-analysis (Hofmann et al. 2016) reviewed studies that have aimed to train ToM in children using different approaches including social cognition and perspective-taking, language and storytelling-based approaches. The authors only included studies that featured a control group and did not exclude clinical conditions such as autism and hearing impairments. Forty-five studies from 32 papers were included in the final analysis. The number of training sessions ranged from 1 to 32. The authors report a moderately strong effect of ToM training, compared with controls, across studies (Hedges' $g = 0.75$, $CI = 0.60–0.89$, $p < .001$). Thus, the overall conclusion is that it is indeed possible to train ToM. However, there are a few important points to note. Firstly, the authors point out that it was not possible to disentangle potential moderator effects, such as the impact of individual differences in language or EF, because many studies do not report these abilities. It also was not possible to parse the training-related or contextual factors that may be most beneficial to training, such as the specific type of training intervention, setting or format of the training. Secondly, none of the included studies provided a long-term measure of effects; therefore, it is unknown whether the effects of ToM training last (the longest interval between training and post-test was 13 days).

Studies That Indirectly Train ToM: Executive Functions

Studies report that children with higher EF skills benefit most from ToM training (Benson et al. 2013; Gao et al. 2020) and that directly training EF leads to improvements in ToM abilities (Kloo and Perner 2003; Fisher and Happé 2005). In one study, training on an EF task, the Dimensional Change Card Sorting task, leads to increases in performance on a false-belief task in children between the ages of 3 and 4.7 (Kloo and Perner 2003). Curiously, training on the false-belief task also leads to improvements on the EF task. The authors suggest that this indicates that both tasks tap into similar processes; however, it does not clarify in what way EF and ToM performance are related.

Another study reports that training autistic children on either EF or ToM abilities leads to improvements in ToM abilities in both groups (Fisher and Happé 2005). Intriguingly, they found that there was a large immediate enhancement of ToM skills in the group trained on ToM abilities, whereas for the group trained on EF, the improvement in ToM skills was most apparent at a later period (measured 6–12 weeks after the intervention). The authors suggest that this may indicate that the ToM training may have a more direct effect, whilst EF training may be indirect and may lead to the development of other skills (such as set shifting) which, at a later timepoint, they then are able to use to their advantage on ToM tasks. Also, the sample size was very small.

A unique study in adults used training to investigate contrasting theories regarding the role of action imitation in socio-cognitive processes (Santiesteban et al. 2012b). They report that training imitation inhibition leads to improvements in performance on a visual perspective-taking task. This contrasted with two other training groups, one in which participants imitated the actions of another person and another where general inhibition skills were trained. This made it possible to determine that the training effects of the imitation inhibition group were specific to social skills and were not a result of inhibitory control training, per se. This finding provided support for the notion that inhibition of imitation is central to perspective-taking (rather than the contrasting theory that imitation facilitates ToM, through activation of the mirror neuron system (Gallese and Goldman 1998)).

Language

Hale and Tager-Flusberg (2003) tested the hypothesis that ToM development is dependent on language in a training study in which 72 children were divided into three groups; one received ToM training, one received training on sentential complements, and a final group (an active control) received training on a different embedded construction – relative clauses. A sentential complement is a grammatical clause which enables 'the relativity of belief and knowledge states' (de Villiers 2000) and has been proposed to be fundamentally involved in ToM skills by enabling us to comprehend how another person may think or feel. The authors report that training in both the ToM and sentential complements groups improved on ToM scores and the control group did not. This shows that, in line with theory, training on sentential complements improved ToM. However, since both groups improved on ToM scores, this does suggest that the linguistic skills are not a *necessary* prerequisite.

Lohmann and Tomasello (2003) pointed out that *all* ToM training studies (including that of Hale and Tager-Flusberg 2003, but also including studies that do not focus on language) include deceptive scenarios. Thus, it cannot be concluded that the effect of training in sentential complements is specifically related to this rather than being exposed to deception. In a well-designed study, they compared multiple training groups in order to directly tease out the influence of several factors believed to contribute to ToM. The training groups included (1) exposure to deceptive experience without language (to test whether deceptive experience per se was the driving force), (2) exposure to sentential complements only (without deception), (3) exposure to perspective-shifting discourse only and (4) a 'full' group which was exposed to both language conditions and deception. They report ToM abilities improved in all groups, apart from the group that did not experience language, suggesting that deceptive experiences alone were not sufficient to lead to ToM improvements. The greatest impact was seen in the 'full' group, suggesting that whilst training on both sentential complements and narrative discourse independently had an impact on ToM abilities, they may have independent effects on ToM and, thus, the most effective training course would include both.

Interim Summary: Training Socio-cognitive Skills

Overall, these studies demonstrate that socio-cognitive skills are plastic and are amenable to a range of different interventions across a broad age range. This is the case in both typically developing children and children with developmental conditions. The most effective training regime may be multifaceted, including multiple elements such as narrative, sentential complements and role-play. Limitations of this area currently are that studies typically include only small sample sizes (Fisher and Happé 2005), rarely provide long-term follow-up and generally do not report the impact of such training on 'real-life' measures reflecting social skills, which may be more ecologically valid. Thus, it is unclear whether training improvements have a lasting effect on children's ToM abilities or whether there is transfer to day-to-day life (Ozonoff and Miller 1995).

Training Socio-affective Skills

Empathy

Studies training empathy have focused on improving understanding of what empathy is and improving the ability to identify emotions or take the perspective of others, through a range of lecture-based (didactic) and experiential approaches (Lam et al. 2011). Experiential approaches include role-play and imagination activities, which aim to foster empathy by providing an insight into the experience of another person. For example, studies have simulated hallucinations to foster empathy for mental health patients (Bunn and Terpstra 2009) and used virtual reality to give participants the experience of visiting a refugee camp, which lead to increased empathy for refugees (Schutte and Stilinović 2017).

Reviews of empathy training conclude that it is possible to train empathy across different populations including physicians, undergraduate medical students, therapists, nurses and couples (Brunero et al. 2010; Lam et al. 2011; Batt-Rawden et al. 2013; Rogge et al. 2013). A meta-analysis of 18 randomised controlled trials supported this conclusion, reporting a moderate overall effect of empathy training (Hedges' $g = 0.63$, CI = 0.39–0.87, $p < .001$) (van Berkhout and Malouff 2016). Training effects were significantly moderated by the type of trainee (training involving health professionals and university students was effective, but for the other groups, it was not), the offer of compensation and the scope and objectivity of outcome measures (objective measures were associated with more effective empathy training compared to self-report measures).

This suggests empathy to be trainable; however, it is unclear what exactly is being trained. Conceptualisations of empathy and how to measure it vary considerably from study to study (Lam et al. 2011). For example, all of the studies reviewed in van Berkhout and Malouff (2016) include a 'cognitive empathy' element, which emphasises being able to take the perspective of another person and may be considered to be a socio-cognitive training approach. This may suggest it is possible, on an intellectual level, to gain understanding of what empathy is and how to take the perspective of another. Less is known about whether it is possible to alter the affective, experience-sharing component of empathy (Lam et al. 2011) or how improvements in empathy understanding relate to behaviour change in more naturalistic settings, since this is also influenced by motivational and contextual factors (Weisz and Zaki 2017).

Compassion

In addition to training empathy, in recent years a number of training studies have specifically targeted compassion. Short-term prosocial effects of compassion induction had been demonstrated previously (Batson et al. 2007); however, the first study to report the potential longer-term (i.e. spanning multiple days) prosocial effects of training compassion was reported by Leiberg et al. (2011). They report increased prosocial behaviours 2–5 days after a day workshop that taught a Buddhist contemplative technique specifically designed to enhance compassion, compared with a group who received a workshop on memory training. Prosocial behaviours were assessed using a novel game to test helping behaviours, the Zurich Prosocial Game, which enables separate investigation of the influence of reciprocity, cost and distress cues on helping behaviour. The number of hours of compassion training related to helping in the no-reciprocity trials but not on the reciprocity trails. This may suggest compassion training specifically enhances prosocial behaviours when there is no opportunity for reciprocity, which contrasts with norm-based (i.e. 'cold', reason-driven) helping (Singer and Steinbeis 2009). Overall, this study was the first to suggest it is possible to impact trait-level compassion.

This finding was further supported by a subsequent study (Weng et al. 2013) which reported that participants who received meditation-based compassion training gave more funds in an anonymous online game, in comparison to a group who received training in emotional reappraisal. The game was an economic decision-making game, in which individuals first saw another individual (the 'dictator') transfer an unfair amount of money to another person. In this way, the game specifically tapped into unfairness norms. Using an ecologically valid design, another study compared compassion and mindfulness training (Condon et al. 2013). After 8 weeks of training, participants visited the lab in order to complete a battery of tests. In the waiting room before entering the lab, an actor appeared with crutches and a walking boot, appearing to be visibly wincing. Without the knowledge of the participants, the experimenters noted whether the participant offered their seat. They found that both meditation groups offered their seat more often than the non-meditation control group and there was no difference between the compassion or mindfulness meditation groups. This suggests that it cannot be determined whether the prosocial effects of compassion training in the early studies of Leiberg et al. (2011) and Weng et al. (2013) are specific to compassion, or a general product of meditation, since they both used non-meditation controls.

More recent studies were designed in order to specifically answer these questions. The ReSource Project is central to this (Singer et al. 2015). This is a large-scale ($N = 323$ healthy adult participants) training study, investigating the specificity of training effects across three distinct training modules which each used different meditation techniques. The training modules were 'presence', 'affect' and 'perspective'. These were based on mindfulness, fostering compassion and developing perspective-taking abilities, respectively. Participants were assigned to one of the three active training groups or to a retest cohort, which acted as a non-active control group. All groups were matched on a range of variables including age, gender, IQ, income, marital status and personality traits. Two of the active training groups experienced the three training modules but in different orders, running consecutively one after another in 3-month blocks. The third active training group received only 'affect' training. Thus, in addition to the non-active control, the training groups were used as active controls for each other. This well-controlled study design enabled direct comparison across the different training types, specifically enabling the investigation of differential effects of training targeting mindfulness, compassion and perspective-taking and their subsequent impact on both behaviour and neural substrates.

The project reported specificity of behavioural outcomes for each module. 'Affect' training leads to increased self-reported compassion, and gains in ToM were achieved by the module targeting perspective-taking (Trautwein et al. 2020). The 'presence' module, based on mindfulness techniques that do not have an affective component, did not lead to increases in either compassion or ToM. There were also differential effects of the distinct training types on subcomponents of prosocial behaviour (Böckler et al. 2018). Altruistic behaviours, measured by a range of behavioural tasks such as donations and helping, were only increased by targeting compassion training ('affect' module), suggesting a specific impact of compassion training on prosocial behaviours. In contrast, self-reported prosociality increased across all active training groups, suggesting meditation training in general leads to increased self-reported prosocial behaviour, despite changes in measured prosociality only occurring in the 'affect' group. This is pertinent in the context of other compassion training studies using self-report measures as their primary outcome.

Neural Effects of Training

A current debate within the field is whether compassion-based meditation training facilitates behavioural effects due to changes in affective or cognitive systems (Dahl et al. 2016; Engen and Singer 2016). Early neuroimaging studies were equivocal, reporting compassion meditation training to be associated with changes in either reward and limbic systems, including the VTA, mOFC, pallidum and putamen (Klimecki et al. 2014), and cognitive regions involved in mentalising, including dlPFC, dmPFC and inferior parietal lobule (Mascaro et al. 2013; Weng et al. 2013).

The ReSource project studies report specificity at a neural level. The behavioural effects of the 'affect' and 'perspective' modules were associated with changes in cortical thickness across the brain across spatially segregated networks (Valk et al. 2017), which overlapped with socio-affective and socio-cognitive networks, respectively, which also mapped onto functional networks. A recent study compared compassion training with two other groups: a placebo group, who were told they were inhaling oxytocin, and a 'familiarity' group. By comparing these different groups, it was possible for the authors to disentangle demand characteristics and placebo effects (Kreplin et al. 2018). They reported increased responses to suffering in the mOFC in a compassion training group compared with two other groups and also in the NAcc compared with the familiarity group. This provides support for the notion and previous studies that have suggested that compassion meditation acts via role for limbic/motivational/affective pathways (Klimecki et al. 2014).

Interim Discussion: Socio-affective Training

Overall, studies suggest it is possible to improve individuals' self-report ratings of empathy and compassion and that prosocial behaviours, including helping and sharing, may increase with compassion training. There may be a dissociation between self-reported measures of increases in prosociality and actual behaviours (Böckler et al. 2018; Ashar et al. 2019). Further, the degree to which individuals actually experienced a change in their own affective experience as opposed to simply learning about it is unclear. This has interesting implications for the potential application of such training paradigms for clinical or applied groups (such as bullying or conduct disorder), as individuals may gain understanding of what empathy and compassion are and rate themselves as more empathetic, but this may not lead to behaviour change.

Overall Conclusions

This chapter summarises the studies that have been conducted targeting socio-cognitive and socio-affective processes. Whilst studies suggest that training may be effective in both of these arenas, the precise methodological factors that facilitate learning are largely unmapped (see also Cochrane and Green, this volume). Additionally, the long-term effects and 'real-world' effects of such training are largely unknown. Further, whilst some efforts have been made to train these skills in developmental populations, more research is needed to see how the effectiveness of training socio-cognitive and socio-affective processes might change with age.

References

Adolphs, R. (2009). The social brain: neural basis of social knowledge. *Annual Review of Psychology, 60*, 693–716.

Ashar, Y. K., Andrews-Hanna, J. R., Halifax, J., Dimidjian, S., & Wager, T. D. (2019). Effects of compassion training on brain responses to suffering others. *bioRxiv*, 616029.

Astington, J. W., & Gopnik, A. (1991). Theoretical explanations of children's understanding of the mind. *British Journal of Developmental Psychology, 9*, 7–31.

Astington, J. W., & Jenkins, J. M. (1999). A longitudinal study of the relation between language and theory-of-mind development. *Developmental Psychology, 35*, 1311–1320.

Baron-Cohen, S., Leslie, A. M., & Frith, U. (1985). Does the autistic child have a "theory of mind"?. *Cognition, 21*, 37–46.

Batson, C. D., Fultz, J., & Schoenrade, P. A. (1987). Distress and empathy: Two qualitatively distinct vicarious emotions with different motivational consequences. *Journal of Personality, 55*, 19–39.

Batson, C. D., Eklund, J. H., Chermok, V. L., Hoyt, J. L., & Ortiz, B. G. (2007). An additional antecedent of empathic concern: Valuing the welfare of the person in need. *Journal of Personality and Social Psychology, 93*, 65–74.

Batt-Rawden, S. A., Chisolm, M. S., Anton, B., & Flickinger, T. E. (2013). Teaching empathy to medical students: an updated, systematic review. *Academic Medicine, 88*, 1171–1177.

Benson, J. E., Sabbagh, M. A., Carlson, S. M., & Zelazo, P. D. (2013). Individual differences in executive functioning predict preschoolers' improvement from theory-of-mind training. *Developmental Psychology, 49*, 1615–1627.

Bischof-Köhler, D. (1994). Self object and interpersonal emotions. Identification of own mirror image, empathy and prosocial behavior in the 2nd year of life. *Zeitschrift fur Psychologie mit Zeitschrift fur angewandte Psychologie, 202*, 349–377.

Blair, C. (2017). Educating executive function. *Wiley Interdisciplinary Reviews: Cognitive Science, 8*, e1403.

Bloom, P. (2017). Empathy and its discontents. *Trends in Cognitive Sciences, 21*, 24–31.

Böckler, A., Tusche, A., Schmidt, P., & Singer, T. (2018). Distinct mental trainings differentially affect altruistically motivated, norm motivated, and self-reported prosocial behaviour. *Scientific Reports, 8*, 1–14.

Broadbent, D. P., Causer, J., Williams, A. M., & Ford, P. R. (2015). Perceptual-cognitive skill training and its transfer to expert performance in the field: Future research directions. *European Journal of Sport Science, 15*, 322–331.

Brunero, S., Lamont, S., & Coates, M. (2010). A review of empathy education in nursing. *Nursing Inquiry, 17*, 65–74.

Bunn, W., & Terpstra, J. (2009). Cultivating empathy for the mentally ill using simulated auditory hallucinations. *Academic Psychiatry, 33*, 457–460.

Carlson, S. M., Mandell, D. J., & Williams, L. (2004). Executive function and theory of mind: Stability and prediction from ages 2 to 3. *Developmental Psychology, 40*, 1105–1122.

Condon, P., Desbordes, G., Miller, W. B., & DeSteno, D. (2013). Meditation increases compassionate responses to suffering. *Psychological Science, 24*, 2125–2127.

Cornwell, E. Y., & Waite, L. J. (2009). Social disconnectedness, perceived isolation, and health among older adults. *Journal of Health and Social Behavior, 50*, 31–48.

Cutting, A. L., & Dunn, J. (1999). Theory of mind, emotion understanding, language, and family background: Individual differences and interrelations. *Child development, 70*, 853–865.

Dahl, C. J., Lutz, A., & Davidson, R. J. (2016). Cognitive processes are central in compassion meditation. *Trends in Cognitive Sciences, 20*, 161–162.

de Villiers, J. (2000). Language and theory of mind: What are the developmental relationships? In S. Baron-Cohen, H. Tager-Flusberg, & D. J. Cohen (Eds.), *Understanding other minds: Perspectives from developmental cognitive neuroscience* (pp. 83–123). New York: Oxford University Press.

Decety, J. (2010). The neurodevelopment of empathy in humans. *Developmental Neuroscience, 32*, 257–267.

Decety, J., & Lamm, C. (2007). The role of the right temporoparietal junction in social interaction: how low-level computational processes contribute to meta-cognition. *The Neuroscientist, 13*, 580–593.

Decety, J., Echols, S., & Correll, J. (2010). The blame game: The effect of responsibility and social stigma on empathy for pain. *Journal of Cognitive Neuroscience, 22*, 985–997.

Devine, R. T., & Hughes, C. (2014). Relations between false belief understanding and executive function in early childhood: A meta-analysis. *Child Development, 85*, 1777–1794.

Eisenberg, N., & Strayer, J. (1987). Critical issues in the study of empathy. In N. Eisenberg & J. Strayer (Eds.), *Cambridge studies in social and emotional development. Empathy and its development* (p. 3–13). Cambridge University Press.

Eisenberg, N., Fabes, R. A., Shepard, S. A., Murphy, B. C., Jones, S., & Guthrie, I. K. (1998). Contemporaneous and longitudinal prediction of children's sympathy from dispositional regulation and emotionality. *Developmental Psychology, 34*, 910–924.

Eisenberg, N., Fabes, R. A., & Spinrad, T. L. (2006). Social, emotional, and personality development. In N.Eisenberg (Ed.), *Handbook of Child Psychology* (p. 129–143). Hoboken: Wiley & Sons.

Engen, H. G., & Singer, T. (2016). Affect and motivation are critical in constructive meditation. *Trends in Cognitive Sciences, 20*, 159–160.

Fan, Y., Duncan, N. W., de Greck, M., & Northoff, G. (2011). Is there a core neural network in empathy? An fMRI based quantitative meta-analysis. *Neuroscience & Biobehavioral Reviews, 35*, 903–911.

Fisher, N., & Happé, F. (2005). A training study of theory of mind and executive function in children with autistic spectrum disorders. *Journal of Autism and Developmental Disorders, 35*, 757–771.

Frith, U., & Frith, C. D. (2003). Development and neurophysiology of mentalizing. *Philosophical Transactions of the Royal Society of London. Series B: Biological Sciences, 358*, 459–473.

Gallese, V., & Goldman, A. (1998). Mirror neurons and the simulation theory of mind-reading. *Trends in cognitive sciences, 2*, 493–501.

Gao, Q., Huang, Q., Zhang, Q., & Chen, W. (2020). Does executive function influence the development of theory of mind in elementary students?. *Current Psychology, 39*, 389–396.

Grazzani, I., & Ornaghi, V. (2012). How do use and comprehension of mental-state language relate to theory of mind in middle childhood?. *Cognitive Development, 27*, 99–111.

Grosse Wiesmann, C., Friederici, A. D., Singer, T., & Steinbeis, N. (2017). Implicit and explicit false belief development in preschool children. *Developmental Science, 20*, e12445.

Hale, C. M., & Tager-Flusberg, H. (2003). The influence of language on theory of mind: A training study. *Developmental Science, 6*, 346–359.

Hein, G., Silani, G., Preuschoff, K., Batson, C. D., & Singer, T. (2010). Neural responses to ingroup and outgroup members' suffering predict individual differences in costly helping. *Neuron, 68*, 149–160.

Heyes, C. (2018). Empathy is not in our genes. *Neuroscience and Biobehavioral Reviews, 95*, 499–507.

Hofmann, S. G., Doan, S. N., Sprung, M., Wilson, A., Ebesutani, C., Andrews, L. A., ... & Harris, P. L. (2016). Training children's theory-of-mind: A meta-analysis of controlled studies. *Cognition, 150*, 200–212.

Hughes, C., & Dunn, J. (1997). "Pretend you didn't know": Preschoolers' talk about mental states in pretend play. *Cognitive Development, 12*, 381–403.

Hutto, D. D. (2007). Narrative and understanding persons. *Royal Institute of Philosophy Supplements, 60*, 1–16.

Jenkins, J. M., & Astington, J. W. (1996). Cognitive factors and family structure associated with theory of mind development in young children. *Developmental Psychology, 32*, 70–78.

Jose, P. E., Ryan, N., & Pryor, J. (2012). Does social connectedness promote a greater sense of well-being in adolescence over time?. *Journal of Research on Adolescence, 22*, 235–251.

Kanske, P., Böckler, A., Trautwein, F. M., & Singer, T. (2015). Dissecting the social brain: Introducing the EmpaToM to reveal distinct neural networks and brain-behavior relations for empathy and theory of mind. *NeuroImage, 122*, 6–19.

Kanske, P., Böckler, A., Trautwein, F. M., Parianen Lesemann, F. H., & Singer, T. (2016). Are strong empathizers better mentalizers? Evidence for independence and interaction between the routes of social cognition. *Social cognitive and affective neuroscience, 11*, 1383–1392.

Klimecki, O. M., Leiberg, S., Ricard, M., & Singer, T. (2014). Differential pattern of functional brain plasticity after compassion and empathy training. *Social Cognitive and Affective Neuroscience, 9*, 873–879.

Kloo, D., & Perner, J. (2003). Training transfer between card sorting and false belief understanding: Helping children apply conflicting descriptions. *Child Development, 74*, 1823–1839.

Kovács, Á. M., Téglás, E., & Endress, A. D. (2010). The social sense: Susceptibility to others' beliefs in human infants and adults. *Science, 330*, 1830–1834.

Kreplin, U., Farias, M., & Brazil, I. A. (2018). The limited prosocial effects of meditation: A systematic review and meta-analysis. *Scientific Reports, 8*, 1–10.

Lam, T. C. M., Kolomitro, K., & Aamparambil, F. C. (2011). Empathy training: Methods, evaluation practices, and validity. *Journal of Multidisciplinary Evaluation, 7*, 162–200.

Lamm, C., Decety, J., & Singer, T. (2011). Meta-analytic evidence for common and distinct neural networks associated with directly experienced pain and empathy for pain. *Neuroimage, 54*, 2492–2502.

Leiberg, S., Klimecki, O., & Singer, T. (2011). Short-term compassion training increases prosocial behavior in a newly developed prosocial game. *PloS one, 6*(3), e17798.

Lillard, A. S. (2013). Playful learning and Montessori education. *American Journal of Play, 5*, 157–186.

Liu, D., Wellman, H. M., Tardif, T., & Sabbagh, M. A. (2008). Theory of mind development in Chinese children: a meta-analysis of false-belief understanding across cultures and languages. *Developmental Psychology, 44*, 523–531.

Lohmann, H., & Tomasello, M. (2003). The role of language in the development of false belief understanding: A training study. *Child Development, 74*, 1130–1144.

Mascaro, J. S., Rilling, J. K., Tenzin Negi, L., & Raison, C. L. (2013). Compassion meditation enhances empathic accuracy and related neural activity. *Social Cognitive and Affective Neuroscience, 8*, 48–55.

McAlister, A., & Peterson, C. (2007). A longitudinal study of child siblings and theory of mind development. *Cognitive Development, 22*, 258–270.

Onishi, K. H., & Baillargeon, R. (2005). Do 15-month-old infants understand false beliefs?. *Science, 308*, 255–258.

Ozonoff, S., & Miller, J. N. (1995). Teaching theory of mind: A new approach to social skills training for individuals with autism. *Journal of Autism and developmental Disorders, 25*, 415–433.
Papale, A. E., & Hooks, B. M. (2018). Circuit changes in motor cortex during motor skill learning. *Neuroscience, 368*, 283–297.
Perner, J., & Lang, B. (1999). Development of theory of mind and executive control. *Trends in Cognitive Sciences, 3*, 337–344.
Perner, J., Ruffman, T., & Leekam, S. R. (1994). Theory of mind is contagious: You catch it from your sibs. *Child Development, 65*, 1228–1238.
Perner, J., Lang, B., & Kloo, D. (2002). Theory of mind and self-control: More than a common problem of inhibition. *Child Development, 73*, 752–767.
Premack, D., & Woodruff, G. (1978). Does the chimpanzee have a theory of mind?. *Behavioral and Brain Sciences, 1*, 515–526.
Rogge, R. D., Cobb, R. J., Lawrence, E., Johnson, M. D., & Bradbury, T. N. (2013). Is skills training necessary for the primary prevention of marital distress and dissolution? A 3-year experimental study of three interventions. *Journal of Consulting and Clinical Psychology, 81*, 949–961.
Russell, J. (1997). How executive disorders can bring about an inadequate "theory of mind". In J. Russell (Ed.), *Autism as an executive disorder* (pp. 256–304). New York: Oxford University Press.
Santiesteban, I., Banissy, M. J., Catmur, C., & Bird, G. (2012a). Enhancing social ability by stimulating right temporoparietal junction. *Current Biology, 22*, 2274–2277.
Santiesteban, I., White, S., Cook, J., Gilbert, S. J., Heyes, C., & Bird, G. (2012b). Training social cognition: From imitation to theory of mind. *Cognition, 122*, 228–235.
Saxe, R., Carey, S., & Kanwisher, N. (2004). Understanding other minds: linking developmental psychology and functional neuroimaging. *Annual Review of Psychology, 55*, 87–124.
Schurz, M., Radua, J., Aichhorn, M., Richlan, F., & Perner, J. (2014). Fractionating theory of mind: a meta-analysis of functional brain imaging studies. *Neuroscience & Biobehavioral Reviews, 42*, 9–34.
Schutte, N. S., & Stilinović, E. J. (2017). Facilitating empathy through virtual reality. *Motivation and Emotion, 41*, 708–712.
Shamay-Tsoory, S. G., Aharon-Peretz, J., & Perry, D. (2009). Two systems for empathy: A double dissociation between emotional and cognitive empathy in inferior frontal gyrus versus ventromedial prefrontal lesions. *Brain, 132*, 617–627.
Siegal, M. (1999). Language and thought: The fundamental significance of conversational awareness for cognitive development. *Developmental Science, 2*, 1–14.
Singer, T. (2006). The neuronal basis and ontogeny of empathy and mind reading: Review of literature and implications for future research. *Neuroscience and Biobehavioral Reviews, 30*, 855–863.
Singer, T., & Klimecki, O. M. (2014). Empathy and compassion. *Current Biology, 24*, R875–R878.
Singer, T., & Steinbeis, N. (2009). Differential roles of fairness- and compassion-based motivations for cooperation, defection, and punishment. *Annals of the New York Academy of Sciences, 1167*, 41–50.
Singer, T., Seymour, B., O'doherty, J., Kaube, H., Dolan, R. J., & Frith, C. D. (2004). Empathy for pain involves the affective but not sensory components of pain. *Science, 303*, 1157–1162.
Singer, T., Kok, B. E., Bornemann, B., Bolz, M., & Bochow, C. (2015). *The ReSource Project: Background, design, samples and measurements* (2nd ed.). Leipzig: Max Planck Institute for Human Cognitive and Brain Sciences.
Slaughter, V., & Gopnik, A. (1996). Conceptual coherence in the child's theory of mind: Training children to understand belief. *Child Development, 67*, 2967–2988.
Steinbeis, N. (2016). The role of self–other distinction in understanding others' mental and emotional states: neurocognitive mechanisms in children and adults. *Philosophical Transactions of the Royal Society B: Biological Sciences, 371*, 20150074.
Trautwein, F. M., Kanske, P., Böckler, A., & Singer, T. (2020). Differential benefits of mental training types for attention, compassion, and theory of mind. *Cognition, 194*, 104039.

Valk, S. L., Bernhardt, B. C., Trautwein, F. M., Böckler, A., Kanske, P., Guizard, N., ... & Singer, T. (2017). Structural plasticity of the social brain: Differential change after socio-affective and cognitive mental training. *Science Advances, 3*, e1700489.

van Berkhout, E. T., & Malouff, J. M. (2016). The efficacy of empathy training: A meta-analysis of randomized controlled trials. *Journal of Counseling Psychology, 63*, 32–41.

Wiesmann, C. G., Schreiber, J., Singer, T., Steinbeis, N., & Friederici, A. D. (2017). White matter maturation is associated with the emergence of theory of mind in early childhood. *Nature Communications, 8*, 14692.

Weisz, E., & Zaki, J. (2017). Empathy building interventions: A review of existing work and suggestions for future directions. In E.M. Seppälä et al. (Eds.), *The Oxford Handbook of Compassion Science* (p. 205–217). New York: Oxford University Press.

Wellman, H. M., & Lagattuta, K. H. (2000). *Developing understandings of mind.* In S. Baron-Cohen, H. Tager-Flusberg, & D. J. Cohen (Eds.), *Understanding other minds: Perspectives from developmental cognitive neuroscience* (p. 21–49). New York: Oxford University Press.

Wellman, H. M., Cross, D., & Watson, J. (2001). Meta-analysis of theory-of-mind development: The truth about false belief. *Child Development, 72*, 655–684.

Weng, H. Y., Fox, A. S., Shackman, A. J., Stodola, D. E., Caldwell, J. Z., Olson, M. C., … & Davidson, R. J. (2013). Compassion training alters altruism and neural responses to suffering. *Psychological Science, 24*, 1171–1180.

Wimmer, H., & Perner, J. (1983). Beliefs about beliefs: Representation and constraining function of wrong beliefs in young children's understanding of deception. *Cognition, 13*, 103–128.

Zaki, J., & Ochsner, K. (2009). The need for a cognitive neuroscience of naturalistic social cognition. *Annals of the New York Academy of Sciences, 1167*, 16–30.

Zaki, J., & Ochsner, K. N. (2012). The neuroscience of empathy: Progress, pitfalls and promise. *Nature Neuroscience, 15*, 675–680.

Commercial Brain Training

Tilo Strobach and Alexandra Kupferberg

Contents

Abstract In the past 10 years, commercial computerized brain trainings became increasingly popular on the digital market. These programs are readily available, easy to administer, motivating, and adaptive. However, their effectiveness based on randomized controlled (RCT) studies with an active control group in healthy populations is hotly debated. Therefore, in this review, we report the characteristics and study outcomes of currently available commercial brain training programs. We critically assess the number and quality of RCT studies evaluating their empirical evidence for enhancing cognitive functioning in healthy adults and discuss their effectiveness.

Tilo Strobach and Alexandra Kupferberg contributed equally with all other contributors.

T. Strobach (✉)
Department Psychology, Medical School Hamburg, Hamburg, Germany
e-mail: tilo.strobach@medicalschool-hamburg.de

A. Kupferberg
University of Bern, Bern, Switzerland

T. Strobach, J. Karbach (eds.), *Cognitive Training*,
https://doi.org/10.1007/978-3-030-39292-5_20

Introduction

In the field of cognitive training, the potentially most hotly debated area from a public perspective is commercial brain trainings, also referred to as brain games. Advertisements of commercial brain trainings claim that such trainings improve real-world performance on tasks that matter in academic, personal, or professional lives. That is, these advertisements claim a transfer from the trained task (e.g., laboratory working-memory training tasks) not only to untrained tasks that tap into the trained domain (e.g., untrained laboratory working-memory tasks) but also to contexts of untrained domains (e.g., tests on intelligence); the former and the latter transfer type is referred to as relatively near and relatively far transfer varying in context and content from the trained tasks (Barnett and Ceci 2002; Simons et al. 2016). However, claims about transfer from commercial brain trainings differ tremendously from those in the scientific context. A group of scientists and therapists published an open letter (www.cognitivetrainingdata.org), claiming that "a substantial and growing body of evidence shows that certain cognitive-training regimens can significantly improve cognitive function, including in ways that generalize to everyday life." Although not focused on commercial brain training exclusively, this letter argued that the literature is replete with "dozens of randomized controlled trials published in peer-reviewed journals that document specific benefits of defined types of cognitive training." In contrast, the Stanford Center on Longevity and the Max Planck Institute for Human Development issued an open letter saying that "… consumers are told that playing brain games will make them smarter, more alert, and able to learn faster and better. … However, … compelling evidence of general and enduring positive effects on the people's minds … has remained elusive" ("A consensus on the brain training industry from the scientific community," 2014). Supporting this position, a large randomized controlled online study with 11,430 participants aged 18–69 years using a commercial brain training program for two to 188 sessions did not show any transfer effect to untrained tasks, even if these tasks were parallel to the trained ones (Owen et al. 2010).

Consistent with similar attempts in previous reviews and meta-analyses (Chiu et al. 2017; Harris et al. 2018; Kelly et al. 2014; Kueider et al. 2012; Lampit et al. 2014; Rabipour and Raz 2012; Rossignoli-Palomeque et al. 2018; Shah et al. 2017; Simons et al. 2016; Tetlow and Edwards 2017), this chapter aims at bringing further light into this debate and its conflicting views. We do so by referring to commercial brain training as computer-based cognitive trainings that, broadly defined, aim to enhance a cognitive skill or general cognitive ability by repeating cognitive tasks over a circumscribed timeframe (Rabipour and Raz 2012). Lately, many commercial trainings take advantage of this idea over the Internet, offering the comfort and privacy of home-based brain exercises (Torous et al. 2016).

Previous Research and Meta-analyses

Reviews of the literature evaluating the efficacy of commercial brain trainings for healthy older adults have shown mixed results when it comes to efficacy and thus the consequences for near and far transfer (Kueider et al. 2012; Lampit et al. 2014). A number of reasons might account for this discrepancy. First, the reviewed studies vary in both methodological approaches (including healthy samples vs. samples with impaired cognitive functions such as cancer, stroke, psychiatric conditions, or traumatic brain injuries). Second, the quality of the studies was not always optimal. Apart from the fact that, until recently, there have been only a few randomized controlled trials (RCTs) using commercial brain trainings in cognitively healthy adults, there has been a lack of active control groups in most studies but the use of simple test – retest/passive/no-contact control groups or even no-control groups. This lack fails to rule out too many confounds and does not allow results to be meaningfully interpreted, because potential changes during training might be the result of numerous cognitive, motivational, or strategic factors (Foroughi et al. 2016). Third, only few studies tested for the maintenance of improvements in various cognitive domains over a longer time period beyond 6 or even 3 months after the end of training (Papp et al. 2009; Rabipour and Raz 2012). Thus, the robustness of potential training and transfer effects is an open issue.

Study Selection

As a consequence of the discrepancy between previous reviews, in this chapter, we exclusively focus on studies that follow the gold standard of active control groups next to treatment groups that receive the training with the commercial brain training (Green et al. 2019; Green et al. 2014; Simons et al. 2016), and we exclusively report studies with samples of healthy participants to provide a good balance between a broad literature overview of this field and a rather consistent set of studies. Importantly, we primarily focus on studies that evaluate the efficiency of commercial brain trainings as a means of a commercial product that is primarily designed for a commercial use and for a commercialization of cognitive training. Thus, the scientific purpose of these products is rather secondary in this definition, saying that we are aware of the fact that this categorization might be intuitive at some point. For instance, this resulted in the exclusion of studies on the CogMed training battery (e.g., Hitchcock and Westwell 2017; Olesen et al. 2004) or Braintwister (Hogrefe et al. 2018; Jaeggi et al. 2008; Studer-Luethi et al. 2012), which have been evaluated scientifically even before the commercialization started (see also de Vries and Geurts, Könen et al., Johann and Karbach, this volume). Furthermore, we included studies on products that provide training of rather cognitive domains and less so on physical domains (see Bherer and Pothier, this volume) or perceptual domains (e.g., the useful field of view task) as well as on commercial video games (also see Bediou et al., Strobach and Schubert, this volume). As a consequence of these criteria, we structure the main section of this chapter by studies on individual products that are offered in the field of commercial brain trainings: Brain Fitness (offered by Posit Science), Lumosity (offered by Lumos Labs), CogniFit (offered by CogniFit), Brain Age and Big Brain Academy (offered by Nintendo), My Brain Trainer (offered by My Brain Trainer), NeuroNation (offered by NeuroNation), and LACE (offered by Neurotone). An overview of study designs and outcomes is illustrated in Table 1.

In the first step, we created this selection of studies by referring to structured searched in reviews and meta-analyses (Chiu et al. 2017; Harris et al. 2018; Kelly et al. 2014; Kueider et al. 2012; Lampit et al. 2014; Rabipour and Raz 2012; Rossignoli-Palomeque et al. 2018; Shah et al. 2017; Simons et al. 2016; Tetlow and Edwards 2017), from which we then selected relevant studies based on our selection criteria listed above. We structure this chapter based on products and not by domains these products aim to train. While the latter would be more relevant from a scientific perspective, most products combine trainings on a varying set of tasks and cognitive domains which makes structuring based on cognitive domains difficult.

The Users of Commercial Brain Training Programs

The launch of Nintendo's Brain Age in 2005 marked a change in the commercialization of cognitive training, since it was the first product marketed to the public on a large scale. The number of users of commercial brain training products increased tremendously in the last 10 years, estimating that 45% of purchases are made by consumers (for themselves personally or for members of their family), while employers, schools, or health providers purchase the rest (Fernandez 2013). Half of the consumers are estimated to be age 50 or over, 30% between 18% and 50%, and 20% younger than age 18. Since commercial brain trainings are mostly computer-based or Web-based, they offer a cost-effective alternative to traditional training programs and can be even used by adults who are homebound or live in an assisted living or nursing home facility.

Lumosity

Lumosity was launched in 2007 by Lumos Labs and involves playing a number of gamified versions of cognitive tasks from scientific laboratory contexts, such as the Eriksen flanker task and the Corsi block-tapping test. The cognitive domains trained by Lumosity include processing speed, working and visuospatial memory, selective and divided attention, inhibitory control, mental flexibility, and reasoning/problem-solving. Lumosity gives users feedback about their brain "fitness" and updates that fitness status as performances on the practiced tasks improve.

Table 1 Overview of the reviewed studies

Company	Software	Publication	*N*	Age (range or mean)	Control intervention	Funded by company	Cognitive training sessions	Duration
Lumos Labs	Lumosity	Ballesteros et al. (2014)	$n = 60$	57–80	Discussing aging and interests	Not funded	1 hour, 20 sessions	10–12 weeks
Lumos Labs	Lumosity	Hardy et al. (2015)	$n = 4715$	18–80	Crossword puzzles	NA	5 min, 50 sessions	10 weeks
Lumos Labs	Lumosity	Mayas et al. (2014)	$n = 27$	57–77	Discussing aging and interests	Not funded	1 hour, 20 sessions	10–12 weeks
Lumos Labs	Lumosity	Kable et al. (2017)	$n = 128$	18–35	Web-based video games	NA	30 minutes, 50 sessions	10 weeks
CogniFit	CogniFit	Peretz et al. (2011)	$n = 155$	>50	Computer games	Yes	20–30 minutes, 36 sessions	3 months
CogniFit	CogniFit	Shatil (2013)	$n = 118$	65–93	Book reading	NA	48 minutes, 32 sessions	4 months
Nintendo	Brain Age	Nouchi et al. (2013)	$n = 32$	69.1	Playing Tetris	Yes	15 minutes, 5 sessions/week	1 month
Nintendo	Big Brain Academy	McLaughlin et al. (2018)	$n = 14$	46–55	Answering a set of randomly generated trivia questions	Not funded	1 hour, 3 sessions/week	6 weeks
My Brain Trainer	My Brain Trainer	Simpson et al. (2012)	$n = 34$	53–75	Playing solitaire	Not funded	20 minutes/ daily	21 days
NeuroNation	NeuroNation	Strobach and Huestegge (2017)	$n = 471$	19–79	Answering trivia questions	Not funded	30 minutes, 21 sessions	36 days
Posit Science	Brain Fitness	Smith et al. (2009)	$n = 487$	65–80	DVD-based educational programs on history, art, and literature	Yes	1 hour, 40 sessions	8–10 weeks
Posit Science	Brain Fitness	Strenziok et al. (2014)	$n = 46$	69.7	Space fortress; the rise of nations	Not funded	1 hour, 6 sessions/week	6 weeks

(continued)

Table 1 (continued)

Company	Software	Publication	N	Age (range or mean)	Control intervention	Funded by company	Cognitive training sessions	Duration
Posit Science	Brain Fitness	Mahncke et al. (2006)	$n = 182$	60–87	Educational DVD	Yes	1 hour, 40 sessions	8–10 weeks
Posit Science	Brain Fitness	Anderson et al. (2013a)	$n = 62$	55–75	Educational DVD	NA	1 hour, 40 sessions	8 weeks
Posit Science	Brain Fitness	Leung et al. (2015)	$n = 209$	<60	Educational video programs	NA	1 hour, 39 sessions	13 weeks
LACE	Neurotone	Saunders et al. (2016)	$n = 279$	68.6	Educational counseling	Not funded	30 minutes, 20 sessions	4 weeks

The effectiveness of the program has been evaluated in four RCT studies (Ballesteros et al. 2014; Hardy et al. 2015; Kable et al. 2017; Mayas et al. 2014) and one 3-month follow-up study (Ballesteros et al. 2015). The first study found significant improvements in the experimental group after training in processing speed, attention, immediate and delayed visual recognition memory, as well as a tendency to improve in some dimensions (affection and assertiveness) of the well-being scale in comparison to a control group undergoing social meetings (Ballesteros et al. 2014). However, the authors failed to show improvements in visuospatial working memory and executive control. The follow-up study showed that only improvements in the subjective well-being domain was maintained while benefits in processing speed, attention, and long-term memory vanished (Ballesteros et al. 2015). This vanishing indicates that booster sessions of brain training may be required for the maintenance of cognitive benefits of this training. A second study used a cross-modal oddball task to measure alertness and distraction after 12 weeks of training with Lumosity and demonstrated significant reduction of distraction and an increase of alertness in the training group but not in the control group (Mayas et al. 2014). Another RCT reported improved processing speed, memory, problem-solving abilities, and concentration after participants completed Lumosity exercises for 15 min, 5 days per week for 10 weeks in comparison to a crossword puzzle control procedure (Hardy et al. 2015).

Across the previous studies, there was good support for the benefits of Lumosity training for near transfer in several cognitive functions, such as speed of processing (Ballesteros et al. 2014) and working memory (Hardy et al. 2015). These results were not in line with a recent study which tested whether training with Lumosity could influence choice behavior and brain responses by enhancing cognitive control, which in turn would lead to better choice behavior in case of immediate and risky rewards (Kable et al. 2017). The study compared 10 weeks of training with Lumosity with Web-based video games that do not specifically target executive function or adapt the level of difficulty throughout training. The authors found no effects of cognitive training on measures of delay discounting or risk sensitivity, while the improvements on the practiced tasks were similar in the training and the control group. In sum, there are mixed findings in studies on immediate effects of Lumosity training in RCT studies so far.

CogniFit

CogniFit initially focused on cognitive training for driving performance and later expanded its product line to form a more complete "brain gym" ("Welcome to the Brain Gym," 2004). This company's first CD-based product for broader brain training, released in 2004, was Mind Fit. That product later was replaced by Web-based and mobile-app-based CogniFit Personal Coach which has been evaluated in two RCT studies using an active control group (Peretz et al. 2011; Shatil 2013).

The first study included 155 healthy older adults (Peretz et al. 2011). It reported that although both the training and an active control condition (playing a video game) improved cognition per se, the training was more effective than playing video games for visuospatial working memory, visuospatial learning, and focused attention. The second study used CogniFit along with a physical activity intervention, using a four-condition design on 118 healthy older adults (Shatil 2013). Participants were randomized into physical activity, brain training, a combination of both, or an active book reading control group. The brain training and combined groups displayed improvements in hand-eye coordination, working memory and long-term memory, speed of processing, visual scanning, and naming. No such improvements were observed in the groups that did not engage in cognitive training. Thus, the authors assumed that the mechanism explaining the improvement in functioning observed in the combined group was likely due to the brain training rather than the physical activity component.

Brain Age and Big Brain Academy

Nintendo's Brain Age (first released in 2005) was one of the first successful mass-marketed commercial brain training products. The program, also known as Dr. Kawashima's Brain Training, was largely based on a book of puzzles and math exercises (e.g., counting currency). Brain Age features mini-games that require players to complete math problems quickly, read aloud, count syllables in phrases, count objects on the screen, or perform other spatial, verbal, and arithmetic tasks. The initial program was followed by a sequel titled Brain Age 2, which was administered via a handheld device and has a touch screen with a pen. Brain Age 2 was evaluated in one RCT (Nouchi et al. 2013). The use of the product for 15 min/day for 5 days/week for 4 weeks resulted in selective improvements in executive functions, working memory, and processing speed. However, no transfer effects were observed in numerous other outcome measures of global cognition or attention. Furthermore, it is critical to mention that this study was conducted with a clear conflict of interest since developers of Brain Age 2 were also authors of this study.

The training package Big Brain Academy was first released by Nintendo in Japan in 2005 and designed to stimulate/train and practice mental abilities such as attention and working memory, perceptual reasoning, and visuospatial skills. It consists of 15 activities grouped into five categories (think, memorize, analyze, compute, and identify), and a recent study examined the feasibility of this training (McLaughlin et al. 2018). In the training group, seven participants were required to complete 18 1-hour training sessions (3 hours/week) for the duration of 6 weeks, and the control group with the same number of participants was required to answer a set of randomly generated questions on different topics such as music, history, science, art, and literature using the Internet. The results showed that relative to the control condition, brain training led to greater improvements on measures of executive attention, as well as reduced improvements in mood. However, in comparison to the control group, participants in the training group did not improve in measures of verbal memory and nonverbal associative learning and memory. Importantly, it is essential to point to the low number of participants and thus the low experimental power in this study. While being essential to investigate the feasibility of training, it has only a limited impact on conclusions about the training's effectiveness.

My Brain Trainer

The program "My Brain Trainer" (Mybraintrainer.com) is based on computerized training in reaction time, inspection time, short-term memory for words, executive function, visuospatial acuity, arithmetic, visuospatial memory, visual scanning/discrimination, and n-back working memory. A RCT investigated the effectiveness of this training on 34 participants between 53 and 75 years of age (Simpson et al. 2012). The participants completed online exercises for 21 days, while an active control group played a solitaire card game. The study reported improved processing speed after training. However, there were no improvements in digit forward and backward tests, spatial working memory, digit symbol substitution, or trail making, showing the limited effectiveness of this "My Brain Trainer" in this study.

NeuroNation

The online platform "NeuroNation" was launched in 2011 and initially served Germany, Austria, and Switzerland, expanding to French, Spanish, Russian, and Portuguese markets 3 years later. Similar to Lumosity, NeuronNation's training includes game-based versions of laboratory tasks, which are categorized into the following domains: logic, perception, memory, reasoning, and maths.

In a recent RCT, the participants were divided into two groups: the training group practiced on NeuroNation working-memory exercises on updating and capacity, whereas the control group conducted verbal knowledge tasks and answered trivia questions (Strobach and Huestegge 2017). The effectiveness of this training was measured by utilizing pre- and posttests in trained tasks and untrained tasks. The trained tasks included trained working-memory updating and capacity. The untrained near-transfer tasks were related to these working-memory domains, while far-transfer tasks were from the domains of processing speed, shifting, inhibition, reasoning, and self-reported cognitive failures. In contrast to the active control group, the training group improved performance in the trained tasks and selective near-transfer tasks on working-memory updating and capacity. Improved performance was also evident in processing speed and shifting tasks (i.e., far-transfer tasks), but these improvements were not as conclusive as those improvements in near-transfer tasks. Further, the number of reported cognitive failures was lower in the training in contrast to the control group at posttest.

Interestingly, performance improvements were more pronounced for high-performing participants, characterizing the efficiency of "NeuroNation" training in more detail.

Brain Fitness

Posit Science's first brain training products were Brain Fitness and InSight, both released as DVDs. In particular, Brain Fitness included auditory discrimination and attention tasks. In 2008, Posit Science added the useful field of view training task used in the ACTIVE trial to train speed of processing (Ball et al. 2002) and incorporated the task into its products. In 2012, Posit Science launched a new Internet-based platform ("Brain HQ") – an online brain training system that includes all the exercises in the Brain Fitness Program and incorporates the games from both auditory and visuospatial training packages. The program operates on the theory that enhancing perception speed and accuracy aiming at the smallest units of auditory and visual processing will lead to general improvement of these processing systems.

Posit Science was evaluated in over ten studies; however, there were only five RCTs with an active control group (Anderson et al. 2013b; Leung et al. 2015; Mahncke et al. 2006; Smith et al. 2009; Strenziok et al. 2014) and two follow-up studies (Zelinski et al. 2014; Zelinski et al. 2011). The first RCT study performed to investigate the effect of Brain Fitness showed significant improvements in memory assessments directly related to the training tasks and significant generalization of improvements to nonrelated standardized neuropsychological measures of auditory memory in the training group even after a 3-month no-contact follow-up period (Mahncke et al. 2006). The second RCT aimed at testing Brain Fitness included 487 participants (Smith et al. 2009). The study also showed improvement of auditory memory and attention as measured by Repeatable Battery for the Assessment of Neuropsychological Status as well as secondary measures (word list total score, word list delayed recall, digits backward, letter-number sequencing). This could be seen as an indication to generalization of performance gains to untrained standardized measures of memory and attention. A follow-up study after 3 months suggested that the training program sustains improvements in cognitive function that endure past the completion of training while decreasing over the 3-month follow-up period (Zelinski et al. 2011). This implies that an ongoing or a repeated use of the program is required to maintain training benefits.

Another study was able to demonstrate transfer effects of Brain Fitness by comparing it with two other cognitive training programs: Space Fortress (aimed at training working memory) and the Rise of Nations (aimed at training strategic reasoning; Strenziok et al. 2014). The authors found transfer of Brain Fitness and Space Fortress to other untrained areas, such as problem resolution of daily life and reasoning. They hypothesized that training in general and independent of the specific product they used produced changes in the attentional networks, leading to improvement in other processes. Therefore, the authors assessed neuroplastic changes: training produced changes in the integrity of gray matter in occipital-temporal areas (associated with improvement in problem-solving of daily life), as well as in the ventral network, while there was no transfer to auditory or visuospatial working memory. This finding of lacking transfer was in contrast to two previous studies showing that Brain Fitness training in healthy older adults transferred to auditory memory (Mahncke et al. 2006; Smith et al. 2009). However, it has to be noticed that in contrast to this younger study, which used the letter-number sequencing subtest of the Wechsler Adult Intelligence Scale III to measure auditory working memory, both older studies used a composite score of the auditory memory subtests of the Repeatable Battery for the Assessment of the Neuropsychological Status, making the comparison difficult.

The benefits of auditory training have been demonstrated in a study aiming to improve the ability to process rapid events that characterize speech, especially in noisy environments (Anderson et al. 2013a). After training, older adults exhibited faster neural timing and experienced gains in memory, attention, speed of processing, and speech-in-noise perception, whereas a matched control group showed no changes in these measures. These results demonstrate that auditory-based cognitive training can partially restore age-related deficits in temporal processing. To examine the long-term maintenance of training gains, the participants were tested 6 months after the completion of training, demonstrating improvements in response peak timing to speech in noise and speed of processing (Anderson et al. 2014).

LACE

LACE (Listening and Communication Enhancement) is an interactive computerized training program which is aimed at training auditory skills and can be used at home (Sweetow and Sabes 2006). It was designed for helping hearing-impaired patients, who's most frequent complaints were the inability to understand speech in noise. While individuals with hearing loss report adequate hearing in quiet environments and for slowly presented speech, they have greater difficulty in noisy situations and when speech is at a rapid rate. The reason for this difficulty could be the detrimental effect of normal aging on central functions and cognitive skills such as speed of processing, working memory, and executive control (Pichora-Fuller 2015). Given the combination of peripheral hearing loss and age-related cognitive decline, individuals which are provided with new hearing aids are still unable to instantly and optimally synthesize the novel and partial degraded auditory signal without experience or training. Thus, some authors proposed that auditory understanding skills might be enhanced with training (Fu et al. 2005; Tremblay 2007).

LACE provides a variety of interactive and adaptive tasks that are divided into three main categories: degraded speech, cognitive skills (auditory memory and speed of processing), and communication strategies such as advices on hearing-conducive seating in a noisy restaurant, telephone use, and communication tips for patients and their friends and loved ones. However, there is only one RTC study which has analyzed the effects of LACE on cognitive abilities. This study included 279 older adults and LACE training on DVD and computer, while a control group performed education counseling; duration of training extended across 20 sessions (30 minutes each) within 4 weeks. However, pre- and posttest comparisons did not result in improved outcomes in any of the cognitive measures (i.e., speech-in-babble, time compression, competing speaker, auditory memory, missing word) over standard-of-care hearing aid intervention alone (Saunders et al. 2016). Since the results were different from the positive outcomes found in a previous study performed with a passive control group (Sweetow and Sabes 2006), the authors suggested the need to examined and evaluate individual differences to assess whether the training has any benefits for particular individuals. Consistently, later performed data analysis of compliance has shown that there were statistically significant improvements in a compliant group, with no statistically significant improvements observed for a noncompliant group (Chisolm et al. 2013). Thus, LACE might have a specific effect under particular conditions.

Limitations

As with any review, our conclusions must be considered within the context of the current search criteria and selection of studies. We included studies which were performed only on healthy populations; hence, different efficacy of trainings in individuals with cognitive decline or mental or physical diseases cannot be ruled out. There are also a large number of excluded studies which use noncommercially or semicommercially available devices. We suggest, however, that a focus on commercial devices was warranted given their growing popularity, easy access, and endorsements and the confusion about their effectiveness for cognitive fitness.

Further, studies published in the brain training industry sometimes can be subject to conflicts of interest that could bias the scientific integrity of the observed findings. Distributors of cognitive exercise programs often fund studies evaluating their product, so that conflicts of interest may impede the objectivity of studies either by provoking the omission of results unfavorable to the desired outcome or mainly reporting of findings that are favorable for the desired outcome (Turner et al. 2008). Authors bound to the industry may overextend the interpretations of their results by emphasizing statistical significance while ignoring relatively small effect sizes that would indicate little or no clinical significance. Thus, conflict of interest was relevant for previous studies and will be for future studies as well.

Future Directions

Improving the quality of experimental designs will be an essential requirement for future studies in order to estimate the extent to which commercial brain trainings promote transfer of skills from the laboratory to real-life contexts. Further research is also needed to scientifically evaluate commercial brain training programs available on the market, through double-blind randomized controlled trials, which include a passive control group and active control group, in addition to proper follow-up assessments.

There are a number of questions to be answered in the future when it comes to the effectiveness of commercial brain trainings. At which age should training start? How much should an individual train? How long should the training last? Do individuals need daily training, how long should the sessions be, and does one need to continue training in order to maintain gains, and if so, in what time intervals? Is it possible to enhance the potential benefits of cognitive training if it is complemented with physical activity or social activities?

Another important question when it comes to commercial brain trainings is the level of compliance, since is strongly related to outcomes. For example, a review of the clinical records of 3000 patients using LACE revealed that only 30% of participants completed 10 or more of the 20 training sessions (Sweetow and Sabes 2006). The closer data analysis of the RCT which evaluated the effectiveness of LACE training (Saunders et al. 2016) indicated that compliant participants had stronger improvements in outcome measures than the noncompliant participants (Chisolm et al. 2013). The authors suggested that the level of compliance can be increased by systematized verbal and written instructions and follow-up via telephone. Indeed, it has been shown that regular face-to-face sessions during Web-based cognitive training were associated with higher training intensities (Cruz et al. 2014). Thus, the combination of classical methods with information technology systems might ensure greater training intensity and therefore a better training outcome. This hypothesis is being tested in an ongoing RCTs with an active control group designed to evaluate the effects of adaptive computerized cognitive training with, among others, the program NeuroNation or the effects of auditory training by KOJ (KOJ auditory training). The latter evaluation uses combinations of face-to-face counseling on cognitive function and speech understanding in hearing impaired individuals wearing hearing aids (Kupferberg et al. 2019) and provides participants with practice on tasks falling into five categories (sound localization; auditory selection, filtering of speech signal from various background noises; auditory separation/dichotomous hearing; auditory differentiation, distinguishing similarly sounding signals from each other; auditory memory). These recent lines of research on compliance provide immediate feedback after each trial of each exercise and should be considered in future studies in general.

Summary

One premise of cognitive training is that training of core cognitive abilities will transfer to other tasks and environments. Therefore, we aimed to provide an overview of the peer-reviewed evidence for these devices in the field of commercial brain training. As such, there was good evidence for training effects and near-transfer effects in most brain training programs included in the current analysis; these trainings likely enhance working memory, processing speed, executive function, and attention in laboratory-based tasks, among others. However, in regard to our stated aims, the reviewed studies showed limited evidence for far transfer effects in general. Five studies have shown evidence of transfer to self-reported measures of everyday function (Ballesteros et al. 2014; Hardy et al. 2015; McLaughlin et al. 2018; Strenziok et al. 2014; Strobach and Huestegge 2017). However, overall evidence is currently weak for real world benefits from commercial trainings in healthy populations.

Only one of the included studies showed effects from Brain Fitness training on the brain connectivity (Strenziok et al. 2014). According to this study, Brain Fitness selectively changed the integrity of occipitotemporal white matter associated with improvement on untrained everyday problem-solving. It therefore seems that mentally challenging activities may trigger brain changes beneficial for enhancing cognition (see also Wenger and Kühn, this volume). However, there are more studies required to replicate, extend, and specify the neuroplastic, and also behavioral, changes as a consequence of commercial brain trainings. Thus, the debate between different positions on commercial brain trainings and their positive effects (www.cognitivetrainingdata.org) and their nonpositive effects ("A consensus on the brain training industry from the scientific community," 2014) is still not solved completely.

Conflicts of Interest Statement Tilo Strobach declares that he has no conflict of interest. Alexandra Kupferberg is partially employed by KOJ.

References

A consensus on the brain training industry from the scientific community. (2014). Retrieved from http://longevity3.stanford.edu/blog/2014/10/15/the-consensus-on-the-brain-training-industry-from-the-scientific-community-2/

Anderson, S., White-Schwoch, T., Parbery-Clark, A., & Kraus, N. (2013a). A dynamic auditory-cognitive system supports speech-in-noise perception in older adults. *Hearing Research, 300*, 18–32.

Anderson, S., White-Schwoch, T., Parbery-Clark, A., & Kraus, N. (2013b). Reversal of age-related neural timing delays with training. *Proceedings of the National Academy of Sciences, 110*, 4357–4362.

Anderson, S., White-Schwoch, T., Choi, H. J., & Kraus, N. (2014). Partial maintenance of auditory-based cognitive training benefits in older adults. *Neuropsychologia, 62*, 286–296.

Ball, K., Berch, D. B., Helmers, K. F., Jobe, J. B., Leveck, M. D., Marsiske, M., & Willis, S. L. (2002). Effects of cognitive training interventions with older adults: A randomized controlled trial. *Journal of the American Medical Association, 288*, 2271–2281.

Ballesteros, S., Prieto, A., Mayas, J., Toril, P., Pita, C., Ponce de León, L., ... & Waterworth, J. (2014). Brain training with non-action video games enhances aspects of cognition in older adults: A randomized controlled trial. *Frontiers in Aging Neuroscience, 6*, 277.

Ballesteros, S., Mayas, J., Prieto, A., Toril, P., Pita, C., Laura, P. D. L., ... & Waterworth, J. A. (2015). A randomized controlled trial of brain training with non-action video games in older adults: Results of the 3-month follow-up. *Frontiers in Aging Neuroscience, 7*, 45.

Barnett, S. M., & Ceci, S. J. (2002). When and where do we apply what we learn? A taxonomy for far transfer. *Psychological Bulletin, 128*, 612–637.

Chisolm, T. H., Saunders, G. H., Frederick, M. T., McArdle, R. A., Smith, S. L., & Wilson, R. H. (2013). Learning to listen again: The role of compliance in auditory training for adults with hearing loss. *American Journal of Audiology, 22*, 339–342.

Chiu, H.-L., Chu, H., Tsai, J.-C., Liu, D., Chen, Y.-R., Yang, H.-L., & Chou, K.-R. (2017). The effect of cognitive-based training for the healthy older people: A meta-analysis of randomized controlled trials. *PLoS One, 12*, e0176742.

Cruz, V. T., Pais, J., Alves, I., Ruano, L., Mateus, C., Barreto, R., ... & Coutinho, P. (2014). Web-based cognitive training: Patient adherence and intensity of treatment in an outpatient memory clinic. *Journal of Medical Internet Research, 16*, e122.

Fernandez, A. (2013). The digital brain health market 2012–2020: Web-based, mobile and biometrics-based technology to assess, monitor, and enhance cognition and brain functioning. *SharpBrains*. http://sharpbrains.com/market-report.

Foroughi, C. K., Monfort, S. S., Paczynski, M., McKnight, P. E., & Greenwood, P. M. (2016). Placebo effects in cognitive training. *Proceedings of the National Academy of Sciences, 113*, 7470–7474.

Fu, Q.-J., Galvin, J., Wang, X., & Nogaki, G. (2005). Moderate auditory training can improve speech performance of adult cochlear implant patients. *Acoustics Research Letters Online, 6*, 106–111.

Green, C. S., Strobach, T., & Schubert, T. (2014). On methodological standards in training and transfer experiments. *Psychological Research, 78*, 756–772.

Green, C. S., Bavelier, D., Kramer, A. F., Vinogradov, S., Ansorge, U., Ball, K. K., ... & Facoetti, A. (2019). Improving methodological standards in behavioral interventions for cognitive enhancement. *Journal of Cognitive Enhancement, 3*, 2–29.

Hardy, J. L., Nelson, R. A., Thomason, M. E., Sternberg, D. A., Katovich, K., Farzin, F., & Scanlon, M. (2015). Enhancing cognitive abilities with comprehensive training: A large, online, randomized, active-controlled trial. *PLoS One, 10*, e0134467.

Harris, D. J., Wilson, M. R., & Vine, S. J. (2018). A systematic review of commercial cognitive training devices: Implications for use in sport. *Frontiers in Psychology, 9*, 709.

Hitchcock, C., & Westwell, M. S. (2017). A cluster-randomised, controlled trial of the impact of Cogmed working memory training on both academic performance and regulation of social, emotional and behavioural challenges. *Journal of Child Psychology and Psychiatry, 58*, 140–150.

Hogrefe, A., Studer-Luethi, B., Kodzhabashev, S., & Perrig, W. J. (2018). Mechanisms underlying n-back training: Response consistency during training influences training outcome. *Journal of Cognitive Enhancement, 1*, 353–357.

Jaeggi, S. M., Buschkuehl, M., Jonides, J., & Perrig, W. J. (2008). Improving fluid intelligence with training on working memory. *Proceedings of the National Academy of Sciences, 105*, 6829–6833.

Kable, J. W., Caulfield, M. K., Falcone, M., McConnell, M., Bernardo, L., Parthasarathi, T., ... & Diefenbach, P. (2017). No effect of commercial cognitive training on brain activity, choice behavior, or cognitive performance. *Journal of Neuroscience, 37*, 7390–7402.

Kelly, M. E., Loughrey, D., Lawlor, B. A., Robertson, I. H., Walsh, C., & Brennan, S. (2014). The impact of cognitive training and mental stimulation on cognitive and everyday functioning of healthy older adults: A systematic review and meta-analysis. *Ageing Research Reviews, 15*, 28–43.

Kueider, A. M., Parisi, J. M., Gross, A. L., & Rebok, G. W. (2012). Computerized cognitive training with older adults: A systematic review. *PLoS One, 7*, e40588.

Kupferberg, A., Koj, A., & Radeloff, A. (2019). Auditorisches Training verbessert Sprachverstehen und kognitive Leistung. *HNO Nachrichten, 49*, 32–37.

Lampit, A., Hallock, H., & Valenzuela, M. (2014). Computerized cognitive training in cognitively healthy older adults: A systematic review and meta-analysis of effect modifiers. *PLoS Medicine, 11*, e1001756.

Leung, N. T., Tam, H. M., Chu, L. W., Kwok, T. C., Chan, F., Lam, L. C., ... & Lee, T. (2015). Neural plastic effects of cognitive training on aging brain. *Neural plasticity*, 535618.

Mahncke, H. W., Connor, B. B., Appelman, J., Ahsanuddin, O. N., Hardy, J. L., Wood, R. A., ... & Merzenich, M. M. (2006). Memory enhancement in healthy older adults using a brain plasticity-based training program: A randomized, controlled study. *Proceedings of the National Academy of Sciences, 103*, 12523–12528.

Mayas, J., Parmentier, F. B. R., Andrés, P., & Ballesteros, S. (2014). Plasticity of attentional functions in older adults after non-action video game training: A randomized controlled trial. *PLoS One, 9*, e92269.

McLaughlin, P. M., Curtis, A. F., Branscombe-Caird, L. M., Comrie, J. K., & Murtha, S. J. E. (2018). The feasibility and potential impact of brain training games on cognitive and emotional functioning in middle-aged adults. *Games For Health Journal, 7*, 67–74.

Nouchi, R., Taki, Y., Takeuchi, H., Hashizume, H., Nozawa, T., Kambara, T., ... & Kawashima, R. (2013). Brain training game boosts executive functions, working memory and processing speed in the young adults: A randomized controlled trial. *PLoS One, 8*, e55518.

Olesen, P. J., Westerberg, H., & Klingberg, T. (2004). Increased prefrontal and parietal activity after training of working memory. *Nature Neuroscience, 7*, 75–79.

Owen, A. M., Hampshire, A., Grahn, J. A., Stenton, R., Dajani, S., Burns, A. S., ... & Ballard, C. G. (2010). Putting brain training to the test. *Nature, 465*, 775–778.

Papp, K. V., Walsh, S. J., & Snyder, P. J. (2009). Immediate and delayed effects of cognitive interventions in healthy elderly: A review of current literature and future directions. *Alzheimer's & Dementia, 5*, 50–60.

Peretz, C., Korczyn, A. D., Shatil, E., Aharonson, V., Birnboim, S., & Giladi, N. (2011). Computer-based, personalized cognitive training versus classical computer games: A randomized double-blind prospective trial of cognitive stimulation. *Neuroepidemiology, 36*, 91–99.

Pichora-Fuller, M. K. (2015). Cognitive decline and hearing health care for older adults. *American Journal of Audiology, 24*, 108–111.

Rabipour, S., & Raz, A. (2012). Training the brain: Fact and fad in cognitive and behavioral remediation. *Brain and Cognition, 79*, 159–179.

Rossignoli-Palomeque, T., Perez-Hernandez, E., & González-Marqués, J. (2018). Brain training in children and adolescents: Is it scientifically valid? *Frontiers in Psychology, 9*, 565.

Saunders, G., Smith, S. L., Chisolm, T. H., Frederick, M. T., McArdle, R. A., & Wilson, R. H. (2016). A randomized control trial: Supplementing hearing aid use with listening and communication enhancement (LACE) auditory training. *Ear and Hearing, 37*, 381–396.

Shah, T. M., Weinborn, M., Verdile, G., Sohrabi, H. R., & Martins, R. N. (2017). Enhancing cognitive functioning in healthy older adults: A systematic review of the clinical significance of commercially available computerized cognitive training in preventing cognitive decline. *Neuropsychology Review, 27*, 62–80.

Shatil, E. (2013). Does combined cognitive training and physical activity training enhance cognitive abilities more than either alone? A four-condition randomized controlled trial among healthy older adults. *Frontiers in Aging Neuroscience, 5*, 8.

Simons, D. J., Boot, W. R., Charness, N., Gathercole, S. E., Chabris, C. F., Hambrick, D. Z., & Stine-Morrow, E. A. L. (2016). Do "Brain-Training" programs work? *Psychological Science in the Public Interest, 17*, 103–186.

Simpson, T., Camfield, D., Pipingas, A., Macpherson, H., & Stough, C. (2012). Improved processing speed: Online computer-based cognitive training in older adults. *Educational Gerontology, 7*, 445–458.

Smith, G. E., Housen, P., Yaffe, K., Ruff, R., Kennison, R. F., Mahncke, H. W., & Zelinski, E. M. (2009). A cognitive training program based on principles of brain plasticity: Results from the Improvement in Memory with Plasticity-based Adaptive Cognitive Training (IMPACT) study. *Journal of the American Geriatrics Society, 57*, 594–603.

Strenziok, M., Parasuraman, R., Clarke, E., Cisler, D. S., Thompson, J. C., & Greenwood, P. M. (2014). Neurocognitive enhancement in older adults: Comparison of three cognitive training tasks to test a hypothesis of training transfer in brain connectivity. *NeuroImage, 85*, 1027–1039.

Strobach, T., & Huestegge, L. (2017). Evaluating the effectiveness of commercial brain game training with working-memory tasks. *Journal of Cognitive Enhancement, 1*, 539–558.

Studer-Luethi, B., Jaeggi, S. M., Buschkuehl, M., & Perrig, W. J. (2012). Influence of neuroticism and conscientiousness on working memory training outcome. *Personality and Individual Differences, 53*, 44–49.

Sweetow, R. W., & Sabes, J. H. (2006). The need for and development of an adaptive Listening and Communication Enhancement (LACE) program. *Journal of the American Academy of Audiology, 17*, 538–558.

Tetlow, A. M., & Edwards, J. D. (2017). Systematic literature review and meta-analysis of commercially available computerized cognitive training among older adults. *Journal of Cognitive Enhancement, 1*, 559–575.

Torous, J., Staples, P., Fenstermacher, E., Dean, J., & Keshavan, M. (2016). Barriers, benefits, and beliefs of brain training smartphone apps: An internet survey of younger US consumers. *Frontiers in Human Neuroscience, 10*, 180.

Tremblay, K. L. (2007). Training-related changes in the brain: Evidence from human auditory-evoked potentials. *Seminars in Hearing, 28*, 120–132.

Turner, E. H., Matthews, A. M., Linardatos, E., Tell, R. A., & Rosenthal, R. (2008). Selective publication of antidepressant trials and its influence on apparent efficacy. *The New England Journal of Medicine, 358*, 252–260.

Zelinski, E. M., Spina, L., Yaffe, K., Ruff, R., Kennison, R., Mahncke, H., & Smith, G. (2011). Improvement in memory with plasticity-based adaptive cognitive training: Results of the 3-month follow-up. *Journal of the American Geriatrics Society, 59*, 258–265.

Zelinski, E. M., Peters, K. D., Hindin, S., Petway, K. T., & Kennison, R. F. (2014). Evaluating the relationship between change in performance on training tasks and on untrained outcomes. *Frontiers in Human Neuroscience, 8*, 617.

Music Training

Swathi Swaminathan and E. Glenn Schellenberg

Contents

Abstract Positive correlations often emerge when researchers ask whether music lessons influence nonmusical cognitive abilities. Experimental studies tend to yield small effects, however, or results that are unlikely to generalize broadly. Here, we review recent empirical studies and suggest that future research could benefit by considering (1) whether transfer effects of music training are domain general or domain specific, (2) mechanisms of transfer, (3) characteristics of the training program, (4) characteristics of the trainee, and (5) the sociocultural context in which the training and research is conducted.

Over the last two decades, researchers have examined whether taking music lessons has a positive influence on nonmusical cognitive abilities. Such an influence would represent a form of *transfer*. The most common design (i.e., correlational) involves comparing musically trained and untrained individuals, which makes it impossible to determine whether music lessons are the cause rather than consequence of improved cognitive performance. Nevertheless, psychologists and neuroscientists routinely but erroneously infer causation from the results of correlational studies

S. Swaminathan · E. G. Schellenberg (✉)
Department of Psychology, University of Toronto Mississauga, Mississauga, ON, Canada

Natural Sciences and Engineering Research Council of Canada, Ottawa, ON, Canada
e-mail: g.schellenberg@utoronto.ca

T. Strobach, J. Karbach (eds.), *Cognitive Training*,
https://doi.org/10.1007/978-3-030-39292-5_21

(Schellenberg 2019; see also Cochrane and Green, this volume), which creates confusion among researchers, the media, and the general public.

True experiments with random assignment are relatively rare because they are costly and because attrition limits the possibility of long-term studies. Experimental studies also tend to yield results that are limited in scope or much smaller effects than the associations reported in correlational studies (for reviews see Schellenberg and Weiss 2013; Swaminathan and Schellenberg 2014). In the present chapter, we review studies published since 2000, with an emphasis on those that inform the issue of causation. We highlight five issues that future research could seek to clarify: (1) whether transfer effects are domain general or domain specific, (2) mechanisms of transfer, (3) characteristics of the music program, (4) characteristics of the trainee, and (5) the sociocultural context.

Domain-General or Domain-Specific Transfer?

One longstanding question asks whether music lessons have putative effects that transfer to *specific* cognitive domains (e.g., visuospatial skills, language abilities) or whether they might enhance domain-*general* cognitive abilities, such as executive functions and intelligence. Correlational evidence documents that musically trained individuals exhibit advantages relative to their untrained counterparts on a wide variety of visuospatial tasks (for review see Schellenberg and Weiss 2013). Longitudinal and experimental results offer a less consistent picture.

For example, one study examined children from families with low socioeconomic status who were having difficulties in school (Portowitz et al. 2009). The children were enrolled in remedial programs at four different after-school centers. Three of these incorporated a 2-year music-enrichment program, which included 2–3 hours per week of music listening, individual instrumental lessons, and group performances. Compared to children at the center without the program, children who received the intervention showed larger improvements in the ability to remember and copy a complex line drawing. Nevertheless, *nonmusical* programs of similar intensity could have a similar effect, and randomization of centers rather than individuals (as in Jaschke et al. 2018) raises the possibility that other differences among centers may have played a role. Moreover, in another study that compared an intensive, 4-week, computerized, music-listening program to a similar program in visual art (Moreno et al. 2011), improvement from pre- to post-test on a visuospatial task (Block Design) did not differ between the two groups of children.

Other scholars argue for specific connections between music training and language skills. Relevant theories suggest that music training fine-tunes listening abilities, which lead to improvements in speech perception in particular, which ultimately have cascading effects that extend to higher-level language abilities such as reading (e.g., Kraus and Chandrasekaran 2010; Patel 2011). This perspective implies that linguistic rather than visuospatial skills are most likely to improve from music training. Supporting evidence indicates that music training is correlated with a wide range of speech skills (for review see Schellenberg and Weiss 2013), including linguistic stress processing, the perception of intonation in speech, speech segmentation, and phonological perception. It is unclear why musicians are better than nonmusicians at perceiving speech in noise in some instances (Parbery-Clark et al. 2009; Tierney et al. 2019) but not in others (Boebinger et al. 2015; Madsen et al. 2019). Musically trained individuals also show advantages on higher-level language tests such as those that measure verbal short-term, long-term, and working memory; vocabulary; reading; and acquisition of a second language (for review see Schellenberg and Weiss 2013).

Nevertheless, associations between music training and language abilities can disappear when music aptitude or IQ is held constant (Swaminathan and Schellenberg 2017; Swaminathan et al. 2018). Convincing evidence for causation—from longitudinal studies with random assignment—is also limited. For example, in one instance, improvements on a brief test of vocabulary (Moreno et al. 2011) were larger among children who took 4 weeks of daily training in music listening compared to children who took a similar amount of training in visual arts. In another instance, 6 months of music or painting training led to larger improvements in pronouncing irregularly spelled words among children taking the music lessons (Moreno et al. 2009). Two other experimental studies found that phonological awareness was enhanced after music training (Degé and Schwarzer 2011; Flaugnacco et al. 2015). Other evidence of positive effects on phonological awareness, auditory memory, or vocabulary came from longitudinal studies *without* random assignment, which allowed self-selection to play a role (Linnavalli et al. 2018; Roden et al. 2012). In sum, associations between music training and language abilities are well documented, and music training could, in principle, play a causal role. Experimental evidence that allows for unambiguous causal inferences is limited, however, to outcome variables that measure very narrow aspects of reading or language use (e.g., phonological awareness).

If music training is associated with both visuospatial *and* language skills, might variance in all three domains (music, visuospatial, and language) be a consequence of general cognitive abilities? Or does music training have widespread transfer effects that influence such abilities, which include intelligence and executive functions? General cognitive improvements could manifest as improvements in specific cognitive abilities whether or not they are attributable to music lessons.

Correlational evidence confirms that musically trained children and adults often have substantially higher IQ scores than their untrained counterparts and that additional music training predicts larger IQ advantages (for review see Schellenberg and Weiss 2013). For example, Canadian children with music training can have IQ scores that are one standard deviation higher than their untrained counterparts (Schellenberg and Mankarious 2012), whereas Finnish adult musicians can have IQs that are one-third of a standard deviation higher than nonmusicians (Criscuolo et al. 2019).

Experimental evidence from three different countries indicates that music lessons may cause small improvements in IQ scores. For example, when Canadian 6-year-olds were randomly assigned to 1 year of music lessons (keyboard or voice) or to control conditions (drama or no lessons at all), larger pre- to post-test improvements in IQ were evident in the two music groups compared to the two control groups (Schellenberg 2004). In studies conducted in Iran and Israel, children who were assigned to a music intervention had larger gains in IQ compared to control groups with no lessons (Kaviani et al. 2014; Portowitz et al. 2009). Although the generality across cultures is reassuring, it is not clear from the Iranian and Israeli results whether the increase in IQ scores was a consequence of *music training* per se, because the control groups had no comparable, nonmusical experience (i.e., there was no "active" control group; Schmiedek, this volume), which means that other aspects of the music programs may have contributed to the findings. In short, good evidence that music training causes small increases in general cognitive ability comes from a single study, yet these results could not be replicated in a large sample of children living in the UK (Haywood et al. 2015). Moreover, a recent meta-analysis reported a negative correlation between quality of design and the size of the effect: the better the design (e.g., random assignment, active control group), the *smaller* the cognitive advantage for children who receive music training, which implies that substantial "effects" are actually the consequence of sub-optimal designs (Sala and Gobet 2017b). Another recent review of the literature found "suggestive" evidence of beneficial by-products of music lessons in childhood, but failed to draw any clear conclusions (Dumont et al., 2017).

Even in correlational studies, music training sometimes has only a marginal or no association with IQ (Schellenberg and Moreno 2010). For example, null or mixed results often occur when highly trained musicians are compared with individuals who have similar amounts of nonmusical training or education (e.g., Brandler and Rammsayer 2003; Helmbold et al. 2005). Moreover, in a recent study, preschool children were assigned to either 6 weeks of group music lessons or no lessons at all (Mehr et al. 2013). The music training had no reliable effects on cognitive abilities. In this instance, however, the children may have been too young for music lessons, or the training may have been too brief (4.5 hours total).

In any event, the available findings make it difficult to attribute most of the effects observed in correlational studies to music lessons, because (1) one would expect such effects to be particularly reliable among individuals with the greatest amount of training and (2) effect sizes from actual experiments are much smaller than those that are typically reported in correlational studies. A simpler explanation is that children who take music lessons, and adults with a history of music training, differ from other individuals in multiple ways, including cognitive abilities, personality, and demographic variables. In some instances, however, music training may exaggerate individual differences that were present before the lessons began.

Mechanisms of Transfer

Although researchers have identified associations between music training and higher-level cognitive abilities, it is unclear why such associations would emerge (Colzato and Hommel, Taatgen, this volume). Indeed, evidence for far transfer—between distantly related domains—is elusive, whether the training focuses on music, chess, or working memory (Sala and Gobet 2017a). Some researchers suggest, however, that music lessons train executive functions, including working memory, which in turn promote general cognitive enhancements (e.g., Schellenberg and Peretz 2008).

Indeed, in some instances, musically trained individuals outperform their untrained counterparts on auditory and non-auditory tests measuring executive functions (Roden et al. 2014; Zuk et al. 2014). Moreover, in one case, the association between music training and IQ appeared to be completely mediated by executive functions (Degé et al. 2011). In another instance, however, music training was associated with IQ but not with executive functions except for working memory (Schellenberg 2011). In a recent longitudinal study of children from underprivileged backgrounds, those who took music lessons after school exhibited an enhanced ability to delay gratification compared to their counterparts who took sports or no after-school program (Hennessy et al. 2019). The effect was weak and transient, however, appearing on only one of two tasks, and evident after 3 years of training but not after 4 years. The music group also improved from 2 to 3 years on a test of response inhibition. Because there was no random assignment and an attrition rate of 32%, the findings might actually suggest that less impulsive children were more likely than other children to take music lessons for years on end. In short, it is still an open question whether the association between music training and general cognitive ability is mediated by executive functions.

Other researchers suggest that music lessons train the auditory brainstem to make high-fidelity copies of auditory stimuli (Kraus and Chandrasekaran 2010). These subcortical changes are often correlated with speech and higher-level language skills including reading and are thought to mediate the language benefits of music training. In line with this hypothesis, musically trained individuals exhibit more precise brainstem responses to speech stimuli (Kraus et al. 2014; Strait et al. 2014). It remains to be seen, however, whether brainstem responses actually mediate any associations between music lessons and language.

A different mechanistic explanation of links between music training and language comes from the OPERA hypothesis (Patel 2011), which posits that music lessons train speech skills when five conditions are met: (1) the speech skill shares a neural overlap (*O*) with a music skill, (2) the music skill involves particularly precise (*P*) auditory processing, (3) the music training has positive emotional (*E*) consequences, (4) the lessons involve repetition (*R*), and (5) the lessons require focused attention (*A*). This theory is largely untested, and it is unclear whether these five conditions are necessary and sufficient for transfer and/or whether transfer is contingent on all five conditions being met.

Another view holds that overlap between language and music abilities occurs primarily in the temporal domain (Goswami 2012; Tallal and Gaab 2006), which implies that *rhythm*-based music interventions are most likely to be effective in training language skills. Evidence consistent with this theory comes from a study of children with dyslexia who were assigned to 6 weeks of auditory rhythm training, to a commercially available phoneme-discrimination intervention, or to a control group (Thomson et al. 2013). Compared to the control group, the rhythm and phoneme groups improved more on tests of phonological processing over the course of the study. In another experimental study of children with dyslexia and an active (painting) control group, 30 weeks of rhythm-based music training improved phonological awareness and pre-reading skills (Flaugnacco et al. 2015). Rhythm-perception abilities are also associated positively with grammatical abilities among typically developing children (Gordon et al. 2015), although the association extends to other tests of language ability (speech perception) and other tests of musical ability (memory for music; Swaminathan and Schellenberg 2019).

Meta-analyses of older adults suggest that music practice may enhance healthy aging by way of specific training mechanisms (i.e., those that are learned during practice), specific compensatory mechanisms (i.e., those that improve specific cognitive problems), and general compensatory mechanisms (i.e., those that improve general cognitive functioning; Román-Caballero et al. 2018), yet it is unknown how much musical expertise is required to predict beneficial effects and whether any benefits continue after interventions have ended (Christie et al. 2017). Future research could focus on evaluating and comparing the different mechanistic explanations of links between music training and nonmusical abilities, as well as on constructing new theories that generate empirically testable hypotheses. Theoretical multiplicity will undoubtedly promote debate and growth in the field.

Characteristics of the Music-Training Program

Private and small-group music lessons emphasize individual accomplishment and skill mastery. Larger, group-based lessons, by contrast, are more likely to emphasize collective outcomes. It is therefore possible that private music training is more effective than group-based lessons at improving scores on tests of cognitive ability, which by definition measure individual ability and accomplishment. Indeed, a recent longitudinal study of group-based music lessons found that advantages emerged only after extended training (Slater et al. 2015). Specifically, after 2 years of lessons, children demonstrated improved performance on a test that measured the ability to perceive speech in the midst of background noise. A separate group of children, who received 1 year of the same lessons, did not show improvement on the same test.

Other experimental studies with individual lessons or lessons taught in small groups have found advantages even with shorter-term interventions, such as when lessons are taught daily for 2 weeks (Moreno et al. 2011), daily for 20 weeks (Degé and

Schwarzer 2011), weekly for 36 weeks (Schellenberg 2004; Thomson et al. 2013), or twice weekly for 30 weeks (Flaugnacco et al. 2015). It is important to note, however, that in the short-term studies with daily training, the lessons focused primarily on music listening rather than learning to play an instrument. In other words, music lessons may be more likely to improve language-related outcomes if the lessons emphasize listening skills. As noted earlier, language benefits could also be more likely if the lessons target rhythm skills (Flaugnacco et al. 2015; Thomson et al. 2013). In any event, many successful music interventions adopted nonstandard pedagogies, which limit the degree to which the findings generalize (Degé and Schwarzer 2011; Flaugnacco et al. 2015; Moreno et al. 2011; Thomson et al. 2013).

Studies of older adults (Guye et al., this volume) and very young children (Rueda et al., this volume) provide converging evidence that characteristics of the music program are an important consideration. For older adults, a recent meta-analysis concluded that the specific focus of instrumental training can differentially affect the consequences of the intervention (Kim and Yoo 2019). For children attending kindergarten, positive associations with language abilities (vocabulary, phonological awareness) emerge after 2 years of music *playschool* (Linnavalli et al. 2018). In short, efficacious interventions need to be age-appropriate and designed specifically for the intended cognitive benefits.

Characteristics of the Trainee

Music training is correlated with cognitive skills in some samples of individuals but not in others (cf. Katz et al., this volume). As noted, highly trained musicians often do not show an IQ advantage compared to equally qualified individuals in nonmusical domains (Brandler and Rammsayer 2003; Helmbold et al. 2005). Thus, the association with general cognitive abilities may emerge primarily when music training is an additional activity rather than an individual's primary focus.

The probability that music training has positive side effects might also increase when the trainee (1) does not come from a privileged background (Barbaroux et al. 2019); (2) experiences atypical developmental trajectories, such as children with dyslexia (Flaugnacco et al. 2015; cf. de Vries and Geurts, this volume); or (3) is very young (Bowmer et al. 2018; cf. Ruede et al., this volume) or very old (Kim and Yoo 2019; cf. Guye et al., this volume). Indeed, two recent reviews reached a similar conclusion: consideration of individual differences is essential for documenting whether music training has actual cognitive benefits (Benz et al. 2016; Costa-Giomi 2015; see Karbach and Kray; Katz et al., this volume).

Other findings suggest that the association between music lessons (or musical involvement) and cognitive ability may be explained by personality factors, particularly the dimension called Openness-to-Experience (Corrigall et al. 2013; Corrigall and Schellenberg 2015; cf. Katz et al., this volume), which is characterized by curiosity, intellectual engagement, and aesthetic sensitivity. These findings imply that musically trained individuals may perform well on intelligence tests because they tend to be particularly interested in learning new things, including music. Moreover, common genetic factors appear to underlie intelligence *and* the propensity to practice music (Mosing et al. 2016).

In short, correlations between music training and cognitive ability may stem from preexisting differences. When considered jointly with evidence for small cognitive benefits of music training (e.g., Schellenberg 2004), it is likely that some individuals benefit more than others from music lessons (cf. Karbach and Kray, this volume). More generally, the study of music training and transfer is well suited to exploring gene-environment interactions (Schellenberg 2015). Future research could consider how preexisting trainee characteristics interact with music training to influence cognitive outcomes.

The Sociocultural Context

The issue of transfer effects from music training to nonmusical cognitive skills has clear practical implications. For example, music interventions may provide an enjoyable way for children with dyslexia to improve reading-related skills (Flaugnacco et al. 2015; Thomson et al. 2013). The study of transfer also has the potential to influence the nature of training and music. Across cultures, music and teaching occupy different places in social life and in their relation to other activities. With a few exceptions (e.g., Kaviani et al. 2014; Swaminathan and Gopinath 2013; Yang et al. 2014), most investigations of transfer have focused on samples of Western individuals learning Western music, which raises the possibility that many findings are Western-specific. Unlike most other cognitive-training programs, music and music training are cultural products that are meaningful in different ways to different individuals (see Colzato and Hommel, this volume).

Music lessons require time, effort, and money. Parents, educators, and policy makers are often motivated to invest in music lessons so that children develop their musical talents and improve their nonmusical skills, such as focus, attention, intelligence, literacy, and school performance. As a result, economic pressures could cause certain types of music programs to be privileged over others. For example, if school-based group lessons are not particularly effective at training nonmusical skills, they could lose financial backing, which would affect who has access to music lessons and what kind of lessons. In sum, because we are dealing with a real-world form of training nested in cultural contexts, the line between the laboratory and real world cannot be neatly defined. It is therefore important that research on music training and transfer becomes a multidisciplinary effort that considers the cultural contexts of producers and consumers of such research and training.

Conclusion

Despite having received much research and media attention, studies of transfer effects of music lessons have predominantly involved correlational designs, which makes it impossible to determine whether music lessons are the cause rather than consequence of improved cognitive performance. Moreover, the relatively small number of experimental and longitudinal studies that exist tends to report small, limited, or mixed effects. As a way forward, future research could examine the extent to which music lessons train general and specific cognitive abilities, the mechanisms by which such transfer occurs, the characteristics of the trainee and training program, and the larger social context in which such training is received.

References

Barbaroux, M., Dittinger, E., & Besson, M. (2019). Music training with De'mos program positively influences cognitive functions in children from low socio-economic backgrounds. *PLoS One, 14*, e0216874.

Benz, S., Sellaro, R., Hommel, B., & Colzato, L. S. (2016). Music makes the world go round: The impact of musical training on non-musical cognitive functions—A review. *Frontiers in Psychology, 6*, 2023.

Boebinger, D., Evans, S., Rosen, S., Lima, C. F., Manly, T., & Scott, S. K. (2015). Musicians and non-musicians are equally adept at perceiving masked speech. *Journal of the Acoustical Society of America, 137*, 378–387.

Bowmer, A., Mason, K., Knight, J., & Welch, G. (2018). Investigating the impact of a musical intervention on preschool children's executive function. *Frontiers in Psychology, 9*, 2389.

Brandler, S., & Rammsayer, T. H. (2003). Differences in mental abilities between musicians and non-musicians. *Psychology of Music, 31*, 123–138.

Christie, G. J., Hamilton, T., Manor, B. D., Farb, N. A. S., Farzan, F., Sixsmith, A., Temprado, J. J., & Moreno, S. (2017). Do lifestyle activities protect against cognitive decline in aging? A review. *Frontiers in Aging Neuroscience, 9*, 381.

Corrigall, K. A., & Schellenberg, E. G. (2015). Predicting who takes music lessons: Parent and child characteristics. *Frontiers in Psychology, 6*, 282.

Corrigall, K. A., Schellenberg, E. G., & Misura, N. M. (2013). Music training, cognition, and personality. *Frontiers in Psychology, 4*, 222.

Costa-Giomi, E. (2015). The long-term effects of childhood music instruction on intelligence and general cognitive abilities. *Update: Applications of Research in Music Education, 33*, 20–26.

Criscuolo, A., Bonetti, L., Särkämö, T., Kliuchko, M., & Brattico, E. (2019). On the association between musical training, intelligence and executive functions in adulthood. *Frontiers in Psychology, 10*, 1704.

Degé, F., & Schwarzer, G. (2011). The effect of a music program on phonological awareness in preschoolers. *Frontiers in Psychology, 2*, 124.

Degé, F., Kubicek, C., & Schwarzer, G. (2011). Music lessons and intelligence: A relation mediated by executive functions. *Music Perception, 29*, 195–201.

Dumont, E., Syurina, E. V., Feron, F. J. M., & van Hooren, S. (2017). Music interventions and child development: A critical review and further directions. *Frontiers in Psychology, 8*, 1694.

Flaugnacco, E., Lopez, L., Terribili, C., Montico, M., Zoia, S., & Schön, D. (2015). Music training increases phonological awareness and reading skills in developmental dyslexia: A randomized control trial. *PLoS One, 10*, e0138715.

Gordon, R. L., Shivers, C. M., Wieland, E. A., Kotz, S. A., Yoder, P. J., & McAuley, J. D. (2015). Musical rhythm discrimination explains individual differences in grammar skills in children. *Developmental Science, 18*, 635–644.

Goswami, U. (2012). Language, music, and children's brains: A rhythmic timing perspective on language and music as cognitive systems. In P. Rebuschat, M. Rohrmeier, J. A. Hawkins, & I. Cross (Eds.), *Language and music as cognitive systems* (pp. 292–301). Oxford, UK: Oxford University Press.

Haywood, S., Griggs, J., Lloyd, C., Morris, S., Kiss, Z., & Skipp, A. (2015). *Creative futures: Act, sing, play. Evaluation report and executive summary*. London: Educational Endowment Foundation.

Helmbold, N., Rammsayer, T., & Altenmüller, E. (2005). Differences in primary mental abilities between musicians and nonmusicians. *Journal of Individual Differences, 26*, 74–85.

Hennessy, S. L., Sachs, M. E., Ilari, B., & Habibi, A. (2019). Effects of music training on inhibitory control and associated neural networks in school-aged children: A longitudinal study. *Frontiers in Neuroscience, 13*, 1080.

Jaschke, A. C., Honing, H., & Scherder, E. J. A. (2018). Longitudinal analysis of music education on executive functions in primary school children. *Frontiers in Neuroscience, 12*, 103.

Kaviani, H., Mirbaha, H., Pournaseh, M., & Sagan, O. (2014). Can music lessons increase the performance of preschool children in IQ tests? *Cognitive Processing, 15*, 77–84.

Kim, S. J., & Yoo, G. E. (2019). Instrument playing as a cognitive intervention task for older adults: A systematic review and meta-analysis. *Frontiers in Psychology, 10*, 151.

Kraus, N., & Chandrasekaran, B. (2010). Music training for the development of auditory skills. *Nature Reviews Neuroscience, 11*, 599–605.

Kraus, N., Slater, J., Thompson, E. C., Hornickel, J., Strait, D. L., Nicol, T., & White-Schwoch, T. (2014). Music enrichment programs improve the neural encoding of speech in at-risk children. *Journal of Neuroscience, 34*, 11913–11918.

Linnavalli, T., Putkinen, V., Lipsanen, J., Huotilainen, M., & Tervaniemi, M. (2018). Music playschool enhances children's linguistic skills. *Scientific Reports, 8*, 8767.

Madsen, S. M. K., Marschall, M., Dau, T., & Oxenham, A. J. (2019). Speech perception is similar for musician and non-musicians across a wide range of conditions. *Scientific Reports, 9*, 10404.

Mehr, S. A., Schachner, A., Katz, R. C., & Spelke, E. S. (2013). Two randomized trials provide no consistent evidence for nonmusical cognitive benefits of brief preschool music enrichment. *PLoS One, 8*, e82007.

Moreno, S., Marques, C., Santos, A., Santos, M., Castro, S. L., & Besson, M. (2009). Musical training influences linguistic abilities in 8-year-old children: More evidence for brain plasticity. *Cerebral Cortex, 19*, 712–723.

Moreno, S., Bialystok, E., Barac, R., Schellenberg, E. G., Cepeda, N. J., & Chau, T. (2011). Short-term music training enhances verbal intelligence and executive function. *Psychological Science, 22*, 1425–1433.

Mosing, M. A., Madison, G., Pedersen, N. L., & Ullén, F. (2016). Investigating cognitive transfer within the framework of music practice: Genetic pleiotropy rather than causality. *Developmental Science, 19*, 504–512.

Parbery-Clark, A., Skoe, E., Lam, C., & Kraus, N. (2009). Musician enhancement for speech-in-noise. *Ear and Hearing, 30*, 653–661.

Patel, A. D. (2011). Why would musical training benefit the neural encoding of speech? The OPERA hypothesis. *Frontiers in Psychology, 2*, 142.

Portowitz, A., Lichtenstein, O., Egorova, L., & Brand, E. (2009). Underlying mechanisms linking music education and cognitive modifiability. *Research Studies in Music Education, 31*, 107–128.

Roden, I., Kreutz, G., & Bongard, S. (2012). Effects of a school-based instrumental music program on verbal and visual memory in primary school children: A longitudinal study. *Frontiers in Psychology, 3*, 572.

Roden, I., Grube, D., Bongard, S., & Kreutz, G. (2014). Does music training enhance working memory performance? Findings from a quasi-experimental longitudinal study. *Psychology of Music, 42*, 284–298.

Román-Caballero, R., Arnedo, M., Triviño, M., & Lupiáñez, J. (2018). Musical practice as an enhancer of cognitive function in healthy aging: A systematic review and meta-analysis. *PLoS One, 13*, e0207957.

Sala, G., & Gobet, F. (2017a). Does far transfer exist? Negative evidence from chess, music, and working memory training. *Current Directions in Psychological Science, 26*, 515–520.

Sala, G., & Gobet, F. (2017b). When the music's over: Does music skill transfer to children's and young adolescents' cognitive and academic skills? A meta-analysis. *Educational Research Review, 20*, 55–67.

Schellenberg, E. G. (2004). Music lessons enhance IQ. *Psychological Science, 15*, 511–514.

Schellenberg, E. G. (2011). Examining the association between music lessons and intelligence. *British Journal of Psychology, 102*, 283–302.

Schellenberg, E. G. (2015). Music training and speech perception: A gene–environment interaction. *Annals of the New York Academy of Sciences, 1337*, 170–177.

Schellenberg, E. G. (2019). Correlation = causation? Music training, psychology, and neuroscience. Advance online publication. *Psychology of Aesthetics, Creativity, and the Arts*. https://doi.org/10.1037/aca0000263.

Schellenberg, E. G., & Mankarious, M. (2012). Music training and emotion comprehension in childhood. *Emotion, 12*, 887–891.

Schellenberg, E. G., & Moreno, S. (2010). Music lessons, pitch processing and g. *Psychology of Music, 38*, 209–221.

Schellenberg, E. G., & Peretz, I. (2008). Music, language and cognition: Unresolved issues. *Trends in Cognitive Sciences, 12*, 45–46.

Schellenberg, E. G., & Weiss, M. W. (2013). Music and cognitive abilities. In D. Deutsch (Ed.), *The psychology of music* (3rd ed., pp. 499–550). Amsterdam: Elsevier.

Slater, J., Skoe, E., Strait, D. L., O'Connell, S., Thompson, E., & Kraus, N. (2015). Music training improves speech-in-noise perception: Longitudinal evidence from a community-based music program. *Behavioural Brain Research, 291*, 244–252.

Strait, D. L., O'Connell, S., Parbery-Clark, A., & Kraus, N. (2014). Musicians' enhanced neural differentiation of speech sounds arises early in life: Developmental evidence from ages 3 to 30. *Cerebral Cortex, 24*, 2512–2521.

Swaminathan, S., & Gopinath, J. K. (2013). Music training and second-language English comprehension and vocabulary skills in Indian children. *Psychological Studies, 58*, 164–170.

Swaminathan, S., & Schellenberg, E. G. (2014). Arts education, academic achievement, and cognitive ability. In P. P. L. Tinio & J. K. Smith (Eds.), *The Cambridge handbook of the psychology of aesthetics and the arts* (pp. 364–384). Cambridge, UK: Cambridge University Press.

Swaminathan, S., & Schellenberg, E. G. (2017). Musical competence and phoneme perception in a foreign language. *Psychonomic Bulletin & Review, 24*, 1929–1934.

Swaminathan, S., & Schellenberg, E. G. (2019). Music ability, music training, and language ability in childhood. Advance online publication. *Journal of Experimental Psychology: Learning, Memory, and Cognition*. https://doi.org/10.1037/xlm0000798.

Swaminathan, S., Schellenberg, E. G., & Venkatesan, K. (2018). Explaining the association between music training and reading in adults. *Journal of Experimental Psychology: Learning, Memory, and Cognition, 44*, 992–999.

Tallal, P., & Gaab, N. (2006). Dynamic auditory processing, musical experience and language development. *Trends in Neurosciences, 29*, 382–390.

Thomson, J. M., Leong, V., & Goswami, U. (2013). Auditory processing interventions and developmental dyslexia: A comparison of phonemic and rhythmic approaches. *Reading and Writing, 26*, 139–161.

Tierney, A., Rosen, S., & Dick, F. (2019). Speech-in-speech perception, nonverbal selective attention, and musical training. Advance online publication. *Journal of Experimental Psychology: Learning, Memory, and Cognition*. https://doi.org/10.1037/xlm0000767.
Yang, H., Ma, W., Gong, D., Hu, J., & Yao, D. (2014). A longitudinal study on children's music training experience and academic development. *Scientific Reports, 4*, 5854.
Zuk, J., Benjamin, C., Kenyon, A., & Gaab, N. (2014). Behavioral and neural correlates of executive functioning in musicians and non-musicians. *PLoS One, 9*, e99868.

Physical Activity and Exercise

Louis Bherer and Kristell Pothier

Contents

Abstract Decades of research have reported that physical activity and exercise training can help improve and maintain cognitive health across the human lifespan. While the first wave of studies in this field was dominated by the positive effect of aerobic training programs, more recent evidence supports the notion that a variety of exercise interventions, from strength training to mind-body exercise, also have a positive impact on cognition, leading to the question of what are the explanatory mechanisms that relate exercise to cognition. This chapter reviews some seminal and recent interventional studies that have attempted to assess whether and how specific physical activities impact cognition. A large number of interventional studies used aerobic and strength programs to investigate the impact of exercise on cognition and showed protective effects on executive functioning, modulated by specific biomarkers. More recently, it was suggested that new training approaches targeting gross motor abilities or using popular soft gymnastics, also designated as mind-body exercises (yoga, dance, tai chi), can also lead to improvements in cognitive functions. Overall, results reviewed here support the notion that physical exercise can contribute to develop and preserve cognitive health across the entire lifespan.

L. Bherer (✉)
Research Center, Montreal Heart Institute, Institut universitaire de gériatrie de Montréal, Department of Medicine, Université de Montréal, Montreal, QC, Canada
e-mail: louis.bherer@umontreal.ca

K. Pothier
EA 2114 – Psychologie des Âges de la Vie et Adaptation (PAVéA), Université de Tours, Tours, France

T. Strobach, J. Karbach (eds.), *Cognitive Training*,
https://doi.org/10.1007/978-3-030-39292-5_22

Introduction

For many years, studies have suggested that lifestyle factors have a significant impact on how well people age (see Kramer et al. 2004). More recently, leading medical organizations such as the Lancet Commission (Livingston et al. 2017), the American Academy of Neurology (Petersen et al. 2018), and the National Academies of Sciences, Engineering, and Medicine of the USA (Downey et al. 2017) have published position papers and recommendations for slowing and halting the development of dementia. All reports emphasized the notion that age-related cognitive decline and potentially dementia can be prevented through better lifestyle choices and management of medical risk factors (see also Guye et al., this volume). They also agreed that further studies are needed to better develop and prescribe lifestyle interventions, as well as to understand the mechanisms by which they enhance cognition and how they interact with other risk factors. In all these reports, physical activity and exercise play a central role and sometimes come first as the main strategy to enhance cognition and prevent age-related cognitive decline. Studies supporting the benefits of physical activity and exercise on cognition are numerous, and methodologies vary greatly with regard to the type of activity and fitness assessment. *Physical activity* is defined as any bodily movement produced by skeletal muscles and that results in energy expenditure. *Exercise* is a subtype of physical activity that is planned, structured, repetitive, and purposive (Caspersen et al. 1985). Physical activity assessment varies from direct observation and self-report questionnaires to pedometers, hip accelerometers, or ActiGraph systems. Physical fitness assessment should ideally involve a submaximal walking test (Kline et al. 1987) or a graded physical exercise test (on a treadmill or a cycle ergometer) to estimate or to provide a direct measure of VO_2max, the ability of the body to transport and use oxygen during intense effort (Betik and Hepple 2008), which is considered as the gold standard for a cardiorespiratory fitness index.

Despite methodological differences among studies, systematic reviews and meta-analyses of longitudinal studies support the association between physical activity and cognitive performances in older adults, with regular practice of physical activity being associated with a reduced risk of cognitive decline and dementia by 20% (Weuve et al. 2004) to 35% (Sofi et al. 2011). In fact, it seems that the risk of cognitive decline is inversely proportional to the amount of physical activity practiced throughout life (Paillard et al. 2015). However, it still remains unclear whether all types of physical activity impact cognition to the same extent or carry the same promise in terms of protection against cognitive decline. Moreover, although some recent reports from the Framingham Heart Study suggest that even light physical activity is associated with total cerebral brain volume in older adults (Spartano et al. 2019), the impact of exercise on biomarkers of cognitive decline and dementia still seems understudied (Jensen et al. 2015), and evidence remains, to date, crucially lacking.

This chapter reviews the literature assessing the impact of physical activity and exercise training as an effective way to improve cognitive performances with a specific focus on neurologically healthy older adult population. It does not claim to be an exhaustive review as the body of literature in this field of investigation that has blossomed at an incredible pace over the last few years. Rather, it aims to provide the reader with an overview of the recent advancement in scientific knowledge, mainly based on interventional studies assessing whether and how physical activity and exercise positively affect cognition. More specifically, we opted to focus on what type of specific physical activity or exercise has an impact on cognitive performances. Although more studies are required to find definitive answers to these important questions, recent interventional studies including various exercises and the use of neuroimaging techniques have opened new research avenues. The following sections report evidence of cognitive improvement in interventional studies using aerobic training, strength or resistance training, gross motor and coordination training, and more innovative exercise such as yoga, dance, and tai chi.

Aerobic Training

Aerobic training usually involves exercises, such as jogging, swimming, or brisk walking, that stimulate and strengthen the heart and lungs, with the goal of improving the body's utilization of oxygen. In a noteworthy study by Kramer et al. (1999), older adults who completed a 6-month walking program showed a greater significant improvement in tasks that tapped into attentional control or executive control functions, compared to other cognitive functions and to a control group (of non-aerobic stretching exercise). This observation was further supported by a meta-analysis by Colcombe and Kramer (2003) of interventional studies (Bherer et al. 2013). Many reports on the link between aerobic fitness and cognitive performances in older adults were recently published. For instance, a recent study (Bherer et al. 2019) investigated the impacts of physical exercise on single- and dual-task performance in younger-old (65–69) and older-old (70–89) adults and to what extent effects were mediated by improvement in cardiorespiratory fitness. One hundred forty-three participants took part in 3 months of aerobic training. In older-old participants, training predicted improved cardiorespiratory fitness and processing speed, and the effect of training on processing speed was fully mediated by change in cardiorespiratory fitness. In younger-old adults, training leads to improved cardiorespiratory fitness and task-set cost, which reflects the working memory load associated with dual-task situation. However, this cognitive improvement was not fully mediated by change in cardiorespiratory fitness, which suggests that other mechanisms induced by exercise come into play to explain the exercise-induced cognitive improvement.

The biological mechanisms by which cognition is enhanced through aerobic exercise remain to be fully elucidated (Etnier and Chang 2019) and likely involve multiple neuroplasticity mechanisms (see Wenger and Kühn, in this volume). In fact, it has been suggested that aerobic exercise can induce angiogenesis, neurogenesis, and synaptogenesis and supports for this are exercise-associated changes in molecular growth factors such as brain-derived neurotrophic factor (BDNF), which plays a crucial role in neuroplasticity and neuroprotection, and increased production of insulin-like growth factor 1 (IGF-1) (Yau et al. 2016). In some studies, these observations paralleled change in brain structures and functions. For instance, in Voss et al. (2010, 2013), 12 months of an aerobic exercise intervention induced improvement in functional connectivity, which was associated with enhanced executive functions, serum level BDNF, and other growth factors. Although reports of functional changes in the brain are not always consistent, the impacts of aerobic exercise on brain volume, at least for some structures like the hippocampus, have often been reported. For instance, Erickson et al. (2009) showed that volume of brain regions that play an important role in aging, such as the frontal lobes and the hippocampus, can be predicted by physical activity level. Indeed, a 1-year aerobic intervention can lead to increased hippocampus volume in older adults (Erickson et al. 2011). While these findings are promising, given the importance of this region in brain aging, others suggest that the level of evidence still calls for more support from randomized controlled trials. In fact, the link between brain marker modifications following exercise programs and cognitive performances is not always clear (Etnier and Chang 2019). Moreover, some studies suggest that the mechanisms by which exercise impacts brain functions remain to be elucidated. In this context, further research is needed to account for the complex role of cerebrovascular function in the link between physical activity and cognition.

In sum, although multiple studies show a link between exercise and cognition, confirmed by improvement in cognitive performance after an exercise intervention, not all studies support that there might be a dose-response relationship between exercise and cognition. In a very recent meta-analysis, Falck et al. (2019) reported a positive correlation between the size of the exercise-induced effect on physical function and on cognitive function, suggesting that both types of improvement are linked, although the mechanism by which they are associated is still under investigation.

Strength or Resistance Training

Strength or resistance training requires the body's musculature to move against an opposing force, usually using some type of training equipment. Recent studies also suggest that strength or resistance training can have beneficial effects on cognition, especially with older adults. Tsai et al. (2015) showed that a 12-month high-intensity resistance exercise intervention could effectively delay the decline in executive functions in healthy elderly males. In contrast, a meta-analysis by Kelly et al. (2014) only partially supported the benefits of resistance training on cognitive function of older adults, with resistance training groups performing better than stretching/toning controls on only 1 out of 3 memory measures and 4 out of 18 executive function tests. Better comparability in behavioral measures of executive functions across studies could help explain discrepant results among studies. A more recent study also supports the effect of resistance training on cognition. In fact, Marston et al. (2019) investigated the effect of 12-week resistance training programs on cognitive function in late middle-aged adults and showed that resistance training increased muscle strength and delayed verbal memory.

Other studies offer more support for the benefit of strength training exercise on brain plasticity. In a neuroimaging study, Liu-Ambrose et al. (2012) showed improved executive scores after a 6-month resistance program in conjunction with increased hemodynamic activity in brain regions associated with response inhibition, suggesting functional plasticity. This plasticity potentially could rely on a molecular pathway. In fact, Cassilhas et al. (2007) showed that both moderate- and high-intensity resistance trainings are associated with increased levels of IGF-1. Similarly, Tsai et al. (2015) showed an increase in serum IGF-1 levels after a long-term high-intensity resistance exercise program, which correlated with changes in reaction times and P3b amplitudes in an oddball task.

Gross Motor/Coordination Training

It has been observed that gross motor training involving coordination, balance, and agility activities can also lead to improvements in older adults' cognitive functions independently of aerobic fitness or resistance training. Voelcker-Rehage et al. (2011) reported improved executive control and perceptual speed after a 12-month coordination-training program in older adults (aged 62–79 years). Forte et al. (2013) also reported improved executive control in adults aged 65–75 years after a 3-month multicomponent training program (including balance and agility). Another study (Berryman et al. 2014) compared aerobic training combined with strength training to gross motor activities training and observed equivalent improvement in an inhibition task across all training programs, highlighting again the beneficial impact of gross motor training on cognition.

Little is known regarding biomarkers of gross motor/coordination training effects. Interestingly, using functional brain imaging approach, Voelcker-Rehage et al. (2011) reported results suggesting that the neurocognitive mechanisms that underlie cognitive changes induced by exercise training could differ depending on training intervention, with cardiovascular training being associated with an increased activation of the sensorimotor network, while coordination training led to an increased activation in the visuospatial network. In a recent study, our group observed that gross motor exercise could induce changes in circulating BDNF in healthy older adults, while aerobic exercise did not (Grégoire et al. 2019). While in appearance surprising, these results find support in a recent meta-analysis showing the effects of different types of exercises, such as of non-aerobic exercise on peripheral blood BNDF concentration (Marinus et al. 2019). Future studies are required to help further understand how gross motor and coordination training impacts specific cognitive functions.

Emerging Training Programs in Mind-Body Exercise: Yoga, Dance, and Tai Chi

There is a growing interest for more ecological-like interventions, based on popular soft gymnastics such as yoga, dance, tai chi, etc. This section mainly reports some interventional studies. Further research will be needed to complete this emerging literature, particularly with regard to potential biomarkers.

Yoga is a widely practiced form of meditation and relaxation with origins in ancient Indian philosophy. It includes postures combined with breathing techniques and is now classified as a form of Complementary and Alternative Medicine by the National Institutes of Health. In a literature review of studies with older adults, Patel et al. (2012) found no improvement in global cognition nor in alertness and attention after yoga training compared to control conditions. However, using a rigorous study design, Gothe et al. (2014) examined the effects of an 8-week Hatha yoga intervention on executive functions in 118 healthy older adults (55–79 years old). Results showed improved performance in measures of working memory, mental set shifting, and flexibility in the yoga group, but not in the stretching-strengthening control group. Gothe and McAuley (2015) reported on 15 randomized controlled trials and 7 acute exposure studies examining the effects of yoga on cognition and reported moderate effects of yoga intervention on attention and processing speed, executive function, and memory, but stronger effects of acute yoga exposure for memory, followed by attention and executive functions. In a more recent narrative review, Rivest-Gadbois and Boudrias (2019) also reported that yoga practice can help improve learning rate, speed, and accuracy, can help increase attentional skills and decrease stress, and seems to have a positive effect on memory. Regarding biomarkers, Pal et al. (2014) found a significant improvement in plasma BDNF level for males aged 20–50 who experimented yoga practice for 3 months, 1 hour per week. The impact of yoga on stress level biomarkers (e.g., cortisol), psychological well-being (e.g., serotonin and dopamine levels), and its potential beneficial effects on reducing the risk for cerebrovascular disease (e.g., reducing lipid profile and lower oxidative stress) need to be further investigated.

Dancing has gained interest in the scientific community. Compared to other activities such as aerobic or resistance training, dancing offers a combination of physical, cognitive, and social activities potentially useful for attenuating age-related decline. A review of the literature (McNeely et al. 2015) showed a strong effect of dance on cognition in older adults. The review included ten studies in which older adults underwent multi-session dance interventions (e.g., salsa, ballroom, contemporary, etc.) between 8 weeks and 18 months, with at least one session per week. Beside positive effects in quality of life (e.g., body pain, physical functioning, or life satisfaction), cognition was improved after dance intervention, with more specific effects on task switching and response speed. However, none of the dance interventions seemed to affect attention, visuospatial memory, or fluid intelligence. Here again, future studies will help clarify and understand the impact of dance on specific cognitive functions in seniors. Predovan et al. (2019) reported on 7 studies representing a total of 429 older adults (70% women), with a mean age of 73 years old, and observed that dance interventions lasting between 10 weeks and 18 months were related to either the maintenance or improvement of cognitive performance (mostly executive functions and memory). Multiple forms of dance interventions exist and might have multiple effects on cognition and health. For instance, while dance movement therapy does not seem to improve cognition in healthy seniors (Esmail et al. 2020), it does seem to improve markers of chronic stress like cortisol levels (Vrinceanu et al. 2019), which was not observed in a group completing an exercise training program. Another study also suggests that dance movement therapy, but not physical exercise, helped improve depression, loneliness, and negative mood, daily functioning, and diurnal cortisol slope in patients with mild cognitive impairment (Ho et al. 2020).

Tai chi, meaning "supreme ultimate boxing," is originally a martial art from China combining the exercise of rhythmic movement and self-defense practice with the use of breathing techniques of yoga. It involves moving from a standing position through a series of postures. Like yoga or dance, tai chi could be described as being a combined intervention, with physical, cognitive, and social characteristics occurring simultaneously. In older adults, Chang et al. (2010) found mixed results with regard to cognitive performance. However, a meta-analysis highlighted the impact of tai chi on attention, processing speed, and working memory in older adults (Kelly et al. 2014). Studies using neuroimaging techniques are still scarce, but Wei et al. (2013) found in middle-aged adults that compared with control participants who never practiced tai chi, those who had practiced tai chi for many years showed thicker cortex in the precentral gyrus, insula sulcus, and middle frontal sulcus in the right hemisphere and superior temporal gyrus, medial occipito-temporal sulcus, and lingual sulcus in the left hemisphere. Although this was a cross-sectional study, it provides partial support to the notion that long-term tai chi practice could induce regional structural changes. In a recent meta-analysis of 32 randomized controlled trials, Wu et al. (2019) reported that mind-body exercise, especially emphasizing tai chi, helps improve global cognition, cognitive flexibility, working memory, verbal fluency, and learning in older adults. However, given the wide variety of these types of interventions, further studies are needed to identify the active ingredient and the dose-response relationship that makes mind-body exercise interventions an efficient way to maintain cognition in older adults (see also Verhaeghen, in this volume).

What Would Be the Best Physical Activity/Exercise?

At this time, we still have insufficient evidence to determine whether specific modalities of exercise have a differential effect on cognitive performance. Over the last decade or so, numerous studies and meta-analyses have reported supporting evidence that aerobic exercise can lead to cognitive improvement. More recently though, strength and resistance training, gross motor training, and other approaches have gained more interest. Consequently, it is hard to provide definitive conclusions regarding the volume and intensity of exercise, as well as the duration of an intervention that would best lead to cognitive improvement. Equally important are the individual differences in fitness level, sedentary lifestyle, and sex differences, which also increasingly seem to become important moderators of fitness intervention effects (see Katz et al., this volume).

With regard to training duration and training modality, a minimum of a 12-week intervention seems to be ideal to improve physical components, depending on which specific components are part of the desired outcome (e.g., from 12 weeks for better balance to 50 weeks for improving muscle morphology, with a minimum of 16 weeks to improve VO_2Max, while some studies including ours show significant improvement after 12 weeks in sedentary seniors; see Predovan et al. 2012). Many of the studies published so far suggest a training frequency of three 1-hour sessions per week to obtain significant physical adaptations or changes. For positive effects on cognition, studies showed more equivocal results that could be linked to variability in study designs, characteristics of the population, and training modalities. In their review of observational studies in healthy older adults and patients with diverse pathologies (prospective cohort studies, case-control studies, and longitudinal studies), Carvalho et al. (2014) observed that study duration varied from 6 months to several years across studies. Future studies should investigate if this could have an impact on cognitive benefits overall and if it could lead to differential gains on various cognitive functions. More recently, Cabral et al. (2019) reported that within 16 weeks of training, referred to as short-term effects, the most global benefits were observed in combined approaches utilizing both aerobic and resistance exercises with at least moderate intensity to improved brain structure, cerebral blood flow, as well as neurotrophic factors such as BDNF and IGF-1. At medium term 16–52 weeks, increased connectivity with aerobic exercise can be observed along with diverse structural changes in white and gray matter volume in frontal and temporal areas, namely, in the hippocampus. Long-term interventions that go beyond 52 weeks, which often include multiple exercise modes like aerobic, coordination, and combined exercise interventions, can lead to increases in white matter integrity and hippocampal volume, whereas long-term resistance training was shown to be effective at increasing IGF-1. Moreover, it seems that different exercise modes favor distinct pathways to improving cognition since BDNF seems to increase after aerobic dance intervention, while IGF-1 increase is observed in individuals that performed moderate to high resistance training. These results support the increasingly accepted notion that multiple modes of intervention are favorable to enhance cognition in seniors due to different biological pathways of improvement.

Conclusion

This chapter is a brief overview of a substantial literature supporting the notion that physical activity and exercise training can positively affect cognitive performances in older adults and perhaps help prevent age-related cognitive decline. In many studies, aerobic training seems to largely impact executive functioning of healthy older adults, and these effects could be, at least in part, induced by improvement in oxygen consumption (e.g., higher VO_2max measures). Moreover, the neuroprotective effect of exercise could be supported by a cascade of molecular mechanisms that involve regulation of specific biomarkers (BDNF, IGF-1). Some of these biomarkers might also be involved in the effects of resistance training on executive control performance. Emerging interventions based on gross motor exercise, yoga, dance, and tai chi have also demonstrated positive effects on cognition. Nevertheless, some important methodological questions remain to be elucidated (see also Cochrane and Green, Könen and Auerswald, Schmiedek, this volume). Among other issues, future studies should help develop more ecological interventions based on individuals' interests and should further investigate short-term benefits and long-term gains of the intervention programs. The question of who would benefit the most among the elderly population should also be raised. Recent interventional studies have shown negative results on cognition (Andrieu et al. 2017), possibly because the included older adults were too robust already to improve intrinsic capacities. A more fragile population (still to define, see Pothier et al. 2019) should take central stage. Future studies should also help uncover potential important moderators for the effects of fitness interventions (e.g., sedentary lifestyle or sex-related differences) on cognition and elucidate biological mechanisms that sustain the positive impacts that exercise induces on brain structure and functions throughout life.

References

Andrieu, S., Guyonnet, S., Coley, N., Cantet, C., Bonnefoy, M., Bordes, S., … & Desclaux, F. (2017). Effect of long-term omega 3 polyunsaturated fatty acid supplementation with or without multidomain intervention on cognitive function in elderly adults with memory complaints (MAPT): A randomised, placebo-controlled trial. *The Lancet Neurology, 16*, 377–389.

Berryman, N., Bherer, L., Nadeau, S., Lauzière, S., Lehr, L., Bobeuf, F., … & Bosquet, L. (2014). Multiple roads lead to Rome: combined high-intensity aerobic and strength training vs. gross motor activities leads to equivalent improvement in executive functions in a cohort of healthy older adults. *Age, 36*, 9710.

Betik, A. C., & Hepple, R. T. (2008). Determinants of VO2 max decline with aging: An integrated perspective. *Applied Physiology, Nutrition, and Metabolism, 33*, 130–140.

Bherer, L., Erickson, K. I., & Liu-Ambrose, T. (2013). A review of the effects of physical activity and exercise on cognitive and brain functions in older adults. *Journal of Aging Research, 2013*, 1–8.

Bherer, L., Langeard, A., Kaushal, N., Vrinceanu, T., Desjardins-Crépeau, L., Langlois, F., & Kramer, A. F. (2019). Physical exercise training effect and mediation through cardiorespiratory fitness on dual-task performances differ in younger–old and older–old adults. Advance Online Publication. The Journals of Gerontology: Series B. https://doi.org/10.1093/geronb/gbz066

Cabral, D. F., Rice, J., Morris, T. P., Rundek, T., Pascual-Leone, A., & Gomes-Osman, J. (2019). Exercise for brain health: An investigation into the underlying mechanisms guided by dose. *Neurotherapeutics, 16*, 580–599.

Carvalho, A., Rea, I. M., Parimon, T., & Cusack, B. J. (2014). Physical activity and cognitive function in individuals over 60 years of age: A systematic review. *Clinical Interventions in Aging, 9*, 661–682.

Caspersen, C. J., Powell, K. E., & Christenson, G. M. (1985). Physical activity, exercise, and physical fitness: Definitions and distinctions for health-related research. *Public Health Reports, 100*, 126–131.

Cassilhas, R. C., Viana, V. A., Grassmann, V., Santos, R. T., Santos, R. F., Tufik, S., & Mello, M. T. (2007). The impact of resistance exercise on the cognitive function of the elderly. *Medicine and Science in Sports and Exercise, 39*, 1401–1407.

Chang, Y. K., Nien, Y. H., Tsai, C. L., & Etnier, J. L. (2010). Physical activity and cognition in older adults: The potential of Tai Chi Chuan. *Journal of Aging and Physical Activity, 18*, 451–472.

Colcombe, S., & Kramer, A. F. (2003). Fitness effects on the cognitive function of older adults: A meta-analytic study. *Psychological Science, 14*, 125–130.

Downey, A., Stroud, C., Landis, S., Leshner, A. I., & National Academies of Sciences, E. (2017). Priorities for future research. In A. Downey, C. Stroud, S. Landis, & A.I. Leshner, *Preventing cognitive decline and dementia: A way forward.* (p. 95–104). Washington, DC: National Academies Press (US).

Erickson, K. I., Prakash, R. S., Voss, M. W., Chaddock, L., Hu, L., Morris, K. S., ... & Kramer, A. F. (2009). Aerobic fitness is associated with hippocampal volume in elderly humans. *Hippocampus, 19*, 1030–1039.

Erickson, K. I., Voss, M. W., Prakash, R. S., Basak, C., Szabo, A., Chaddock, L., ... & Wojcicki, T. R. (2011). Exercise training increases size of hippocampus and improves memory. *Proceedings of the National Academy of Sciences, 108*, 3017–3022.

Esmail, A., Vrinceanu, T., Lussier, M., Predovan, D., Berryman, N., Houle, J., ... & Bherer, L. (2020). Effects of dance/movement training vs. aerobic exercise training on cognition, physical fitness and quality of life in older adults: A randomized controlled trial. *Journal of Bodywork and Movement Therapies, 24*, 212–220.

Etnier, J. L., & Chang, Y.-K. (2019). Exercise, cognitive function, and the brain: Advancing our understanding of complex relationships. *Journal of Sport and Health Science, 8*, 299–300.

Falck, R. S., Davis, J. C., Best, J. R., Crockett, R. A., & Liu-Ambrose, T. (2019). Impact of exercise training on physical and cognitive function among older adults: a systematic review and meta-analysis. *Neurobiology of Aging, 79*, 119–130.

Forte, R., Boreham, C. A., Leite, J. C., De Vito, G., Brennan, L., Gibney, E. R., & Pesce, C. (2013). Enhancing cognitive functioning in the elderly: Multicomponent vs resistance training. *Clinical Interventions in Aging, 8*, 19–27.

Gothe, N. P., & McAuley, E. (2015). Yoga and cognition: A meta-analysis of chronic and acute effects. *Psychosomatic Medicine, 77*, 784–797.

Gothe, N. P., Kramer, A. F., & McAuley, E. (2014). The effects of an 8-week Hatha yoga intervention on executive function in older adults. *Journals of Gerontology Series A: Biomedical Sciences and Medical Sciences, 69*, 1109–1116.

Grégoire, C.-A., Berryman, N., St-Onge, F., Vu, T. T. M., Bosquet, L., Arbour, N., & Bherer, L. (2019). Gross motor skills training leads to increased brain-derived neurotrophic factor levels in healthy older adults: A pilot study. *Frontiers in Physiology, 10*, 410.

Ho, R. T., Fong, T. C., Chan, W. C., Kwan, J. S., Chiu, P. K., Yau, J. C., & Lam, L. C. (2020). Psychophysiological effects of dance movement therapy and physical exercise on older adults with mild dementia: A randomized controlled trial. *The Journals of Gerontology: Series B, 75*, 560–570.

Jensen, C. S., Hasselbalch, S. G., Waldemar, G., & Simonsen, A. H. (2015). Biochemical markers of physical exercise on mild cognitive impairment and dementia: Systematic review and perspectives. *Frontiers in Neurology, 6*, 187.

Kelly, M. E., Loughrey, D., Lawlor, B. A., Robertson, I. H., Walsh, C., & Brennan, S. (2014). The impact of exercise on the cognitive functioning of healthy older adults: A systematic review and meta-analysis. *Ageing Research Reviews, 16*, 12–31.

Kline, C. J., Porcari, J. P., Hintermeister, R., Freedson, P. S., Ward, A., McCarron, R. F., & Rippe, J. (1987). Estimation of from a one-mile track walk, gender, age and body weight. *Medicine & Science in Sports & Exercise, 19*, 253–259.

Kramer, A. F., Hahn, S., & Gopher, D. (1999). Task coordination and aging: Explorations of executive control processes in the task switching paradigm. *Acta Psychologica, 101*, 339–378.

Kramer, A. F., Bherer, L., Colcombe, S. J., Dong, W., & Greenough, W. T. (2004). Environmental influences on cognitive and brain plasticity during aging. *The Journals of Gerontology Series A: Biological Sciences and Medical Sciences, 59*, M940–M957.

Liu-Ambrose, T., Nagamatsu, L. S., Voss, M. W., Khan, K. M., & Handy, T. C. (2012). Resistance training and functional plasticity of the aging brain: A 12-month randomized controlled trial. *Neurobiology of Aging, 33*, 1690–1698.

Livingston, G., Sommerlad, A., Orgeta, V., Costafreda, S. G., Huntley, J., Ames, D., ... & Cooper, C. (2017). Dementia prevention, intervention, and care. *The Lancet, 390*, 2673–2734.

Marinus, N., Hansen, D., Feys, P., Meesen, R., Timmermans, A., & Spildooren, J. (2019). The impact of different types of exercise training on peripheral blood brain-derived neurotrophic factor concentrations in older adults: A meta-analysis. *Sports Medicine, 49*, 1–18.

Marston, K. J., Peiffer, J. J., Rainey-Smith, S. R., Gordon, N., Teo, S. Y., Laws, S. M., ... & Brown, B. M. (2019). Resistance training enhances delayed memory in healthy middle-aged and older adults: A randomised controlled trial. *Journal of Science and Medicine in Sport, 22*, 1226–1231.

McNeely, M. E., Duncan, R. P., & Earhart, G. M. (2015). Impacts of dance on non-motor symptoms, participation, and quality of life in Parkinson disease and healthy older adults. *Maturitas, 82*, 336–341.

Paillard, T., Rolland, Y., & de Souto Barreto, P. (2015). Protective effects of physical exercise in Alzheimer's disease and Parkinson's disease: A narrative review. *Journal of Clinical Neurology, 11*, 212–219.

Pal, R., Singh, S. N., Chatterjee, A., & Saha, M. (2014). Age-related changes in cardiovascular system, autonomic functions, and levels of BDNF of healthy active males: Role of yogic practice. *Age, 36*, 9683.

Patel, N. K., Newstead, A. H., & Ferrer, R. L. (2012). The effects of yoga on physical functioning and health related quality of life in older adults: A systematic review and meta-analysis. *The Journal of Alternative and Complementary Medicine, 18*, 902–917.

Petersen, R. C., Lopez, O., Armstrong, M. J., Getchius, T. S., Ganguli, M., Gloss, D., ... & Sager, M. (2018). Practice guideline update summary: Mild cognitive impairment: Report of the Guideline Development, Dissemination, and Implementation Subcommittee of the American Academy of Neurology. *Neurology, 90*, 126–135.

Pothier, K., Saint-Aubert, L., Hooper, C., Delrieu, J., Payoux, P., de Souto Barreto, P., ... & MAPT/DSA Study Group. (2019). Cognitive changes of older adults with an equivocal amyloid load. *Journal of Neurology, 266*, 835–843.

Predovan, D., Fraser, S. A., Renaud, M., & Bherer, L. (2012). The effect of three months of aerobic training on stroop performance in older adults. *Journal of Aging Research*, 2012, 269815.

Predovan, D., Julien, A., Esmail, A., & Bherer, L. (2019). Effects of dancing on cognition in healthy older adults: A systematic review. *Journal of Cognitive Enhancement, 3*, 161–167.

Rivest-Gadbois, E., & Boudrias, M. H. (2019). What are the known effects of yoga on the brain in relation to motor performances, body awareness and pain? A narrative review. *Complementary therapies in Medicine, 44*, 129–142.

Sofi, F., Valecchi, D., Bacci, D., Abbate, R., Gensini, G. F., Casini, A., & Macchi, C. (2011). Physical activity and risk of cognitive decline: A meta-analysis of prospective studies. *Journal of Internal Medicine, 269*, 107–117.

Spartano, N. L., Davis-Plourde, K. L., Himali, J. J., Andersson, C., Pase, M. P., Maillard, P., ... & Seshadri, S. (2019). Association of accelerometer-measured light-intensity physical activity with brain volume: The Framingham heart study. *JAMA Network Open, 2*, e192745–e192745.

Tsai, C.-L., Wang, C.-H., Pan, C.-Y., & Chen, F.-C. (2015). The effects of long-term resistance exercise on the relationship between neurocognitive performance and GH, IGF-1, and homocysteine levels in the elderly. *Frontiers in Behavioral Neuroscience, 9*, 23.

Voelcker-Rehage, C., Godde, B., & Staudinger, U. M. (2011). Cardiovascular and coordination training differentially improve cognitive performance and neural processing in older adults. *Frontiers in Human Neuroscience, 5*, 26.

Voss, M. W., Prakash, R. S., Erickson, K. I., Basak, C., Chaddock, L., Kim, J. S., ... & Wójcicki, T. R. (2010). Plasticity of brain networks in a randomized intervention trial of exercise training in older adults. *Frontiers in Aging Neuroscience, 2*, 32.

Voss, M. W., Erickson, K. I., Prakash, R. S., Chaddock, L., Kim, J. S., Alves, H., ... & Olson, E. A. (2013). Neurobiological markers of exercise-related brain plasticity in older adults. *Brain, Behavior, and Immunity, 28*, 90–99.

Vrinceanu, T., Esmail, A., Berryman, N., Predovan, D., Vu, T. T. M., Villalpando, J. M., ... & Bherer, L. (2019). Dance your stress away: Comparing the effect of dance/movement training to aerobic exercise training on the cortisol awakening response in healthy older adults. *Stress, 22*, 687–695.

Wei, G. X., Xu, T., Fan, F. M., Dong, H. M., Jiang, L. L., Li, H. J., ... & Zuo, X. N. (2013). Can Taichi reshape the brain? A brain morphometry study. *PLoS One, 8*, e61038.

Weuve, J., Kang, J. H., Manson, J. E., Breteler, M. M., Ware, J. H., & Grodstein, F. (2004). Physical activity, including walking, and cognitive function in older women. *JAMA, 292*, 1454–1461.

Wu, C., Yi, Q., Zheng, X., Cui, S., Chen, B., Lu, L., & Tang, C. (2019). Effects of mind-body exercises on cognitive function in older adults: A meta-analysis. *Journal of the American Geriatrics Society, 67*, 749–758.

Yau, S. Y., Li, A., Sun, X., Fontaine, C. J., Christie, B. R., & So, K.-F. (2016). Potential biomarkers for physical exercise-induced brain health. In M.Wang & F.A. Witzman (Eds.), *Role of biomarkers in medicine* (pp. 169–191). London, UK: IntechOpen.

Part V
Cognitive Training in Applied Contexts

Educational Application of Cognitive Training

Verena E. Johann and Julia Karbach

Contents

Abstract There are studies indicating that executive functions (EF) such as working memory (WM), inhibition, and flexibility are related to academic abilities and that deficits in reading and mathematical abilities are associated with WM deficits. Hence, it can be assumed that academic abilities may be enhanced by means of EF training. In this chapter, we review the effects of cognitive training on academic abilities in children. We first focus on transfer of WM training and training programs targeting other EF on academic abilities in children with learning difficulties or attention deficits and typically developing children. Despite many promising results, existing findings are heterogeneous. We hypothesize that these inconsistencies are caused by several factors, which vary between studies such as the trained WM domain and the tasks applied to measure transfer effects. We also discuss two more factors assumed to modify training and transfer effects: presentation format and personality factors.

V. E. Johann (✉) · J. Karbach
University of Koblenz-Landau, Department of Psychology, Landau in der Pfalz, Germany

Center for Research on Individual Development and Adaptive Education of Children at Risk (IDeA), Frankfurt am Main, Germany
e-mail: johann@uni-landau.de

T. Strobach, J. Karbach (eds.), *Cognitive Training*,
https://doi.org/10.1007/978-3-030-39292-5_23

Recent research indicates that game-based training tasks might be more effective than standard training tasks and that the personality factors neuroticism and effortful control might modulate transfer gains. As compared to domain-specific reading or mathematical training programs, domain-general WM training seems to evoke either equal or even greater improvement in academic abilities. However, a combination of WM training and domain-specific academic training might be more effective as separate training programs targeting WM or academic abilities. Further research is necessary to determine how EF training has to be designed to elicit transfer to reading and mathematical abilities in children with different needs.

Introduction

Academic abilities, such as reading and mathematical abilities, are involved in many daily activities, and academic achievement in these subjects is predictive for various outcomes such as vocational success (Dyer 1987; Rabiner et al. 2016). Therefore, improvements on academic abilities are the aim of different training programs. Most of the evaluated training programs focused on working memory (WM: see also Könen et al., this volume); however, recent studies also investigated the effects of training of other executive functions (EF) such as inhibition or cognitive flexibility (see also Karbach and Kray; Strobach and Schubert, this volume). Significant associations between EF and academic abilities form the basis for the idea that training of EF transfers to scholastic skills. We first review briefly the literature regarding relations between EF and academic abilities. In the following paragraph, we summarize research results regarding transfer of WM training and training programs targeting other cognitive abilities to reading and mathematical abilities in children with cognitive or learning deficits and typically developing children (see also deVries and Geurts, Rueda et al., this volume). Furthermore, we evaluate the usefulness of cognitive training interventions as compared to domain-specific reading or mathematical training programs. Despite many promising results, there are still studies in which no transfer effects to academic relevant abilities were found. In the last paragraph, we discuss factors which might modulate training and transfer gains, namely, motivational aspects and personality traits.

The Relations of EF to Academic Abilities

There is a large amount of research suggesting that WM is involved in reading comprehension and reading efficiency (for a review, see Titz and Karbach 2014). In a recent meta-analysis, Peng et al. (2018) found a significant relation between reading and WM. Moreover, different types of reading performance (phonological coding, decoding, vocabulary, and comprehension) were associated with WM to a similar degree. Mathematical abilities have also been linked to WM (for a review, see Titz and Karbach 2014). However, the specific type of math task and the age of the sample modulate this association (for a meta-analysis, see Peng et al. 2016). Not only WM but also inhibition is associated with reading and mathematical abilities. However, the results are more inconsistent and the relations depend on the specific inhibition component which is being measured. Whereas resistance to proactive interference seems to be related to reading comprehension (Borella et al. 2010; Carretti et al. 2009), response inhibition might be involved in reading speed (Protopapas et al. 2007; Savage et al. 2006). Moreover, there is evidence for significant associations of measures of cognitive flexibility with reading and mathematical abilities (for a meta-analysis, see Yeniad et al. 2013). However, other studies failed to find associations of inhibition and cognitive flexibility with reading and mathematical abilities or only found WM to be related to academic abilities (Agostino et al. 2010; Cartwright et al. 2010; Colé et al. 2014; Lee et al. 2009). In our own study, we found that children's WM span and inhibition ability were related to reading speed, whereas shifting ability was positively related to reading comprehension (Johann et al. 2019). Taken together, there is strong evidence for an involvement of EF in academic abilities.

Effects of Cognitive Training on Academic Abilities in Children with Cognitive or Learning Deficits

Considering the significant relationship between EF and academic abilities, it seems conceivable that training of EF might be an effective way to elicit transfer to academic abilities. Even though there are large methodological differences between study designs and results are heterogeneous, there are some promising outcomes that we discuss in detail below.

Training Programs Targeting WM

Many training studies applied the Cogmed Working Memory Training program (http://www.cogmed.com, Klingberg et al. 2005) consisting of a variety of visuospatial and verbal short-term memory (STM) and WM tasks (see also deVries and Geurts, this volume). Dahlin (2011) examined the effects of the Cogmed training battery on reading abilities (reading comprehension, word decoding, and orthographical knowledge) in 9–12-year-old children with special needs. After 25 training sessions, the training group showed greater performance improvements in terms of WM and reading comprehension as compared to a passive control group. These effects were still present 6 months after the training. This result is in line with the findings from Holmes and Gathercole (2014) who investigated the effects of Cogmed training on WM, English, and mathematical abilities in children with low academic abilities (9–11 years of age). The authors reported greater progress across the academic year in English and math attainment in the training group as compared to a control group. Other training studies in which the Cogmed training program was applied also reported transfer to mathematical abilities in children with attention deficits and special educational needs (Dahlin 2013) and in children with WM deficits (Bergman-Nutley and Klingberg 2014).

Jungle Memory is another program, which advertises that WM training can improve cognitive abilities in children (Alloway et al. 2013). In contrast to the Cogmed Working Memory Training, this program aims at training WM in the context of reading, math, and letter recognition. The program also includes features thought to increase motivation such as positive verbal feedback, a display of the children's best scores, and the number of "super monkeys" earned as a reward for completing the training levels. Transfer effects of this program were investigated in children with learning difficulties in a pilot study with the result that training benefitted WM, vocabulary, and mathematical abilities (Alloway 2012). In a second study with a larger sample (Alloway et al. 2013), children with learning difficulties were allocated into a training group, active control group, or passive control group. Participants in the training group trained four times a week, whereas participants in the active control group only trained once a week over an 8-week period. Transfer effects on WM, verbal and nonverbal IQ, and standardized measures of academic attainment were investigated immediately after training as well as 8 months later. There were near-transfer effects to WM as well as far-transfer effects to verbal and nonverbal IQ, and spelling, but not math, in the training group. Most of these effects were still present 8 months later.

There is also evidence for transfer of WM training to mathematical abilities in developmental dyscalculia. The WM training in a study by Layes et al. (2018) comprised 11 tasks focusing on manipulating and maintaining arithmetic information. Participants with dyscalculia were randomized into a training group or a control group and were assessed on measures of WM, mathematic abilities, and nonverbal ability before and after training. The authors reported that the training group showed greater gains in WM and mathematic abilities than the control group.

Transfer to mathematical abilities was also found in Chinese children with learning difficulties. Chen et al. (2018) investigated the effects of a WM training on WM, IQ, and academic abilities in Chinese children (mean age = 10 years) with learning difficulties. Training included three forms of an updating task with animals, letter, and locations. Participants in the training group trained five times a week over a period of 20 days and were compared to a passive control group. The authors reported immediate transfer to WM and IQ in the training group. Moreover, the math scores but not Chinese language scores of the training group improved significantly by 6 months after the training. The lack of transfer effects to language can be explained with the characteristics of the test material and the training tasks. The Chinese test included reading comprehension and writing work. The authors assume that WM training may not improve writing skills. Moreover, the training material contained letters rather than Chinese characters, and it is possible that the training improved the participants' sensitivity to English letters but not Chinese characters. Zhang et al. (2018) applied the same training tasks in children with learning difficulties (10–13 years of age) and investigated transfer to academic performance and effects on children's brain activity. They found that the training group exhibited greater performance improvements in terms of WM and mathematical scores as compared to a control group with learning difficulties. Additionally, the amplitudes of N160 (representative of visual recognition) and P300 (representative of updating processing) increased from pretest to posttest indicating training-related positive changes in brain activity.

Whereas most of the reviewed studies focused on elementary school children, evidence for training-related gains in adolescents is rare. Van der Molen et al. (2010) examined the effects of an adaptive or nonadaptive WM training based on the principle of the Odd-One-Out test by Henry (2001) on i.a. scholastic abilities and the recall of stories in adolescents (13–16 years of age) with mild to borderline intellectual disabilities. Transfer measures were assessed before and after 5 weeks of training as well as 10 weeks after the training was finished. At follow-up, both the adaptive and nonadaptive WM training showed higher improvements on STM, arithmetic, and story recall as compared to a control group.

In contrast, there are also studies in which WM training did not benefit academic abilities. In a study from Dunning et al. (2013) conducted on children with WM deficits (7–9 years of age), Cogmed training did not improve children's reading and mathematical abilities neither immediately after training nor 1 year later. Gray et al. (2012) investigated children with coexisting learning disabilities and attention deficit hyperactivity (12–17 years of age). Participants were randomized into a WM training group or a math training group. Adolescents in the WM training group showed greater improvement in a subset of WM criterion measures compared with those in the math training group, but there were no transfer effects to cognitive or academic performance. However, in this study transfer was only assessed 3 weeks after completing the training. Especially in children with learning disabilities, attention deficits, or low WM, it might take time until improved WM skills can be used efficiently in more complex activities. A study from Holmes et al. (2009) in children with low WM supports this assumption. Whereas transfer to WM and STM was found immediately after the training, far transfer to mathematical abilities was only found 6 month following training (see also van der Molen et al. 2010; Zhang et al. 2018).

Training Programs Targeting Other Cognitive Abilities

Although there is a significant association between inhibition and flexibility with reading and mathematical abilities (see above), there is a lack of studies investigating the effects of training programs targeting other cognitive abilities such as inhibition or cognitive flexibility (see also Karbach and Kray; Strobach and Schubert, this volume). Wang et al. (2019) compared the performance of two low-achieving training groups, targeting WM or inhibition with that of a matched low-achieving control group and a normal-achieving control group. Children's scores on the school-administered tests of Chinese language and math as well as performance on a matrices test were collected before training, after training, and 2 months after completing training. The normal-achieving control group outperformed the other three groups in language and math before training. After completing training, the normal-achieving control group still performed better than the other three groups in language and math. At follow-up, the low-achieving WM and inhibition training groups showed greater performance improvements in language than the low-achieving control group. Moreover, the difference in language between the inhibition training group and the normal-achieving control group was no longer significant, and the difference between the WM training group and the normal-achieving control group was markedly reduced. Whereas most of the research examining the effects of WM training on language or reading skills focused on alphabet-based languages, this study is one of the first focusing on Chinese language-related skills and illustrates that WM could be extremely relevant for managing the challenges of this logographic-based language. The finding that training-related improvements in language skills only emerged at the follow-up assessment supports the assumption that it might take time for children to implement their newly acquired skills in other situations (Wang et al. 2019).

Effects of Cognitive Training on Academic Abilities in Typically Developing Children

There is also evidence that cognitive training benefits academic abilities in typically developing children. Again, most training studies focused on WM, but there is also one study in which the effects of WM, inhibition, and flexibility training on academic abilities were compared (Johann and Karbach 2019).

Training Programs Targeting WM

There are three studies in which the effects of the Braintwister WM training (Buschkuehl et al. 2008) or single tasks of this battery on scholastic performance were investigated (Karbach et al. 2015; Loosli et al. 2012; Studer-Luethi et al. 2016). The Braintwister WM training is a battery comprising verbal and visuospatial simple and complex span tasks as well as an auditory, visual, or dual n-back task. Two of the verbal complex span tasks are specially designed for children, including child-friendly pictorial stimuli (animals set in a farm or safari context) as well as appealing performance feedback at the end of each trial.

Karbach et al. (2015) examined the effects of this WM training program on WM, inhibition, task switching, reading, and mathematical abilities in typically developing children (7–9 years of age). Participants were randomized into an adaptive training group or a nonadaptive control group. Training consisted of two adaptive or nonadaptive verbal complex span tasks in which participants had to remember the sequence of animal pictures against a secondary processing task. After 14 training sessions, there were transfer effects on the untrained WM task and reading ability, but not on task switching, inhibition, or mathematical ability. The result regarding reading is in line with the findings from Loosli et al. (2012). In their study, children (9–11 years of age) performed 10 sessions of an adaptive verbal complex span task from the Braintwister training. Before and after 2 weeks of training, participants were tested on measures of reading and problem-solving. Compared to a passive control group, WM training benefitted text reading, but not word reading.

The finding that WM training benefits reading abilities in children fits the results from the meta-analysis from Peng et al. (2018) who found that WM and reading abilities were moderately related. Moreover, they concluded that the domain-general central executive of WM was more relevant in early reading acquisition stages, whereas verbal WM (and therefore a domain-specific WM subsystem) was more important for reading after the children became more experienced readers. Thus, in less experiences readers, the trained WM domain should not be crucial for transfer effects on reading provided that the training tasks involve central executive demands. In contrast, training of verbal complex span tasks is assumed to be the most efficient way to improve reading abilities in experienced readers. Participants in the studies from Karbach et al. (2015) and Loosli et al. (2012) can be considered experienced readers, and, therefore, verbal complex span task elicited transfer to reading abilities.

The results from another study support this assumption (Studer-Luethi et al. 2016). In this study, training comprised the same complex span task as applied in the studies from Karbach et al. (2015) and Loosli et al. (2012) and additionally a visuospatial n-back task. Children (mean age = 8;3 years) were assigned to a WM training group, reading training group, or a passive control group. Before and after training, children performed an untrained WM task, a stroop task, a vocabulary task, and the same standardized reading ability test and math test as in the study from Karbach et al. (2015). The authors reported near-transfer effects of WM training on WM and far transfer on a vocabulary task. However, there was only a trend toward greater performance improvements on reading and math. Participants were comparable in terms of age with those in the study from Karbach et al. (2015) and Loosli et al. (2012). In contrast, training comprised not only a verbal complex span task but also a visuospatial n-back task. According to the result from the meta-analysis from Peng et al. (2018), it may be assumed that a pure verbal WM training could have led to more pronounced transfer effects on reading ability.

Henry et al. (2014) investigated the effects of WM training to untrained WM tasks, reading, and mathematical abilities in younger children (5–8 years of age). WM training comprised a spatial WM task (odd one out span) and a complex span task (listening recall). At posttest and a 6-month follow-up, there were greater performance improvements on the untrained WM tasks, but not on word reading or number skills in the WM training groups as compared to an active control group. However, 12 months after training, the WM training group showed larger gains in reading comprehension as compared to the active control group. In contrast to the study from Studer-Luethi (2016), a WM training consisting of a visuospatial and a verbal WM task benefitted reading abilities. However, participants in this study were younger and less experienced, supporting the assumption that at this skill level, the trained WM domain is not crucial for transfer effects as long as training tasks involve central executive demands (Peng et al. 2018).

Given that all components of WM have been related to mathematical abilities (see above), it seems likely that WM training also benefits mathematics in typically developing children. Indeed, there are some studies supporting this assumption. Rode et al. (2014) investigated the effects of an adaptive WM training program on WM and academic abilities in elementary school students (third grade) as compared to a control group, which participated in regular classroom activities. The WM training program consisted of a task including numerical material. Participants were instructed to remember the sequence of numbers against a secondary processing task in which simple arithmetic tasks had to be solved. The authors reported significant but small far-transfer effects on mathematical abilities, but not reading abilities. In a study from Kuhn and Holling (2014), children (mean age = 9 years) participated in either an adaptive training of number sense or WM or served as a control group. The WM training program consisted of a spatial n-back task, a Corsi block task, and a letter span task, which was assumed to tap spatial attention due to its requirements to memorize stimuli in a specific order. There were significant higher gains in terms of mathematical abilities in the WM training group and the number sense training group as compared to the control group (see also Sánchez-Pérez et al. 2018).

However, there are also studies in which transfer to reading but not math was found (Henry et al. 2014; Karbach et al. 2015). These inconsistencies could be due to the different training tasks applied in the studies. As described, the WM training programs in the study from Rode et al. (2014) and Kuhn and Holling (2014) comprised visuospatial WM tasks or mathematical processing tasks. In contrast, participants in the studies from Karbach et al. (2015) and Henry et al. (2014) were trained on verbal complex span tasks or visuospatial tasks without numerical content, and thus, it can be hypothesized that the arithmetic demands of the training task in the study from Rode et al. (2014) initiated transfer to mathematical abilities. This assumption is however contradicted by the finding from Peng et al. (2016), who reported in their meta-analysis that mathematics showed comparable association with verbal WM, numerical WM, and visuospatial WM. In contrast, the strength of the relation between WM and different mathematical skills was not invariant, and word-problem-solving and whole-number calculations showed the strongest relation with WM, whereas geometry showed the weakest relation with WM. In the studies from Karbach et al. (2015) and Henry et al. (2014), composite scores of mathematical abilities were used as dependent variables, which also might explain the lack of transfer effects. Finally, Peng et al. (2016) demonstrated that the relation between WM and mathematics is stronger in individuals with mathematics difficulties as compared with typically developing individuals. This finding implies that WM training might be more effective in children with mathematics difficulties than in typically developing children. Given that WM training transferred to mathematical abilities in most of the studies investigating children with cognitive or leaning deficits (Alloway 2012; Dahlin 2013; Holmes and Gathercole 2014; Ramani et al. 2017, see above), we assume that WM training might also be more effective in children with these deficits. Moreover, there is evidence that EF training results in compensation effects with low-performing individuals benefitting more from interventions. Karbach et al. (2015) found that WM performance at baseline and transfer gains in WM at posttest and follow-up were significantly correlated. Furthermore, reading ability at baseline and transfer gains at posttest were also related, indicating that poorer performance at pretest was consistently associated with larger transfer gains (see also Dahlin 2011). One might argue that compensation effects can be explained by regression to the mean. However, a recent study (Karbach et al. 2017) applied latent-change modeling and showed that compensation effects were significantly higher in the training group than in an active control group, indicating that they were more likely based on the effects of EF training than on regression to the mean or retest effects.

Beside these positive findings regarding transfer of WM training to academic abilities, there are also two meta-analyses with less promising results. Schwaighofer et al. (2015) investigated the effects of WM training on WM, verbal and nonverbal abilities, word coding, and mathematical abilities. The authors concluded that there was no evidence for transfer to word coding and mathematics. Although different moderators such as age, training dose, and training modality were included in the analyses, the authors did not differentiate between participants with learning disabilities or attention deficits and typically developing children. However, it can be assumed that different learning abilities modulate training and transfer gains (see above). In another meta-analysis from Sala and Gobet (2017), this issue was considered, and only studies with typically developing children were included. They investigated far transfer of WM training to mathematics, literacy/word decoding, fluid intelligence, cognitive control, crystallized intelligence, and science. Although there was a significant overall effect size for mathematics and a marginally significant effect size for literacy/word decoding, the authors concluded that WM is not effective at improving children's cognitive or academic abilities. A limitation of this study is that the training domain was not taken into account as moderator. As already described, transfer effects depend on the training domain and the transfer measure. Regarding these complex interactions, it is not surprising that meta-analyses which did not simultaneously take age, training modality, and different outcome measures into account failed to reveal far-transfer effects to academic abilities.

Training Programs Targeting Other Cognitive Abilities

There is a lack of studies investigating training programs targeting other cognitive abilities such as inhibition or cognitive flexibility in typically developing children. Still, there is ample evidence for transfer to structurally similar tasks and transfer to other domains of EF and, in some cases, reasoning in children and adolescents (e.g.,

Karbach and Kray 2009; Thorell et al. 2009; see deVries and Geurts, Karbach and Kray; Rueda et al., Schaeffner et al., this volume). Moreover, there is one study in which the effects of a combined WM, planning, and inhibition training on academic abilities were examined (Goldin et al. 2014). The authors reported near transfer to attention and inhibitory control and far transfer to language and math scores. However, it remains unclear which component of the training led to the transfer effects. In our own study (Johann and Karbach 2019), we investigated systematically the effects of game-based and standard training regimens targeting WM, inhibition, or cognitive flexibility on cognitive and academic abilities in elementary school children. Participants were randomized into a game-based WM training group, a standard WM training group, a game-based inhibition training group, a standard inhibition training group, a game-based flexibility training group, a standard flexibility training group, or a passive control group. Standardized reading and math tests, as well as performance on untrained WM, inhibition, and cognitive flexibility tasks, were assessed before training, immediately after training, and 6 weeks after the training was finished. We found greater performance improvements on reading in the game-based flexibility training group and the game-based inhibition training group as compared to the control group. Transfer effects were still present at follow-up. These findings provide first evidence for transfer effects of inhibition and flexibility training on academic abilities.

The Effects of Cognitive Training as Compared to Domain-Specific Academic Abilities Training

Compared to domain-specific training programs directly targeting reading or mathematical strategies, cognitive training might be useful in educational environments, if two criteria are met. First, domain-general cognitive training should evoke comparable improvements in academic abilities, and second, cognitive training should transfer simultaneously to different academic abilities. As described above, there is evidence that WM training enhanced both reading and mathematical abilities, even though transfer effects vary as a function of the task design (see above). However, there are only few studies in which the effects of a domain-specific reading or mathematical training were systematically compared to the effects of a domain-general basic cognitive training.

Nevo and Breznitz (2014) examined the effects of a reading acceleration training program and combined reading acceleration and WM training programs on reading skills and WM abilities. Participants (mean age = 8;6 years) were divided into three training groups and a passive control group. The three training groups received only reading acceleration training, WM followed by reading acceleration training, or reading acceleration followed by WM training. There were larger improvements regarding word accuracy in the reading acceleration only group as compared to the control group. Both training groups, in which reading acceleration and WM were trained, improved more strongly in word and pseudoword fluency. Furthermore, phonological complex memory was enhanced in the reading acceleration only training group and the reading acceleration followed by WM training group. The authors conclude that a combination of an intensive reading acceleration program followed by a WM training program might have the greatest effects on reading and WM skills. In an already mentioned study from Studer-Luethi et al. (2016), the effects of a WM training on WM and academic abilities were compared to those of a computer-based reading training in elementary school students (mean age = 8;3 years). Children who took part in the WM training showed greater improvements on a WM task and a vocabulary test as children in the reading training group and children in a no-contact control group. Regarding scholastic abilities, the WM training group demonstrated greater improvements in mathematical and reading abilities by trend as compared to the reading training group and the passive control group. These studies suggest that a WM training program or a combination of WM and reading training might have greater impact on reading abilities than a pure reading training.

Similar results exist regarding mathematical abilities: Passolunghi and Costa (2014) compared the effects of a WM and a domain-specific training program targeting early numeracy in preschool children. After 5 weeks, the early numeracy intervention specifically improved early numeracy abilities, whereas the WM training intervention improved not only WM but also early numeracy abilities. Importantly, the gains regarding early numeracy abilities in the WM training group did not differ significantly from the gains obtained in the early numeracy training group. These results show that a domain-general training targeting basic cognitive abilities can be as effective as a domain-specific training targeting numerical competence in preschoolers. This finding is line with the results from Ramani et al. (2017) who used two approaches to improve numerical knowledge in kindergarten children from low-income backgrounds. Children were allocated to a training group targeting conceptual knowledge, or a training group targeting underlying cognitive processes such as WM. After ten training sessions, children in the numerical knowledge training group and those in the WM training group showed greater performance improvements in numerical magnitude knowledge as compared to a no-contact control group. This result demonstrated that both domain-specific and domain-general interventions facilitate mathematical learning equally. Sánchez-Pérez et al. (2018) even conclude that training of WM had greater impact on mathematical abilities than training of mathematics tasks. In their study, the training program comprised a combination of basic mathematical tasks as well as three different WM tasks. Children were allocated to the combined training group or an active control group. The authors reported greater performance improvements on nonverbal IQ, inhibition, mathematical, and reading abilities in the training group as compared to the control group. Moreover, the contribution of each specific component of the training (WM and math) on transfer effects was analyzed. Transfer effects on nonverbal IQ, inhibition, and reading abilities were more related to the WM activities than to the math exercises. However, improvements on mathematical performance were associated with performance on the whole training program and not related to either WM or math activities separately. The authors conclude that WM training in addition to math exercise is the most effective way to improve mathematical abilities.

In sum, there is evidence that a domain-general WM training evoked either equal or even greater improvement in reading or mathematical abilities than domain-specific reading or mathematical training programs. However, the studies by Nevo and Breznitz (2014) and Sánchez-Pérez et al. (2018) suggest that a combination of WM training and domain-specific reading or mathematical training might be more effective as separated training programs targeting WM or academic abilities. Further research is necessary to determine how training of basic cognitive functions such as WM, inhibition, or flexibility training should be combined with domain-specific training tasks to maximize transfer to academic abilities.

Moderating Variables in Training and Transfer Effects

Findings regarding transfer of cognitive training to academic abilities in children are quite inconsistent. This heterogeneity may be due to different variables moderating training outcomes. Possible moderating variables are training-specific features such as training type, training modality, duration of training, frequency of training, or instructional support (Sala and Gobet 2017; Schwaighofer et al. 2015; for a review, see von Bastian and Oberauer 2014; see also Cochrane and Green, Karbach and Kray, this volume). In children, especially the presentation format might play an important role because many training programs using psychometric cognitive tasks are relatively monotonous and require participants to perform a single training task repeatedly, which could adversely affect motivation in children. Furthermore, individual differences in age and personality are assumed to influence training and transfer gains (see Katz et al., this volume). The following chapter addresses the possibly moderating effects of the presentation format and personality in children.

Presentation Format and Motivational Effects

Especially in children, motivational aspects have been discussed as important predictors for training gains (see Katz et al., this volume). Locke and Braver (2010) assumed that motivation modulates the effort an organism is willing to invest to achieve goals, thereby aligning goal-directed behavior. Therefore, it seems likely that training willingness and possibly training gains are affected by motivational factors. Adding game elements to the training tasks could be one effective way to enhance motivation and training willingness. Even though many training studies used partially game-based training batteries such as the Cogmed training program, surprisingly few have systematically compared game-based training tasks to standard training tasks.

Prins et al. (2011) examined the benefits of adding game elements to WM training in children with ADHD (7–12 years of age). The game-based WM training enhanced motivation (assessed by the time voluntarily spend on training), training performance, and transfer to an untrained WM task as compared to the standard training setting. Dörrenbächer et al. (2014) investigated the effects of a task-switching training in a high-motivational setting with game elements and a low-motivational setting without game elements in middle-aged children (8–11 years of age). They found that training willingness and near transfer in switching costs but not far transfer were enhanced in the high-motivational setting as compared to the low-motivational setting. In our study (Johann and Karbach 2019), we relied on the concept of intrinsic interest and the self-determination theory (SDT; Ryan and Deci 2000) and developed a WM training, an inhibition training, and a flexibility training in a game-based and a standard version, respectively. The game-based versions featured experimental manipulations designed to satisfy the three basic psychological needs: relatedness (feeling connected and involved with others and having a sense of belonging), autonomy (need to experience one's behavior as self-determined), and competency (feeling effective in one's interactions with the environment). SDT suggests that fulfilling these needs facilitates intrinsic motivation, which may increase training motivation and training-induced performance gains. In the game-based version, tasks were framed by a cover story to enhance the feeling of relatedness (see Fig. 1a–d). In each task, participants could earn magic power points that made the protagonist stronger to enhance the feeling of perceived competence (see Fig. 1e). In order to increase the feeling of autonomy, there were pseudo-choices providing participants the opportunity to decide which route to take. All training tasks were adaptive (i.e., task difficulty was continuously adapted to individual performance across 7 levels), and a progress bar turned green after responses that were correct and provided in time (see Fig. 1f) and red after responses that were incorrect or too slow. The difficulty level was increased after a certain number of correct reactions and decreased after a certain number of incorrect reactions. Whereas the presentation format (game-based, standard) did not modulate training effects, there were differences regarding far transfer on academic abilities. We found greater performance improvements in sentence comprehension and reading speed in the game-based inhibition and game-based flexibility training group as compared to the control group. Those transfer effects were not found in the standard inhibition or standard flexibility training group. This result cannot be explained by higher training gains since training performance increased equally in the game-based and the standard training groups. However, there is some evidence that isolated training of specific cognitive functions, such as executive functions in a narrow task context, may constrain transfer to dissimilar activities in complex activity contexts (Greeno et al. 1993; Schwaighofer et al. 2015). According to this view, adding game elements to executive control training tasks may enhance the complexity of the training context and therefore facilitate transfer to academic abilities which are also acquired and practiced in a complex context.

Fig. 1 Selected pictures of the cover story and the training tasks. The map of the kingdom Asfallon (**a**) where the king and the queen ruled until the evil wizard Ansgar destroys the magic stone and takes the control of the kingdom (**b**); The protagonists Edvin, Bragi, and Finja (**c**); Edvin fighting against Ansgar at the end of the story (**d**) Feedback in terms of magic power points in the game-based setting (**e**); example for an inhibition training task (**f**)

Personality

Since EF and academic abilities are associated with personality factors (Neuenschwander et al. 2013; Poropat 2009), it seems likely that personality also modulates training and transfer gains (see Katz et al., this volume). Studer-Luethi et al. (2016) investigated the moderating effect of the personality traits neuroticism and effortful control on WM training outcomes in children (mean age = 8;3 years). As already described, participants were allocated to a WM training group, reading training group, or a no-contact control group. There were greater improvements on visual WM, vocabulary, and academic abilities by trend in the WM training group as compared to the other groups. Moreover, they found a moderation effect of neuroticism and effortful control on transfer gains. The WM training program predicted higher post-training gains compared to the reading training group and the control group only in children with high effortful control or low neuroticism. The authors conclude that sufficient self-regulative abilities and emotional stability are necessary for WM training to be effective.

Conclusion

To summarize, recent findings indicate that WM may indeed benefit academic abilities in children with learning disabilities or attention deficits (Alloway 2012; Alloway et al. 2013; Bergman-Nutley and Klingberg 2014; Dahlin 2011, 2013; Holmes et al. 2009; Holmes and Gathercole 2014; see also deVries and Geurts, this volume) and typically developing children (Henry et al. 2014; Karbach et al. 2015; Kuhn and Holling 2014; Loosli et al. 2012; Rode et al. 2014; Sánchez-Pérez et al. 2018; see also Schaeffner et al., this volume). Moreover, there is the first evidence for benefits of inhibition and flexibility training on academic abilities (Goldin et al. 2014; Johann and Karbach 2019; Wang et al. 2019). Further research is necessary to investigate how different training modalities can be combined to maximize training and transfer effects. So far, most cognitive training programs have a one-size-fits-all design. This could be the reason why training benefits in several studies did not generalize to untrained tasks and why meta-analyses failed to detect performance benefits after cognitive training. Current studies demonstrated that there are different factors such as the training modality, the presentation format, and personality factors which modulate training and transfer gains. Further research should focus on tailoring training programs to individual abilities and needs (see also Cochrane and Green, Colzato and Hommel, Guye et al., Karbach and Kray; Könen and Auerswald, Könen et al., this volume). Moreover, it is necessary to investigate how cognitive training could be implemented in school life to enhance academic abilities efficiently.

Acknowledgment This work was supported by a grant from the German Research Foundation awarded to the last author (KA 3216/3-1).

References

Agostino, A., Johnson, J., & Pascual-Leone, J. (2010). Executive functions underlying multiplicative reasoning: Problem type matters. *Journal of Experimental Child Psychology, 105*, 286–305.

Alloway, T. (2012). Can interactive working memory training improving learning? *Journal of Interactive Learning Research, 23*, 197–207.

Alloway, T., Bibile, V., & Lau, G. (2013). Computerized working memory training: Can it lead to gains in cognitive skills in students? *Computers in Human Behavior, 29*, 632–638.

Bergman-Nutley, S., & Klingberg, T. (2014). Effect of working memory training on working memory, arithmetic and following instructions. *Psychological Research, 78*, 869–877.

Borella, E., Carretti, B., & Pelgrina, S. (2010). The specific role of inhibition in reading comprehension in good and poor comprehenders. *Journal of Learning Disabilities, 43*, 541–552.

Buschkuehl, M., Jaeggi, S., Kobel, A., & Perrig, W. J. (2008). *Braintwister-Aufgabensammlung für kognitives Training [Braintwister training battery for cognitive training]*. Bern: Universität Bern, Institut für Psychologie.

Carretti, B., Borella, E., Cornoldi, C., & De Beni, R. (2009). Role of working memory in explaining the performance of individuals with specific reading comprehension difficulties: A meta-analysis. *Learning and Individual Differences, 19*, 246–251.

Cartwright, K. B., Marshall, T. R., Dandy, K. L., & Isaac, M. C. (2010). The development of graphophonological-semantic cognitive flexibility and its contribution to reading comprehension in beginning readers. *Journal of Cognition and Development, 11*, 61–85.

Chen, X., Ye, M., Chang, L., Chen, W., & Zhou, R. (2018). Effect of working memory updating training on retrieving symptoms of children with learning disabilities. *Journal of Learning Disabilities, 51*, 507–519.

Colé, P., Duncan, L. G., & Blaye, A. (2014). Cognitive flexibility predicts early reading skills. *Frontiers in Psychology: Cognitive Science, 5*, 565.

Dahlin, K. I. E. (2011). Effects of working memory training on reading in children with special needs. *Reading and Writing, 24*, 479–491.

Dahlin, K. (2013). Working memory training and the effect on mathematical achievement in children with attention deficits and special needs. *Journal of Education and Learning, 2*, 118–133.

Dörrenbächer, S., Müller, P. M., Tröger, J., & Kray, J. (2014). Dissociable effects of game elements on motivation and cognition in a task-switching training in middle childhood. *Frontiers in Psychology, 5*, 1275.

Dunning, D. L., Holmes, J., & Gathercole, S. E. (2013). Does working memory training lead to generalized improvements in children with low working memory? A randomized controlled trial. *Developmental Science, 16*, 915–925.

Dyer, E. D. (1987). Can university success and first-year job performance be predicted from academic achievement, vocational interest, personality and biographical measures? *Psychological Reports, 61*, 655–671.

Goldin, A. P., Hermida, M. J., Shalom, D. E., Costa, M. E., Lopez-Rosenfeld, M., Segretin, M. S., ... & Sigman, M. (2014). Far transfer to language and math of a short software-based gaming intervention. *Proceedings of the National Academy of Sciences, 111*, 6443–6448.

Gray, S. A., Chaban, P., Martinussen, R., Goldberg, R., Gotlieb, H., Kronitz, R., Hockenberry, M., & Tannock, R. (2012). Effects of a computerized working memory training program on working memory, attention, and academics in adolescents with severe LD and comorbid ADHD: A randomized controlled trial. *Journal of Child Psychology and Psychiatry, 53*, 1277–1284.

Greeno, J. G., Moore, J. L., & Smith, D. R. (1993). Transfer of situated learning. In D. K. Detterman & R. J. Sternberg (Eds.), *Transfer on trial: Intelligence, cognition, and instruction* (pp. 99–167). Ablex Publishing.

Henry, L. A. (2001). How does the severity of a learning disability affect working memory performance? *Memory, 9*, 233–247.

Henry, L. A., Messer, D. J., & Nash, G. (2014). Testing for near and far transfer effects with a short, face-to-face adaptive working memory training intervention in typical children. *Infant and Child Development, 23*, 84–103.

Holmes, J., & Gathercole, S. E. (2014). Taking working memory training from the laboratory into schools. *Educational Psychology, 34*, 440–450.

Holmes, J., Gathercole, S. E., & Dunning, D. L. (2009). Adaptive training leads to sustained enhancement of poor working memory in children. *Developmental Science, 12*, F9–F15.

Johann V. E., & Karbach, J. (2019). Effects of game-based and standard executive control training on cognitive and academic abilities in elementary school children. *Developmental Science*, e12866.

Johann, V. E., Könen, T., & Karbach, J. (2019). The unique contribution of working memory, inhibition, cognitive flexibility, and intelligence to reading comprehension and reading speed. *Child Neuropsychology, 26*, 324–344.

Karbach, J., & Kray, J. (2009). How useful is executive control training? Age differences in near and far transfer of task-switching training. *Developmental Science, 12*, 976–990.

Karbach, J., Strobach, T., & Schubert, T. (2015). Adaptive working memory training benefits reading, but not mathematics in middle childhood. *Child Neuropsychology, 21*, 285–301.

Karbach, J., Könen, T., & Spengler, M. (2017). Who Benefits the Most? Individual Differences in the Transfer of Executive Control Training Across the Lifespan. *Journal of Cognitive Enhancement, 1*, 394–405.

Klingberg, T., Fernell, E., Olesen, P. J., Johnson, M., Gustafsson, P., Dahlström, K., ... & Westerberg, H. (2005). Computerized training of working memory in children with ADHD-a randomized, controlled trial. *Journal of the American Academy of Child & Adolescent Psychiatry, 44*, 177–186.

Kuhn, J.-T., & Holling, H. (2014). Number sense or working memory? The effect of two computer-based trainings on mathematical skills in elementary school. *Advances in Cognitive Psychology, 10*, 59–67.

Layes, S., Lalonde, R., Bouakkaz, Y., & Rebai, M. (2018). Effectiveness of working memory training among children with dyscalculia: Evidence for transfer effects on mathematical achievement – A pilot study. *Cognitive Processing, 19*, 375–385.

Lee, K., Ng, E. L., & Ng, S. F. (2009). The contributions of working memory and executive functioning to problem representation and solution generation in algebraic word problems. *Journal of Educational Psychology, 101*, 373–387.

Locke, H. S., & Braver, T. S. (2010). Motivational influences on cognitive control: A cognitive neuroscience perspective. In R. Hassin, K. Ochsner, & Y. Trope (Eds.), *Self control in society, mind, and brain* (pp. 114–140). New York: Oxford University Press.

Loosli, S. V., Buschkuehl, M., Perrig, W. J., & Jaeggi, S. M. (2012). Working memory training improves reading processes in typically developing children. *Child Neuropsychology, 18*, 62–78.

Neuenschwander, R., Cimeli, P., Röthlisberger, M., & Roebers, C. M. (2013). Personality factors in elementary school children: Contributions to academic performance over and above executive functions. *Learning and Individual Differences, 25*, 118–125.

Nevo, E., & Breznitz, Z. (2014). Effects of working memory and reading acceleration training on improving working memory abilities and reading skills among third graders. *Child Neuropsychology, 20*, 752–765.

Passolunghi, M. C., & Costa, H. M. (2014). Working memory and early numeracy training in preschool children. *Child Neuropsychology, 22*, 81–98.

Peng, P., Namkung, J., Barnes, M., & Sun, C. Y. (2016). A meta-analysis of mathematics and working memory: Moderating effects of working memory domain, type of mathematics skill, and sample characteristics. *Journal of Educational Psychology, 108*, 455–473.

Peng, P., Barnes, M., Wang, C., Wang, W., Li, S., Swanson, H. L., Dardick, W., & Tao, S. (2018). A meta-analysis on the relation between reading and working memory. *Psychological Bulletin, 144*, 48–76.

Poropat, A. E. (2009). A meta-analysis of the five-factor model of personality and academic performance. *Psychological Bulletin, 135*, 322–338.

Prins, P. J., Dovis, S., Ponsioen, A., Ten Brink, E., & Van der Oord, S. (2011). Does computerized working memory training with game elements enhance motivation and training efficacy in children with ADHD? *Cyberpsychology, Behavior and Social Networking, 14*, 115–122.

Protopapas, A., Archonti, A., & Skaloumbakas, C. (2007). Reading ability is negatively related to stroop interference. *Cognitive Psychology, 54*, 251–282.

Rabiner, D. L., Godwin, J., & Dodge, K. A. (2016). Predicting academic achievement and attainment: The contribution of early academic skills, attention difficulties, and social competence. *School Psychology Review, 45*, 250–267.

Ramani, G. B., Jaeggi, S. M., Daubert, E. N., & Buschkuehl, M. (2017). Domain-specific and domain-general training to improve kindergarten children's mathematics. *Journal of Numerical Cognition, 3*, 468–495.

Rode, C., Robson, R., Purviance, A., Geary, D. C., & Mayr, U. (2014). Is working memory training effective? A study in a school setting. *PLoS One, 9*, e104796.

Ryan, R. M., & Deci, E. L. (2000). Intrinsic and extrinsic motivations: Classic definitions and new directions. *Contemporary Educational Psychology, 25*, 54–67.

Sala, G., & Gobet, F. (2017). Working memory training in typically developing children: A meta-analysis of the available evidence. *Developmental Psychology, 53*, 671–685.

Sánchez-Pérez, N., Castillo, A., López-López, J. A., Pina, V., Puga, J. L., Campoy, G., González-Salinas, C., & Fuentes, L. J. (2018). Computer-based training in math and working memory improves cognitive skills and academic achievement in primary school children: Behavioral results. *Frontiers in Psychology, 8*, 2327.

Savage, R., Cornish, K., Manly, T., & Hollis, C. (2006). Cognitive processes in children's reading and attention: The role of working memory, divided attention, and response inhibition. *British Journal of Psychology, 97*, 365–385.

Schwaighofer, M., Fischer, F., & Bühner, M. (2015). Does working memory training transfer? A meta-analysis including training conditions as moderators. *Educational Psychologist, 50*, 138–166.

Studer-Luethi, B., Bauer, C., & Perrig, W. J. (2016). Working memory training in children: Effectiveness depends on temperament. *Memory & Cognition, 44*, 171–186.

Thorell, L. B., Lindqvist, S., Bergman-Nutley, S., Bohlin, G., & Klingberg, T. (2009). Training and transfer effects of executive functions in preschool children. *Developmental Science, 12*, 106–133.

Titz, C., & Karbach, J. (2014). Working memory and executive functions: Effects of training on academic achievement. *Psychological Research, 78*, 852–868.

Van der Molen, M. J., Van Luit, J. E. H., Van der Molen, M. W., Klugkist, I., & Jongmans, M. J. (2010). Effectiveness of a computerised working memory training in adolescents with mild to borderline intellectual disabilities. *Journal of Intellectual Disability Research, 54*, 433–447.

von Bastian, C. C., & Oberauer, K. (2014). Effects and mechanisms of working memory training: A review. *Psychological Research, 78*, 803–820.

Wang, C., Jaeggi, S. M., Yang, L., Zhang, T., He, X., Buschkuehl, M., & Zhang, Q. (2019). Narrowing the achievement gap in low-achieving children by targeted executive function training. *Journal of Applied Developmental Psychology, 63*, 87–95.

Yeniad, N., Malda, M., Mesman, J., van Ijzendoorn, M. H., & Pieper, S. (2013). Shifting ability predicts math and reading performance in children: A meta-analytical study. *Learning and Individual Differences, 23*, 1–9.

Zhang, H., Chang, L., Chen, X., Ma, L., & Zhou, R. (2018). Working memory updating training improves mathematics performance in middle school students with learning difficulties. *Frontiers in Human Neuroscience, 12*, 154.

Cognitive Training in Children with Neurodevelopmental Conditions

Marieke de Vries, Lauren Kenworthy, Sebastiaan Dovis, and Hilde M. Geurts

Contents

M. de Vries (✉)
School of Psychology, University of Nottingham Malaysia, Semenyih, Malaysia
e-mail: marieke.deVries@nottingham.edu.my

L. Kenworthy
George Washington University School of Medicine & Health Sciences, Washington, DC, USA

Division of Pediatric Neuropsychology, Children's National Health System, Washington, DC, USA

S. Dovis
Department of Psychology, University of Amsterdam, Amsterdam, The Netherlands

Youz Child and Adolescent Psychiatry Unit, Zaandam, The Netherlands

H. M. Geurts
Department of Psychology, University of Amsterdam, Amsterdam, The Netherlands

Dr. Leo Kannerhuis Amsterdam, Amsterdam, The Netherlands

T. Strobach, J. Karbach (eds.), *Cognitive Training*,
https://doi.org/10.1007/978-3-030-39292-5_24

Abstract Neurodevelopmental conditions and associated disabilities such as autism spectrum disorders (ASD), attention deficit hyperactivity disorder (ADHD), and learning disorders (LD) become apparent in childhood. These conditions often come with difficulties in cognitive functions, e.g., executive functions (EFs). Targeting EFs in an intervention might benefit these children. The child's brain is malleable, hence susceptible for cognitive training. In this chapter we give an overview of the state of knowledge about the effectiveness of cognitive training for children with ASD, ADHD, and LD. Additionally, we shed some light on cognitive training for pediatric conditions with similar cognitive problems: prematurity, brain tumors, and sickle cell disease. Despite the first promising results from process-based training, transfer to broader cognitive functions and daily life remains challenging. Strategy-based training seems more promising when combined with extensive opportunities for practice. Several factors might influence the effectiveness of cognitive training for children with neurodevelopmental conditions: the type of training, the training level (adaptive), and the targeted behavior. Training multiple functions in a broad variety and focusing on generalization appears most effective.

Cognitive control or executive functioning (EF) is an important predictor for positive life outcome. Children with well-developed EFs show better academic performance, quality of life, and social abilities (e.g., Moffitt et al. 2011). Improving EFs might lead to improvement in several aspects of daily functioning; hence EFs are a popular intervention target. Children with so-called neurodevelopmental conditions often show difficulties in cognitive functioning; hence training cognitive functioning in these children might be particularly promising (see also Johann and Karbach, Rueda et al., Schaeffner et al., this volume).

For the last decades, several studies have focused on active treatment to improve cognitive functions such as working memory (WM, keeping in mind and updating information), cognitive flexibility (the ability to flexibly switch between different tasks and behaviors), and inhibition (stopping an initiated response). EF relies on a broad neural network including, among others, the fronto-striatal network that develops throughout childhood (Gogtay et al. 2004) and hence might be amenable for cognitive training (Wass et al. 2012). This gave rise to the idea that training the brain "like a muscle" (Shipstead et al. 2010) might lead to more effective use in daily life. Initial results appeared promising (Klingberg 2010).

This chapter describes the use of cognitive training for neurodevelopmental conditions. After a description of some neurodevelopmental conditions, process- and strategy-based cognitive training are discussed and their effectiveness for different conditions.

Neurodevelopmental Conditions

Neurodevelopmental conditions are conditions that most often develop during childhood (American Psychiatric Association 2013). An overarching difficulty in these conditions are cognitive functioning problems, although the specifically affected cognitive function differs between and within conditions. The DSM 5 (American Psychiatric Association 2013) describes the following neurodevelopmental conditions: intellectual disabilities (ID), communication disorders, autism spectrum disorders (ASD), attention deficit hyperactivity disorder (ADHD), specific learning disorder (LD), and motor disorder (American Psychiatric Association 2013). In the current chapter, we will focus on the cognitive challenges that children with an ADHD, ASD, and/or LD diagnosis experience, as these are most often targeted in cognitive training. ID and communication and motor disorders are also often included in training studies, given the overlap in symptomatology. For example, a review on cognitive training for children with ID included many studies on children with ADHD (Kirk et al. 2015). Additionally, we will describe pediatric conditions with similar cognitive difficulties (prematurity, brain tumor, and sickle cell disease).

ADHD

Children with ADHD have difficulties with focused attention and may be hyperactive and impulsive (American Psychiatric Association 2013). Theories of ADHD suggest that EF problems are at the core of the ADHD syndrome and play a pivotal role in explaining the difficulties children with ADHD encounter in daily life (e.g., Barkley 2014; Nigg 2006). EF impairments appear related to problems in attention, hyperactivity, and impulsivity (e.g., Crosbie et al. 2013; Sarver et al. 2015) and to associated problems such as deficient academic and social functioning (Titz and Karbach 2014; Kofler et al. 2019; Kofler et al. 2018). More specifically, WM and to lesser extent inhibition and cognitive flexibility are impaired in individuals with ADHD (Martinussen et al. 2005; Willcutt et al. 2005).

ASD

Autistic children[1] experience challenges in social interactions and communication, difficulties in dealing with unpredictability, and sensory sensitivities (i.e., social and communication difficulties and restricted and repetitive behavior (American

[1] The term "autistic children" refers to children with an ASC diagnosis. Although preferences with respect to identity-first versus person-first language use are heterogeneous, identity-first language appears to be mostly preferred by autistic adults (Kenny et al. 2016).

Psychiatric Association 2013). EF difficulties are considered a primary problem in autistic children. There is accumulating evidence from meta-analyses that besides specific problems in flexibility and planning (Hill 2004), EF is more globally impaired in ASD, including problems in working memory, flexibility, inhibition, generativity, and organization (Demetriou et al. 2018; Lai et al. 2017), when accounting for cognitive ability and co-occurring ADHD (Lai et al. 2017). Hence, EF impairments can be considered a core problem in autistic children.

Learning Disorders

Children with LD have difficulties in learning and academic skills (American Psychiatric Association 2013), related to reading and/or writing (i.e., dyslexia), or mathematics (i.e., dyscalculia) that do not result from ID. Children with LDs show general cognitive or EF difficulties, and specific problems are reported in verbal and mathematical LD (Moll et al. 2016; Cirino et al. 2015). Children with verbal and mathematical LD show deficits in WM, processing speed, and verbal comprehension (Willcutt et al. 2013), but those with verbal LD show deficits in naming speed and phoneme awareness, and those with mathematical LD show deficits in flexibility (Willcutt et al. 2013). Since EFs appear largely related to academic performance, and children with LD have specific difficulties in academic performance, training EFs might be particularly fruitful in this population.

Pediatric Conditions

Children with some pediatric diagnoses (e.g., prematurity, brain tumors, and sickle cell disease) experience similar cognitive problems as children with neurodevelopmental conditions and might be similarly susceptible to cognitive training. Luckily, the survival rate of preterm children (born before 37 weeks of pregnancy) increases. However, this comes with a cost; many preterm children and children with very low birth weight experience cognitive difficulties early and later in life, such as learning difficulties (maths, reading, and spelling), and EF deficits (fluency, WM, and cognitive flexibility), with larger problems with lower gestational age (Aarnoudse-Moens et al. 2009).

Children with acquired brain injury, with internal (e.g., brain tumor) or external (e.g., trauma) causes, experience cognitive problems related to the affected area, which might diffuse to other areas. Moreover, the affected brain areas might develop differently, leading to difficulties later in development. Treatments, particularly chemotherapy and radiation therapy for brain tumor, additionally negatively influence cognitive functioning (de Ruiter et al. 2013; Araujo et al. 2017). In brain tumor survivors, general cognitive functioning (i.e., intelligence), attention (de Ruiter et al. 2013), and WM and verbal memory (Margelisch et al. 2015) are affected, though not all survivors show these difficulties (Araujo et al. 2017).

Sickle cell disease results from a genetic deficit, causing a change in the shape of the red blood cells. This influences the blood flow and the production of new blood cells. Children with sickle cell disease suffer from chronic anemia and are at a constant risk of blockage of the blood vessels by the sickle-shaped blood cells, which can lead to tissue and organ damage. Blocked blood flow to the brain or brain areas can lead to brain damage and related cognitive problems. Children with sickle cell disease show neuropsychological deficits, such as visual motor functioning (van der Land et al. 2015), and visuospatial WM (Hijmans et al. 2011). The severity of the deficits depends on the presence and location of cortical infarcts (DeBaun et al. 2012) and white matter hyperintensities, which is related to lower intelligence, processing speed, and EF difficulties (van der Land et al. 2015).

Cognitive Training

There are roughly two types or cognitive training programs: process-based and strategy-based training (Jolles and Crone 2012). Process-based training consists of performing a cognitive task repeatedly, ideally with increasing difficulty. The goal is to improve the targeted and related functions and apply these functions better in daily life. Strategy-based training teaches procedures and strategies to improve cognitive functions (e.g., improve WM by imagery, rehearsal, chunking, or story formation). Instructions focus on specific behavior and daily life.

Cognitive training has been studied extensively, and studies differ largely with respect to study design, training program, target group, and outcome measures. The study design is essential to study effectiveness. Firstly, the number of participants who participate *and* complete a training should be sufficient; motivation is essential. Secondly, to test whether improvement in task performance reflects improvement in the underlying mechanism, "transfer" to other tasks should be evaluated. Thirdly, a multiple baseline study can show whether a training leads to improvement in the targeted functions. A randomized controlled trial (RCT, including an active control group and random group assignment) can confirm whether pre- to post-training improvements are actual training effects, accounting for expectancy, Hawthorne, and test-retest effects (see Cochrane and Green, this volume).

Near and Far Transfer

Effectiveness of cognitive training is generally measured on two levels: near and far transfer. Near transfer refers to improvement on different tasks measuring the same function (e.g., does WM training improve performance on other WM tasks). Far transfer refers to generalization to other (related) functions and daily life. EFs are interrelated; hence training one EF could additionally improve other EFs (e.g., training WM might also improve inhibition). Far transfer to daily life indicates that the trained function improves in daily life (e.g., remembering and following up on instructions better after WM training). Long-term effects are additionally important; training effects might diminish upon ceasing the training, as with physical exercise. However, if brain networks are structurally altered, or cognition or strategies are better applied, effects might last longer (see Wenger and Kühn, this volume). In short, while near transfer is needed as proof of the pudding, far transfer to daily life and long-term effects seem essential to confirm training effectiveness.

Process-Based Cognitive Training

Cognitive training programs mainly focused on WM, and cognitive flexibility and inhibition training were also studied (see Karbach and Kray, Könen et al., Strobach and Schubert, this volume). The training consists of repeatedly performing a cognitive task over a fixed period (e.g., 30 minutes/day for 5 weeks). Most training programs are adaptive; i.e., the task increases in difficulty, ideally adaptive to individual performance, to ensure continuous training at the top of one's ability (Diamond and Ling 2016).

WM training has been studied in different modalities (i.e., audio and visual) and forms (i.e., spatial and verbal). An example of a visuospatial WM training is blocks lighting up one by one in a grid, to be repeated by mouse clicks. An example of a visual verbal WM task is the N-back task, in which a sequence of words is displayed and participants have to state whether the current word is similar to N words earlier. Increasing the number of stimuli to remember increases the difficulty (number of blocks that light up or number of words (N) to keep in mind; see Könen et al., this volume).

An extensively studied WM training protocol is the Cogmed WM training (Roche and Johnson 2014, e.g., Klingberg 2010). The basic Cogmed WM training consists of 25 sessions (30–45 minutes), each including eight exercises to be performed five times per week for 5 weeks. This training has been studied in various populations, and various outcome measures were studied, such as WM, condition-specific difficulties (e.g., attention in ADHD), and general factors (e.g., intelligence, academic performance) (Shipstead et al. 2012). In their review, Chacko et al. (2013) described Cogmed WM training as "possibly efficacious." Cogmed seems to improve performance on trained tasks; there are indications of improvement on untrained cognitive tasks, but findings on improvement of behavior were mixed (See also Shipstead et al. 2012).

There are alternative computerized WM training programs available, such as Jungle Memory, Cognifit (Melby-Lervåg and Hulme 2013), BrainTwister verbal WM training (Buschkuehl et al. 2008), and Braingame Brian (Prins et al. 2013). The latter includes a WM, cognitive flexibility, and inhibition training. Although Cogmed is studied most extensively, reviews and meta-analyses report no large differences between programs (Klingberg 2010; Kassai et al. 2019). Melby-Lervag and Hulme (2013) reported that Cogmed was only more effective to improve visuospatial WM (not in other outcome measures) than other programs, though not when compared to other commercially available programs (Cognifit and Memory booster). Hence, although Cogmed is the most studied WM training program, it does not seem more effective than other (commercially available) programs.

Cognitive flexibility training has been studied with RCTs in ASD (de Vries et al. 2015) and ADHD (Kray et al. 2012; Dovis et al. 2015b). Cognitive flexibility is mostly trained with switch tasks; stimuli have to be sorted on alternating rules (e.g., color or form) (Monsell 2003). These switch tasks are performed repeatedly and adaptively; speed increases when performance improves. A meta-analysis showed that near-transfer effects of cognitive flexibility training were small though significant. However, the far-transfer effect was very small and not significant (Kassai et al. 2019). The number of studies that focused on cognitive flexibility is limited, and effects are less promising compared to WM training. However, alternating between different tasks during cognitive training does seem to enhance training effectiveness (Buitenweg et al. 2012), and training cognitive flexibility might be more effective when several EFs are trained simultaneously (Dovis et al. 2015b).

Training inhibition can be done with Stroop-like tasks (Stroop 1935), flanker tasks (Eriksen and Eriksen 1974), stop tasks, or go-no go tasks. Tasks become more challenging (adaptive to performance) by increasing speed or decreasing the time to inhibit a response. For example, in a stop task one has to respond to a stimulus, but inhibit the response when a (visual or auditory) stop sign appears. Inhibiting the response is more challenging when the time between stimulus and stop time increases. Training inhibition appears challenging compared to WM and cognitive flexibility, with small though significant near-transfer effects and very small and nonsignificant far-transfer effects (Kassai et al. 2019).

Strategy-Based Cognitive Training

Reviews of treatment research in ADHD and ASD have called for expansion of cognitive interventions that "directly target neuropsychological processes," or EF (Sonuga-Barke et al. 2013), including strategy-based training. Organization training was classified as a "well-established" strategy-based intervention to improve EF, based on its efficacy in a number of trials for children and young adolescents (See review: Evans et al. 2014). These interventions focus on planning, organizing, and time management strategies, and some include extensive practice until the strategies become habits, a key component of effective EF strategy training interventions (Diamond and Ling 2016).

Various strategy-based training methods have been developed, varying from a digital game to teach time management, planning, and organizing skills to children, with extensive practice (Bul et al. 2016), to a clinic-based cognitive training for adolescents, augmented by parent training with little practice of the skills (Sprich et al. 2016; Boyer et al. 2015).

Ylvisaker and Feeny (1998, 2009) created a strategy-based cognitive training model, incorporating self-regulation, EF flexibility, and meta-cognitive training for pediatric traumatic injury (see also Schaeffner et al., this volume). In this "coaching" methodology, adults model good EF strategies, scaffold and support children as they try to use the EF strategies, and only fade support when the child has lots of practice and success at initiating, completing, and generalizing the use of the skill to multiple settings. Cognitive and behavioral self-regulation strategies are taught through "self-regulatory scripts" (words/phrases that support good EF, like "Big Deal/Little Deal") and routines that guide behavior when executive demands are high (Ylvisaker and Feeney 2009). The Unstuck and On Target curriculum (Kenworthy et al. 2014a; Cannon et al. 2011) builds on this model to develop a strategy-based intervention to improve flexibility, goal setting, and planning in children with an ASD and ADHD diagnosis. Evidence-based teaching techniques from ASD and ADHD interventions are included, such as use of visual cues and positive reinforcement (Wong et al. 2015). Unstuck and On Target is delivered in a small groups, teaching with self-regulatory scripts what flexibility, planning, and organization are, the usefulness, and how to be more flexible, planful, and organized. School-based sessions introduce and practice skills using games, vignettes/movies, consistent visuals, and extensive modeling. Parent training is provided so that the same skills and self-regulatory vocabulary is modeled and reinforced at home and at school.

Stichter and colleagues' strategy building social competence intervention addresses social-emotional skills and includes an EF component of problem-solving. The intervention, delivered in a clinic setting, included instruction, modeling, and practice of skills, with a strong emphasis on a "scaffolded approach by which the concepts and skills learned in previous units were incorporated into latter units" (P. 1071: Stichter et al. 2010).

Important Considerations

Initial level of cognitive functioning might influence training effects as lower functioning children might have more room for improvement (Jolles and Crone 2012). This so-called compensation effect is found in process-based EF training (Karbach and Unger 2014). On the other hand, initially higher performing children might be predisposed to perform better, reach a higher optimal level, and thus improve more. This so-called magnification effect has been found in strategy-based cognitive training (Karbach and Unger 2014). Moreover, despite positive training effects, an individual might not reach a "normal" level of cognitive functioning (Jolles and Crone 2012); hence managing expectancies is important. It is important to focus on the most pressing problem, as, besides the obvious necessity, this might increase essential motivation.

Training might be relatively more effective in younger children, despite the shorter training times (Wass et al. 2012). This might result from the larger plasticity of the young brain, although the underdeveloped brain has a limited optimal reachable performance (Jolles and Crone 2012). Moreover, it is challenging to keep young children motivated and focused, hence the shorter training times.

Motivation is essential, as cognitive training acquires sustained attention and endurance (see Katz et al., this volume). This is particularly challenging for children with neurodevelopmental conditions with attention and concentration difficulties. Incorporating rewards in the training improves motivation. However, reward sensitivity and the effectiveness of reward types vary. Autistic children appear less sensitive to social and more sensitive to tangible rewards (Dichter et al. 2012; Demurie et al. 2012). Children with ADHD respond better to immediate than postponed rewards (Tripp and Alsop 2001) and better to tangible than implicit rewards (Kohls et al. 2009). Besides its importance for training completion, motivation and reward sensitivity might increase training-induced improvement (Keshavan et al. 2014). Rewards hence seem essential to consider to improve training commitment and effectiveness.

Finally, it is important that a treatment is intensive, adaptive, and varying (Buitenweg et al. 2012; Diamond and Ling 2016). Training at the border of one's abilities is essential for improvement (Diamond and Ling 2016), and variation is important for motivation and effectiveness.

Training Effects

Neurodevelopmental conditions often co-occur, and there is overlap in cognitive problems between conditions; hence studies that focus on a specific condition often include a heterogeneous population. This comes to light in reviews, e.g., a meta-analytic review about LD (Peijnenborgh et al. 2016) and a review about training effectiveness in ID (Kirk et al. 2015), including many studies on children with ADHD. Cognitive training has been studied extensively in ADHD, while for other conditions studies are limited, but effectiveness of cognitive training might not be very condition specific.

ADHD

Meta-analyses suggest that process-based EF training for children with ADHD mainly improve on measures of near transfer, but have very limited far-transfer effects (Dovis et al. 2015a; Hodgson et al. 2014; Rapport et al. 2013; Sonuga-Barke et al. 2013; Chacko et al. 2013). In most placebo-controlled EF training studies transfer to measures of untrained EF has been limited at best, and effects on parent- or teacher-rated behavior (e.g., ADHD or EF) are generally not found (Dovis et al. 2015b).

ADHD is a heterogeneous condition, and not all children with ADHD have EF deficits (e.g., Dovis et al. 2015c; Fair et al. 2012; Nigg et al. 2005). Possibly, the meta-analytic findings might have been more positive if only children with ADHD with EF impairments were selected for training. EF-impaired children might benefit from EF training, because they have more room for improvement (Diamond 2012). Moreover, in EF-unimpaired children with ADHD, symptoms might not originate from impairments in EF, and training would probably have less impact on ADHD symptoms. However, a recent placebo-controlled moderation study (Dovis et al. 2019) did not support this suggestion; training only those children with EF impairments with ADHD will probably not change the conclusions of the earlier meta-analyses. In short, in children with ADHD, EF training can produce changes in EF performance, but in its current form, it seems not more effective than a placebo training in improving symptoms of ADHD or EF behavior in everyday life.

Strategy-based training seems to improve parent and adolescent reported symptoms of ADHD, but there is limited evidence of reduced impairment in EF (Boyer et al. 2015; Boyer et al. 2016b; Evans et al. 2014). The long-term effects of a planning focused intervention appeared more promising for adolescents with less depressive, but more anxiety, symptoms than a solution-focused therapy, which seemed more promising for those who had more depressive symptoms (Boyer et al. 2016a).

ASD

Process-based cognitive training appears feasible in ASD, but children with ASD and/or ID might need more time to complete a training (Benyakorn et al. 2018). Weckstein et al. (2017) reported subjective improvement of Cogmed training in attention, impulsivity, emotional reactivity, and in academics and social interaction and awareness. Although promising, this study did not include a control group. A blind RCT of a WM and cognitive flexibility training for autistic children with Braingame Brian (de Vries et al. 2015) reported limited effects. Both the WM and flexibility training induced near-transfer effects, but improvements did not transfer to other EFs nor daily life. However, a pilot study showed that attention training in a school-based setting did induce some improvement in cognitive and academic measures in autistic children, as compared to an active control group (Spaniol et al. 2018). In short, process-based cognitive training for autistic children seems feasible, but effectiveness needs to be confirmed.

The strategy-based Unstuck and On Target curriculum was evaluated in two random effectiveness trials in public and low-income schools. The first trial compared Unstuck and On Target to an equal intensity social skills curriculum, following the same teaching format to introduce social communication skills lessons (Baker 2003). The participants (7–11 year, ASD without ID) improved in both interventions. The interventions had equivalent impacts on social skills. However, Unstuck and On Target resulted in greater improvements than the social skills intervention on treatment-blind EF measures (flexibility, planning, organization) and problem-solving in a variety of modalities, including parent/teacher report and treatment-masked performance-based tasks and observed classroom behavior (Kenworthy et al. 2014b). The second trial (children with ASD, ADHD, and flexibility problems) compared the effectiveness of Unstuck and On Target and an adaptation of contingency behavioral management (parents and teachers supporting students). Both interventions targeted EF, Unstuck and On Target with cognitive training techniques, and parents and teachers supporting students with behavior techniques. Unstuck and On Target resulted in more improvement on treatment-masked EF tasks and EF-related classroom behaviors for autistic children than parents and teachers supporting students. Children with ADHD improved with either intervention (Anthony et al. 2019).

Several cognitive social skills interventions for autistic children include an EF training component. Uncontrolled single group studies reveal improvement on parent reported EF behaviors and EF/problem-solving task performance (See Wallace et al. 2016 for a review). Fisher and Happé (2005) compared a strategy-based theory of mind and EF training program for autistic children. Both training groups improved in theory of mind task performance compared to the control groups, but none improved in EF task performance. In a cognitive intervention, Solomon et al. (2004) evaluated the social adjustment enhancement curriculum, which addressed social understanding and EFs, "with a special emphasis on individual and group problem solving." Lessons focused on specific EF skills (identifying relevant information, prioritizing problems, flexibly generating alternative solutions to problems). The younger (8–10 years old) children improved in problem-solving compared to a waitlist control group.

There is some evidence that strategy-based EF training is effective in people with autism or ADHD. Extensive practice is necessary, and there are few well-controlled RCTs of these interventions. Limited data appears to indicate that EF strategy training interventions can be implemented across difference neurodevelopmental conditions, while behavioral interventions may require condition-specific interventions.

Learning Disorders

In their meta-analysis, Peijnenborgh et al. (2016) state that up until then no studies specifically focused on effectiveness of WM training for verbal and nonverbal LDs. This meta-analysis (13 RCTs: 10 ADHD, 1 ADHD+LD, 2 LD unspecified) reported promising effects. Short-term near-transfer effects on verbal and visuospatial WM and far transfer on decoding (correctly pronouncing written words) were small to medium and lasted until 8 months follow-up (Peijnenborgh et al. 2016).

Pediatric Conditions

Cogmed might induce improvement in WM and verbal learning task performance in preterm adolescents (Løhaugen et al. 2011) and, in WM, language and attention in preterm children (Grunewaldt et al. 2013). The studies were no RCTs, but Løhaugen et al. (2011) included a non-active typically developing control group which did not improve, and Grunewald et al. (2013) used a crossover design. Memory strategy training is possibly more effective for preterm children than WM training (Everts et al. 2019) as compared to a waitlist control group. Moreover, a pilot study suggests that a combined WM, cognitive flexibility, and inhibition training with Braingame Brian might be feasible in preterm children with attention difficulties (Aarnoudse-Moens et al. 2018), leading to improved WM and speed, but not inhibition, cognitive flexibility, and daily life attention.

A recent review of cognitive training for childhood cancer survivors (four computerized and nine strategy-based interventions) described no adverse effects of cognitive training for cancer survivors (Olson and Sands 2016). Attention seems to be particularly susceptible for improvement, and WM and math showed mixed results. EF and reading seemed challenging to improve. Pre-intervention IQ positively predicted the effectiveness.

Cogmed training might improve WM in children with sickle cell disease, as compared to a waitlist control group (Hardy et al. 2016b). However, completing the training was very challenging (dropout rate ±50%) (Hardy et al. 2016a), and there seemed to be a dose effect (Hardy et al. 2016b). Feasibility and effectiveness of cognitive training are yet to be confirmed in sickle cell disease. Cognitive training effectiveness for the abovementioned pediatric conditions should be studied thoroughly with RCTs before firm conclusions can be drawn.

Discussion

Many children with neurodevelopmental conditions have cognitive problems; hence cognitive training might be fruitful to improve daily life functioning. The abundance of research on the topic shows that both process-based and strategy-based cognitive training might improve the trained cognitive function, but that far-transfer effects remain a challenge for primarily the process-based cognitive training programs (Diamond and Lee 2011; Diamond and Ling 2016; Karbach and Unger 2014; Melby-Lervåg and Hulme 2013).

A recent meta-analysis including 38 process-based cognitive training studies showed that there are significant medium and heterogeneous near-transfer effects of EF training. The effect sizes for WM training are higher than for cognitive flexibility and inhibition training. However, far-transfer effects are small, nonsignificant, and not heterogenous (Kassai et al. 2019). Inhibition appears difficult to improve. This might be particularly challenging for children with ADHD, who show inhibition difficulties, and might hence need improvement in inhibition most urgently.

Strategy training programs seem (more) promising, especially far-transfer effects. However, there are far less studies focusing on this specific type of interventions for neurodevelopmental conditions as compared to the process-based cognitive training programs. These studies are often less methodologically rigorous compared to the recent process-based training studies, though similar to the earlier studies. It is promising for clinical practice that direct community-based studies are run with these cognitive strategy training programs given the gap between the outcome of efficacy trials and outcomes observed in the community (Nahmias et al. 2019).

The metaphor of training EF like training a "muscle" (Shipstead et al. 2010) does not do the truth just. While improving specific task performance can be useful in certain circumstances (e.g., exam training; the sheer repetition of the content increases consolidation of content), it is not useful for EF. Moreover, it is theoretically challenging that performing a task repeatedly improves task performance, but not the underlying construct (see Könen and Auerswald, Schmiedek, this volume). This raises concerns regarding the construct validity of the task used to measure and train a specific construct and about the working mechanism: What exactly are we training?

The effectiveness of cognitive training possibly largely results from the support during the training: parent support, clinician's involvement, and personal attention (Roche and Johnson 2014). Cognitive training might not be a replacement for regular treatment for children with neurodevelopmental conditions, but could be a promising supplement to support other treatment methods. The best time to induce a training is yet unknown and might differ per condition and child. Training at a younger age might be better (Wass et al. 2012), but the child's brain is still in development with limited plasticity (Jolles and Crone 2012).

Different conditions come with specific challenges. For children with ADHD it might be challenging to improve inhibition, and for ASD and sickle cell disease the high dropout rates form a serious challenge. Preterm children show many similarities with children with neurodevelopmental conditions and hence experience similar challenges. For brain tumor survivors the difficulties and training effects largely depend on the type and location of the tumor, the presence of hydrocephaly, and the received treatment. For children with sickle cell disease, the risk of new vain blockages is always present; hence training effects might be temporary. Moreover, other concerns, such as pain control, might be more pressing.

Future opportunities to improve far transfer might be to increase ecological validity of process-based training, and generalization tasks should be incorporated within the training, to ensure better applications in daily life (see Colzato and Hommel, this volume). It would be worthwhile to study whether combined process and strategy training would induce near and far transfer on short and long term.

References

Aarnoudse-Moens, C. S. H., Weisglas-Kuperus, N., van Goudoever, J. B., & Oosterlaan, J. (2009). Meta-analysis of neurobehavioral outcomes in very preterm and/or very low birth weight children. *Pediatrics, 124*, 717–728.

Aarnoudse-Moens, C. S., Twilhaar, E. S., Oosterlaan, J., Van Veen, H. G., Prins, P. J., Van Kaam, A. H., & van Wassenaer-Leemhuis, A. G. (2018). Executive function computerized training in very preterm-born children: A pilot study. *Games for Health Journal, 7*, 175–181.

American Psychiatric Association. (2013). *Diagnostic and statistical manual of mental disorders* (5th ed.). Arlington: American Psychiatric Association.

Anthony, L., Anthony, B., & Kenworthy, L. (2019). *A community-based executive function intervention for children from low income schools with ADHD and ASD*. Final Research Report, Patient-Centered Outcomes Research Institute (Peer and Institute Reviewed).

Araujo, G. C., Antonini, T. N., Anderson, V., Vannatta, K. A., Salley, C. G., Bigler, E. D., & Lo, W. (2017). Profiles of executive function across children with distinct brain disorders: Traumatic brain injury, stroke, and brain tumor. *Journal of the International Neuropsychological Society, 23*, 529–538.

Baker, J. (2003). *Social skills training for children and adolescents with Asperger syndrome and social communications problems* (1st ed.). Shawnee Mission: Autism Asperger Publishing.

Barkley, R. A. (2014). *Attention-deficit hyperactivity disorder: A handbook for diagnosis and treatment*. New York: Guilford Publications.

Benyakorn, S., Calub, C. A., Riley, S. J., Schneider, A., Iosif, A. M., Solomon, M., … & Schweitzer, J. B. (2018). Computerized cognitive training in children with autism and intellectual disabilities: Feasibility and satisfaction study. *JMIR Mental Health, 5*, e40.

Boyer, B. E., Geurts, H. M., Prins, P. J., & Van der Oord, S. (2015). Two novel CBTs for adolescents with ADHD: The value of planning skills. *European Child & Adolescent Psychiatry, 24*, 1075–1090.

Boyer, B. E., Doove, L. L., Geurts, H. M., Prins, P. J., Van Mechelen, I., & Van der Oord, S. (2016a). Qualitative treatment-subgroup interactions in a randomized clinical trial of treatments for adolescents with ADHD: Exploring what cognitive-behavioral treatment works for whom. *PLoS One, 11*, e0150698.

Boyer, B. E., Geurts, H. M., Prins, P. J., & Van der Oord, S. (2016b). One-year follow-up of two novel CBTs for adolescents with ADHD. *European Child & Adolescent Psychiatry, 25*, 333–337.

Buitenweg, J. I., Murre, J. M., & Ridderinkhof, K. R. (2012). Brain training in progress: A review of trainability in healthy seniors. *Frontiers in Human Neuroscience, 6*, 183.

Bul, K. C., Kato, P. M., Van der Oord, S., Danckaerts, M., Vreeke, L. J., Willems, A., … & Franken, I. H. (2016). Behavioral outcome effects of serious gaming as an adjunct to treatment for children with attention-deficit/hyperactivity disorder: A randomized controlled trial. *Journal of Medical Internet Research, 18*, e26.

Buschkuehl, M., Jaeggi, S., Kobel, A., Perrig, W. J. (2008). *Braintwister–Aufgabensammlung für kognitives Training*. Universität Bern: Institut für Psychologie.

Cannon, L., Kenworthy, L., Alexander, K. C., Werner, M. A., & Anthony, L. (2011). *Unstuck and on target! Teaching flexibility and goal directed behavior to children with ASD* (2nd ed.). Baltimore: Brookes Publishing Company.

Chacko, A., Feirsen, N., Bedard, A. C., Marks, D., Uderman, J. Z., & Chimiklis, A. (2013). Cogmed working memory training for youth with ADHD: A closer examination of efficacy utilizing evidence-based criteria. *Journal of Clinical Child & Adolescent Psychology, 42*, 769–783.

Cirino, P. T., Fuchs, L. S., Elias, J. T., Powell, S. R., & Schumacher, R. F. (2015). Cognitive and mathematical profiles for different forms of learning difficulties. *Journal of Learning Disabilities, 48*, 156–175.

Crosbie, J., Arnold, P., Paterson, A., Swanson, J., Dupuis, A., Li, X., … & Schachar, R. J. (2013). Response inhibition and ADHD traits: Correlates and heritability in a community sample. *Journal of Abnormal Child Psychology, 41*, 497–507.

De Ruiter, M. A., Van Mourik, R., Schouten-Van Meeteren, A. Y., Grootenhuis, M. A., & Oosterlaan, J. (2013). Neurocognitive consequences of a paediatric brain tumour and its treatment: A meta-analysis. *Developmental Medicine & Child Neurology, 55*, 408–417.

de Vries, M., Prins, P. J., Schmand, B. A., & Geurts, H. M. (2015). Working memory and cognitive flexibility-training for children with an autism spectrum disorder: A randomized controlled trial. *Journal of Child Psychology and Psychiatry, 56*, 566–576.

DeBaun, M. R., Armstrong, F. D., McKinstry, R. C., Ware, R. E., Vichinsky, E., & Kirkham, F. J. (2012). Silent cerebral infarcts: A review on a prevalent and progressive cause of neurologic injury in sickle cell anemia. *Blood, 119*, 4587–4596.

Demetriou, E. A., Lampit, A., Quintana, D. S., Naismith, S. L., Song, Y. J. C., Pye, J. E., ... & Guastella, A. J. (2018). Autism spectrum disorders: A meta-analysis of executive function. *Molecular Psychiatry, 23*, 1198–1204.

Demurie, E., Roeyers, H., Baeyens, D., & Sonuga-Barke, E. (2012). Temporal discounting of monetary rewards in children and adolescents with ADHD and autism spectrum disorders. *Developmental Science, 15*, 791–800.

Diamond A. (2012). Activities and programs that improve children's executive functions. *Current Directions in Psychological Science, 21*, 335–341.

Diamond, A., & Lee, K. (2011). Interventions shown to aid executive function development in children 4 to 12 years old. *Science, 333*, 959–964.

Diamond, A., & Ling, D. S. (2016). Conclusions about interventions, programs, and approaches for improving executive functions that appear justified and those that, despite much hype, do not. *Developmental Cognitive Neuroscience, 18*, 34–48.

Dichter, G. S., Damiano, C. A., & Allen, J. A. (2012). Reward circuitry dysfunction in psychiatric and neurodevelopmental disorders and genetic syndromes: Animal models and clinical findings. *Journal of Neurodevelopmental Disorders, 4*, 1–43.

Dovis, S., Agelink van Rentergem, J., & Huizenga, H. M. (2015a). Does cogmed working memory training really improve inattention in daily life? A reanalysis. *PLoS One, 10*, e0119522.

Dovis, S., Van der Oord, S., Wiers, R. W., & Prins, P. J. (2015b). Improving executive functioning in children with ADHD: Training multiple executive functions within the context of a computer game. A randomized double-blind placebo controlled trial. *PLoS One, 10*, e0121651.

Dovis, S., Van der Oord, S., Huizenga, H. M., Wiers, R. W., & Prins, P. J. (2015c). Prevalence and diagnostic validity of motivational impairments and deficits in visuospatial short-term memory and working memory in ADHD subtypes. *European Child & Adolescent Psychiatry, 24*, 575–590.

Dovis, S., Maric, M., Prins, P. J., & Van der Oord, S. (2019). Does executive function capacity moderate the outcome of executive function training in children with ADHD?. *ADHD Attention Deficit and Hyperactivity Disorders, 11*, 445–460.

Eriksen, B. A., & Eriksen, C. W. (1974). Effects of noise letters upon the identification of a target letter in a nonsearch task. *Percept Psychophys, 16*, 143–149.

Evans, S. W., Owens, J. S., & Bunford, N. (2014). Evidence-based psychosocial treatments for children and adolescents with attention-deficit/hyperactivity disorder. *Journal of Clinical Child & Adolescent Psychology, 43*, 527–551.

Everts, R., Schöne, C. G., Mürner-Lavanchy, I., & Steinlin, M. (2019). Development of executive functions from childhood to adolescence in very preterm-born individuals-A longitudinal study. *Early Human Development, 129*, 45–51.

Fair, D. A., Bathula, D., Nikolas, M. A., & Nigg, J. T. (2012). Distinct neuropsychological subgroups in typically developing youth inform heterogeneity in children with ADHD. *Proceedings of the National Academy of Sciences, 109*, 6769–6774.

Fisher, N., & Happé, F. (2005). A training study of theory of mind and executive function in children with autistic spectrum disorders. *Journal of Autism and Developmental Disorders, 35*, 757–771.

Gogtay, N., Giedd, J. N., Lusk, L., Hayashi, K. M., Greenstein, D., Vaituzis, A. C., … & Rapoport, J. L. (2004). Dynamic mapping of human cortical development during childhood through early adulthood. *Proceedings of the National Academy of Sciences, 101*, 8174–8179.

Grunewaldt, K. H., Løhaugen, G. C. C., Austeng, D., Brubakk, A. M., & Skranes, J. (2013). Working memory training improves cognitive function in VLBW preschoolers. *Pediatrics, 131*, e747–e754.

Hardy, S. J., Hardy, K. K., Schatz, J. C., Thompson, A. L., & Meier, E. R. (2016a). Feasibility of home-based computerized working memory training with children and adolescents with sickle cell disease. *Pediatric Blood & Cancer, 63*, 1578–1585.

Hardy, S. J., Hardy, K. K., Wise, S. M., Olson, K. J., Thompson, A. L., & Meier, E. R. (2016b). Initial results of a randomized controlled trial of computerized working memory training in pediatric sickle cell disease. *Blood, 128*, 247.

Hijmans, C. T., Fijnvandraat, K., Grootenhuis, M. A., van Geloven, N., Heijboer, H., Peters, M., & Oosterlaan, J. (2011). Neurocognitive deficits in children with sickle cell disease: A comprehensive profile. *Pediatric Blood & Cancer, 56*, 783–788.

Hill, E. L. (2004). Evaluating the theory of executive dysfunction in autism. *Developmental Review, 24*, 189–233.

Hodgson, K., Hutchinson, A. D., & Denson, L. (2014). Nonpharmacological treatments for ADHD: A meta-analytic review. *Journal of Attention Disorders, 18*, 275–282.

Jolles, D., & Crone, E. A. (2012). Training the developing brain: A neurocognitive perspective. *Frontiers in Human Neuroscience, 6*, 76.

Karbach, J., & Unger, K. (2014). Executive control training from middle childhood to adolescence. *Frontiers in Psychology, 5*, 390.

Kassai, R., Futo, J., Demetrovics, Z., & Takacs, Z. K. (2019). A meta-analysis of the experimental evidence on the near-and far-transfer effects among children's executive function skills. *Psychological Bulletin, 145*, 165–188.

Kenny, L., Hattersley, C., Molins, B., Buckley, C., Povey, C., & Pellicano, E. (2016). Which terms should be used to describe autism? Perspectives from the UK autism community. *Autism, 20*, 442–462.

Kenworthy, L., Anthony, L., & Alexander, K. (2014a). *Solving executive function challenges: Simple ways to get kids with autism unstuck and on target*. Baltimore: Brookes Publishing.

Kenworthy, L., Anthony, L. G., Naiman, D. Q., Cannon, L., Wills, M. C., Luong-Tran, C., ... & Sokoloff, J. L. (2014b). Randomized controlled effectiveness trial of executive function intervention for children on the autism spectrum. *Journal of Child Psychology and Psychiatry, 55*, 374–383.

Keshavan, M. S., Vinogradov, S., Rumsey, J., Sherrill, J., & Wagner, A. (2014). Cognitive training in mental disorders: Update and future directions. *American Journal of Psychiatry, 171*, 510–522.

Kirk, H. E., Gray, K., Riby, D. M., & Cornish, K. M. (2015). Cognitive training as a resolution for early executive function difficulties in children with intellectual disabilities. *Research in Developmental Disabilities, 38*, 145–160.

Klingberg, T. (2010). Training and plasticity of working memory. *Trends in Cognitive Sciences (Regul Ed), 14*, 317–324.

Kofler, M. J., Harmon, S. L., Aduen, P. A., Day, T. N., Austin, K. E., Spiegel, J. A., … & Sarver, D. E. (2018). Neurocognitive and behavioral predictors of social problems in ADHD: A Bayesian framework. *Neuropsychology, 32*, 344–355.

Kofler, M. J., Spiegel, J. A., Soto, E. F., Irwin, L. N., Wells, E. L., & Austin, K. E. (2019). Do working memory deficits underlie reading problems in attention-deficit/hyperactivity disorder (ADHD)?. *Journal of Abnormal Child Psychology, 47*, 433–446.

Kohls, G., Herpertz-Dahlmann, B., & Konrad, K. (2009). Hyperresponsiveness to social rewards in children and adolescents with attention-deficit/hyperactivity disorder (ADHD). *Behavioral and Brain Functions, 5*, 20.

Kray, J., Karbach, J., Haenig, S., & Freitag, C. (2012). Can task-switching training enhance executive control functioning in children with attention deficit/-hyperactivity disorder?. *Frontiers in Human Neuroscience, 5*, 180.

Lai, C. L. E., Lau, Z., Lui, S. S., Lok, E., Tam, V., Chan, Q., ... & Cheung, E. F. (2017). Meta-analysis of neuropsychological measures of executive functioning in children and adolescents with high-functioning autism spectrum disorder. *Autism Research, 10*, 911–939.

Løhaugen, G. C., Antonsen, I., Håberg, A., Gramstad, A., Vik, T., Brubakk, A. M., & Skranes, J. (2011). Computerized working memory training improves function in adolescents born at extremely low birth weight. *The Journal of Pediatrics, 158*, 555–561.

Margelisch, K., Studer, M., Ritter, B. C., Steinlin, M., Leibundgut, K., & Heinks, T. (2015). Cognitive dysfunction in children with brain tumors at diagnosis. *Pediatric Blood & Cancer, 62*, 1805–1812.

Martinussen, R., Hayden, J., Hogg-Johnson, S., & Tannock, R. (2005). A meta-analysis of working memory impairments in children with attention-deficit/hyperactivity disorder. *Journal of the American Academy of Child & Adolescent Psychiatry, 44*, 377–384.

Melby-Lervåg, M., & Hulme, C. (2013). Is working memory training effective? A meta-analytic review. *Developmental Psychology, 49*, 270–291.

Moffitt, T. E., Arseneault, L., Belsky, D., Dickson, N., Hancox, R. J., Harrington, H., ... & Sears, M. R. (2011). A gradient of childhood self-control predicts health, wealth, and public safety. *Proceedings of the National Academy of Sciences, 108*, 2693–2698.

Moll, K., Göbel, S. M., Gooch, D., Landerl, K., & Snowling, M. J. (2016). Cognitive risk factors for specific learning disorder: Processing speed, temporal processing, and working memory. *Journal of Learning Disabilities, 49*, 272–281.

Monsell, S. (2003). Task switching. *Trends in Cognitive Sciences, 7*, 134–140.

Nahmias, A. S., Pellecchia, M., Stahmer, A. C., & Mandell, D. S. (2019). Effectiveness of community-based early intervention for children with autism spectrum disorder: A meta-analysis. *Journal of Child Psychology and Psychiatry, 60*, 1200–1209.

Nigg, J. T. (2006). *What causes ADHD?: Understanding what goes wrong and why*. New York: Guilford Press.

Nigg, J. T., Willcutt, E. G., Doyle, A. E., & Sonuga-Barke, E. J. (2005). Causal heterogeneity in attention-deficit/hyperactivity disorder: Do we need neuropsychologically impaired subtypes?. *Biological Psychiatry, 57*, 1224–1230.

Olson, K., & Sands, S. A. (2016). Cognitive training programs for childhood cancer patients and survivors: A critical review and future directions. *Child Neuropsychology, 22*, 509–536.

Peijnenborgh, J. C., Hurks, P. M., Aldenkamp, A. P., Vles, J. S., & Hendriksen, J. G. (2016). Efficacy of working memory training in children and adolescents with learning disabilities: A review study and meta-analysis. *Neuropsychological Rehabilitation, 26*, 645–672.

Prins, P. J., Brink, E. T., Dovis, S., Ponsioen, A., Geurts, H. M., De Vries, M., & Van Der Oord, S. (2013). "Braingame Brian": Toward an executive function training program with game elements for children with ADHD and cognitive control problems. *Games for Health Journal, 2*, 44–49.

Rapport, M. D., Orban, S. A., Kofler, M. J., & Friedman, L. M. (2013). Do programs designed to train working memory, other executive functions, and attention benefit children with ADHD? A meta-analytic review of cognitive, academic, and behavioral outcomes. *Clinical Psychology Review, 33*, 1237–1252.

Roche, J. D., & Johnson, B. D. (2014). Cogmed working memory training product review. *Journal of Attention Disorders, 18*, 379–384.

Sarver, D. E., Rapport, M. D., Kofler, M. J., Raiker, J. S., & Friedman, L. M. (2015). Hyperactivity in attention-deficit/hyperactivity disorder (ADHD): Impairing deficit or compensatory behavior?. *Journal of Abnormal Child Psychology, 43*, 1219–1232.

Shipstead, Z., Redick, T. S., & Engle, R. W. (2010). Does working memory training generalize? *Psychologica Belgica, 50*, 245–276.

Shipstead, Z., Hicks, K. L., & Engle, R. W. (2012). Cogmed working memory training: Does the evidence support the claims? *Journal of Applied Research in Memory and Cognition, 1*, 185–193.

Solomon, M., Goodlin-Jones, B. L., & Anders, T. F. (2004). A social adjustment enhancement intervention for high functioning autism, Asperger's syndrome, and pervasive developmental disorder NOS. *Journal of Autism and Developmental Disorders, 34*, 649–668.

Sonuga-Barke, E. J., Brandeis, D., Cortese, S., Daley, D., Ferrin, M., Holtmann, M., … & Dittmann, R. W. (2013). Nonpharmacological interventions for ADHD: Systematic review and meta-analyses of randomized controlled trials of dietary and psychological treatments. *American Journal of Psychiatry, 170*, 275–289.

Spaniol, M. M., Shalev, L., Kossyvaki, L., & Mevorach, C. (2018). Attention training in autism as a potential approach to improving academic performance: A school-based pilot study. *Journal of Autism and Developmental Disorders, 48*, 592–610.

Sprich, S. E., Safren, S. A., Finkelstein, D., Remmert, J. E., & Hammerness, P. (2016). A randomized controlled trial of cognitive behavioral therapy for ADHD in medication-treated adolescents. *Journal of Child Psychology and Psychiatry, 57*, 1218–1226.

Stichter, J. P., Herzog, M. J., Visovsky, K., Schmidt, C., Randolph, J., Schultz, T., & Gage, N. (2010). Social competence intervention for youth with Asperger syndrome and high-functioning autism: An initial investigation. *Journal of Autism and Developmental Disorders, 40*, 1067–1079.

Stroop, J. R. (1935). Studies of interference in serial verbal reactions. *Journal of Experimental Psychology, 18*, 643–662.

Titz, C., & Karbach, J. (2014). Working memory and executive functions: Effects of training on academic achievement. *Psychological Research, 78*, 852–868.

Tripp, G., & Alsop, B. (2001). Sensitivity to reward delay in children with attention deficit hyperactivity disorder (ADHD). *Journal of Child Psychology and Psychiatry, 42*, 691–698.

van der Land, V., Hijmans, C. T., de Ruiter, M., Mutsaerts, H. J., Cnossen, M. H., Engelen, M., … & Fijnvandraat, K. (2015). Volume of white matter hyperintensities is an independent predictor of intelligence quotient and processing speed in children with sickle cell disease. *British Journal of Haematology, 168*, 553–556.

Wallace, G. L., Yerys, B. E., Peng, C., Dlugi, E., Anthony, L. G., & Kenworthy, L. (2016). Assessment and treatment of executive function impairments in autism spectrum disorder: An update. *International Review of Research in Developmental Disabilities, 51*, 85–122

Wass, S. V., Scerif, G., & Johnson, M. H. (2012). Training attentional control and working memory–Is younger, better? *Developmental Review, 32*, 360–387.

Weckstein, S. M., Weckstein, E. J., Parker, C. D., & Westerman, M. W. (2017). A retrospective chart analysis with follow-up of cogmed working memory training in children and adolescents with autism spectrum disorder. *Medical Science Monitor Basic Research, 23*, 336–343.

Willcutt, E. G., Doyle, A. E., Nigg, J. T., Faraone, S. V., & Pennington, B. F. (2005). Validity of the executive function theory of attention-deficit/hyperactivity disorder: A meta-analytic review. *Biological Psychiatry, 57*, 1336–1346.

Willcutt, E. G., Petrill, S. A., Wu, S., Boada, R., DeFries, J. C., Olson, R. K., & Pennington, B. F. (2013). Comorbidity between reading disability and math disability: Concurrent psychopathology, functional impairment, and neuropsychological functioning. *Journal of Learning Disabilities, 46*, 500–516.

Wong, C., Odom, S. L., Hume, K. A., Cox, A. W., Fettig, A., Kucharczyk, S., ... & Schultz, T. R. (2015). Evidence-based practices for children, youth, and young adults with autism spectrum disorder: A comprehensive review. *Journal of Autism and Developmental Disorders, 45*, 1951–1966.

Ylvisaker, M., Feeney, T. J. (1998). *Collaborative brain injury intervention: Positive everyday routines*. Singular Publishing Group.

Ylvisaker, M., & Feeney, T. (2009). Apprenticeship in self-regulation: Supports and interventions for individuals with self-regulatory impairments. *Developmental Neurorehabilitation, 12*, 370–379.

Changes of Electrical Brain Activity Due to Cognitive Training in Old Adults and Older Industrial Workers

Michael Falkenstein and Patrick D. Gajewski

Contents

Abstract Cognitive training (CT) has effects on performance as well as brain measures. In most CT studies measures of brain structure and perfusion have been used, while methods based on the electroencephalogram (EEG) have been almost neglected. The first part of the chapter provides an overview about CT studies in older adults using EEG-based methods. The results generally reveal enhancements of specific EEG frequency bands or ERP components after CT while timing was not affected. This suggests an enhancement but no acceleration of the underlying processes due to CT. The second part presents some results of a CT study with older industrial workers who showed specific cognitive deficits. CT led to an improvement of the affected functions which was seen in an increase of performance accuracy and enhancement of specific ERP components. In the final part the advantages of brain-related, and particular EEG-based, measures for CT research are outlined and recommendations for their use in future CT studies with older adults and particularly older workers are given.

M. Falkenstein (✉)
Institute for Working Learning Ageing (ALA), Bochum, Germany
e-mail: falkenstein@ala-institut.de

P. D. Gajewski
Leibniz Research Centre for Working Environment and Human Factors at TU Dortmund (IfADo), Dortmund, Germany
e-mail: gajewski@ifado.de

T. Strobach, J. Karbach (eds.), *Cognitive Training*,
https://doi.org/10.1007/978-3-030-39292-5_25

Introduction

The goal of cognitive training (CT) for older adults is to improve their performance on cognitive skills that usually deteriorate with age, but are important for everyday life performance. In most CT studies test-like tasks or games which target different cognitive functions were trained via PC (computerized cognitive training, CCT) for an extended time. Recent meta-analyses suggest that CCT leads to improvements of various cognitive functions and also transfers to untrained cognitive tasks or even everyday situations (far transfer) in healthy older adults (Ballesteros et al. 2015; Bediou et al. this volume; Karbach and Kray this volume; Karbach and Verhaeghen 2014; Kelly et al. 2014; Lampit et al. 2014).

Usually the outcome of CT or CCT is the performance in psychometric or neuropsychological tests. Measures of brain structure or activity go beyond performance in that they reveal the underlying mechanisms which accompany possible changes in test or everyday performance. Among the recent reviews Brehmer et al. (2014) and Bamidis et al. (2014) emphasized brain measures (cf. also Wenger et al. this volume).

Most of the measures used in CT studies are based on anatomic and structural scans obtained by magnet resonance imaging (MRI) and functional scans during task performance by functional magnet resonance imaging (fMRI). fMRI has a high spatial but a poor time resolution. Brehmer et al. (2014) provided a comprehensive review on such changes due to different types of CT in healthy older adults. The electroencephalogram (EEG) records and measures this activity from the human scalp during task performance. Complementary to fMRI, the EEG has a high time resolution that allows detailed analysis of distinct cognitive processing steps affected by CT. Surprisingly there are only a few studies that used the EEG to evaluate effects of CT in healthy older adults, and they are almost neglected in the current reviews (but see Wenger et al. this volume). Thus, the first part of the present chapter aims at filling this gap by providing a comprehensive overview about EEG-based CT studies in healthy older subjects.

In the second part results of an EEG-based CT study with older industrial workers are presented. There is converging evidence that long-term cognitively undemanding jobs increase the age-related cognitive decline (e.g., Finkel et al. 2009). Such unchallenging jobs are much more frequent than cognitively demanding ones. The possibility to use CT to improve cognitive abilities in older workers with cognitively undemanding jobs is of high relevance to enhance their mental fitness and health and thus also their employability.

CT Studies with Older Adults Using EEG-Based Methods

The EEG records the neural activity of cortical and some subcortical sources with electrodes placed on the scalp. Functional activity is either reflected in oscillations at specific frequencies and locations, such as frontal theta activity which has been related to top-down control processes (Onton et al. 2005) and which is usually reduced in older subjects (Cummins and Finnigan 2007).

The other type of functional EEG activity are so-called event-related potentials (ERPs) which can be extracted by averaging EEG segments that are time-locked to specific events, such as stimuli or responses. The ERP components are related to sensory, cognitive, or central motor processes. The brain sources of ERPs can be estimated with source analysis software like LORETA. In older adults some cognitive ERP components are attenuated or delayed (e.g. Gajewski et al. 2010b) while ERPs after irrelevant stimuli are often enhanced (Hahn et al. 2011). In the following the existing CT studies using EEG-based measures are shortly reported.

Falkenstein, Gajewski and colleagues were the first to use ERPs in a CT study. In their "Dortmund training study" 152 healthy older adults were randomized to three different group interventions: CT, physical training, stretching and relaxation training (active control group), and a no-contact control group (Gajewski et al. 2010a; Gajewski and Falkenstein 2012, 2018; Küper et al. 2017; Wild-Wall et al. 2012). Training duration was equal for the three active groups (two sessions of 90 min per week for 16 weeks). For the CT group paper- and PC-based trainings were selected from different commercial products so that each targeted cognitive function was trained with several different tasks. The difficulty of the tasks was adapted and adjusted to the current individual performance of the trainees. A battery of paper- and PC-based psychometric tests that address cognitive control functions was administered before and after the training. During the eight PC-based tasks the EEG was recorded and ERPs were computed. Some of the tasks were insensitive to CT (e.g., antisaccade task, auditory distraction task). However, behavioral and ERP effects due to CT were seen in other executive control tasks mainly for the CT group (visual search, Wild-Wall et al. 2012), task switching (Gajewski and Falkenstein 2012), Stroop (Küper et al. 2017), and 2-back (Gajewski et al. 2018).

The general pattern of the ERP results was an amplitude increase of ERP components related to cognitive processes such as preparation (CNV), task-set retrieval (P2), response selection (N2), resource allocation and working memory (P3b), and error detection (Ne/ERN). In contrast, early sensory ERPs were not affected. However, the ERP effects due to CT differed across the tests. For example, in the switch task increases of the N2, P3b, and Ne were observed which were accompanied by a reduction of mixing costs in accuracy and intraindividual RT variability (Gajewski and Falkenstein 2012). Similar effects of CT were observed using the Stroop task (Küper et al. 2017). In the visual search task the P2 was enhanced which was accompanied by an increased target detection rate. Source analysis (LORETA) located this effect in lingual and parahippocampal brain areas which are linked to visual feature processing (Wild-Wall et al. 2012). In the n-back task the frontal P3a was enhanced in trials preceding a correctly detected target, suggesting enhanced maintenance of information due to CT (Gajewski and Falkenstein, 2018). The enhancement of the CNV after CT in the switch task is of particular interest because it reflects a compensatory mechanism. A strong preparation (i.e., a large CNV) is related to low error rates (Hohnsbein et al. 1998). Hence, the enhancement of the CNV is likely one reason for the reduction of the error rate.

Other studies reported similar findings. For example Anguera et al. (2013) had their subjects play a dual task video game which consisted of tracking while responding to traffic signs with adaptive difficulty for 12 sessions. An active control group only trained the subtasks, while a third group received no training. Only after the dual task training, multitasking costs in a nonadaptive version of the game were reduced, with gains persisting for 6 months. Further, working memory and sustained attention were enhanced. In the EEG, frontal theta power and frontal-posterior theta coherence which were attenuated before the training were found to be enhanced but only after the dual task training. Notably, the dual task training led to changes in the neural processing of signs that reached a level comparable to neural activity patterns observed in younger adults.

O'Brien et al. (2013) trained 11 old subjects with a speed of processing (SOP) training aimed at enhancing perceptual processing of visual stimuli in visual attention tasks. SOP training primarily involves practice of perceptual processing with exercise difficulty adapted to the individual user. Before and after the training a visual search task was administered and ERPs determined. Selective attention to a target was enhanced after the CT compared to a no training control group. In the ERPs the amplitudes of the N2pc and P3b were increased after training, reflecting enhanced focusing of attentional and processing capacity allocation due to the training. The authors conclude that SOP training may be successful in counteracting age-related declines in selective attention.

As mentioned above, older adults have deficits in ignoring distracting irrelevant information. Mishra et al. (2014) conducted two parallel experiments with older rats and humans. They administered a cognitive training with adaptive difficulty of distinguishing between auditory targets and distractors. Training resulted in enhanced discrimination abilities in both species. After CT, neural responses to distractors in the auditory cortex were selectively reduced in both species, mimicking the behavioral effects. Training gains generalized to group and individual level benefits in aspects of working memory and sustained attention. Moreover, frontal theta measures of top-down engagement with distractors were selectively restrained in trained humans. This study shows converging cross-species evidence for training-induced improvement of distractor control.

Rose et al. (2015) administered a CT aimed at improving prospective memory (PM) for 12 sessions. One group received music training and a no-contact group served as active and passive control. Performance and ERPs during a lab-based PM task, real-world PM, and instrumental activities of daily living were assessed before and after training. Importantly, the CT produced far transfer to performance on real-world PM and activities of daily living. The ERPs revealed a reduction of a negative ERP component which is likely related to the processing of PM cues, which suggests more automatic PM retrieval.

In a recent study Gaál and Czigler (2018) administered a task-switching training to young and old subjects in which the difficulty was individually adjusted. The training led to strong performance increases in the old subjects up to the level of the young subjects. Moreover, this performance improvement was maintained 1 year later. The training-related gains were accompanied by an increase in the N2 and P3 amplitude after training that remained larger compared to the pretest in a 1-year follow-up study, while no changes were found in the control groups. Olfers and Band (2018) administered video game training in young participants for 6 weeks and found benefits in task performance and an enhanced N2 after training relative to controls, corroborating the findings in older and middle-aged adults.

In summary the EEG-based CT studies showed enhanced performance accompanied by increased ERPs in executive control tasks (CNV, P2, N2, P3a, P3b, Ne/ERN). Additionally, the increase of frontal theta activity after a complex dual task training suggests improvement of control. In contrast, frontal theta activity was reduced after an auditory discrimination training suggesting less need for control. This reduction of control was paralleled by a selective reduction of distractor processing, as revealed in auditory ERPs. Generally (with one exception), the ERP-based studies showed increases of "cognitive" components while latencies remained unaffected. This was usually paralleled by a reduction of error rate but not a speeding of responses.

EEG-Based Studies with Older Industrial Workers

Challenging work represents an important cognitive stimulation to protect age-related decline or to enhance compensatory mechanisms (Wild-Wall et al. 2009) which may reduce the risk of cognitive decline in older age (Andel et al. 2005). In a longitudinal study with about 3000 workers Marquié et al. (2010) showed that the more cognitively stimulating the work the higher the performance in tests of episodic memory, attention, and speed of processing, and the more favorable the change of these functions over a 10-year follow-up.

The first part of the project PFIFF, a program for improving cognitive abilities in older employees, aimed at investigating cognitive functions in older industrial workers with low vs. high stimulating work characteristics in a cross-sectional design (Gajewski et al. 2010b). Four groups of industrial workers ($n = 91$) of a big car factory participated in the study. One group consisted of older workers (mean age 52) and the other of younger workers (mean age 22). Both groups were again divided: one group worked flexibly in areas such as service and maintenance while the other worked in the repetitive assembly line production. In other aspects the groups were well matched for age, education, and health status. The workers were administered a series of EEG-supported tests as in the "Dortmund training study." The most difficult task was a memory-based task-switching paradigm. The comparison between both groups of the old adults showed that flexibly working older employees responded faster and produced lower error rates than the older assembly line employees. This was accompanied by enhancements of the CNV, the P3b, and the Ne/ERN. These results support the findings in the literature reported above, suggesting a positive influence of long-term flexible and challenging job characteristics on performance in a difficult novel task. The ERP changes reveal in detail the cognitive functions that differed between employees with repetitive and flexible work, namely, preparation, working memory, and error processing.

In a subsequent longitudinal intervention study CT was administered to 120 middle-aged industrial workers (mean age 47 years) with long-term repetitive work from the same factory (Gajewski et al. 2017). The sample was subdivided into two groups with 60 workers each which completed the study successively. The subgroups were randomly assigned either to immediate CT or a waiting control group that received CT later. Participants were trained with a similar supervised variable CT and CCT as used in the "Dortmund training study" for 3 months (two sessions of 90 minutes per week). The training was evaluated with paper-based tests and the EEG-supported memory-based task-switching task as administered in the preceding cross-sectional study.

Compared to the waiting control group the trainees improved their performance in a number of psychometric paper and pencil tests. The results of the task switching showed also improvements of cognitive performance after CT, particularly a decrease of error rates, suggesting improved maintenance of a complex task sequence in working memory. The behavioral benefits were accompanied by a number of electrophysiological changes, in particular an increase of the N2 and the Ne/ERN due to the CT. This pattern of results persisted even 3 months after the training was finished. In contrast no changes in the waiting control group were found at this time point. However, after this group had also received CT, the same performance improvements and ERP changes were observed as in the cognitive training group before corroborating these effects in two independent samples (Gajewski et al. 2017).

The findings suggest that job and/or aging-related deficits in certain cognitive processes can be ameliorated by CT. However, positive effects of a cognitive training might depend on the age or the a priori cognitive state of the trainees. For example, a recent study conducted by Borness et al. (2013) did not find any positive impact of CT on cognition and well-being among white collar office workers. In contrast, in our study (Gajewski et al. 2017) only elderly blue-collar workers were trained who showed clear deficits in the preceding study (Gajewski et al. 2010).

General Discussion and Conclusion

Summary of the Results

In the EEG-based CT studies enhancements of processes have been consistently reported. With respect to EEG oscillations, frontal theta activity and frontal-parietal coherence were enhanced after a dual task CT, which suggests an increase of control. In contrast, theta and hence cognitive control was reduced after a simple discrimination training which led to a decrease of the ERP after irrelevant stimuli. This also suggests that the processing of distracting stimuli, which is usually enhanced in older individuals, can be attenuated due to CT.

Generally, ERPs after task-relevant stimuli show increased amplitudes of cognitive ERP components after CT, while latency reductions have not been observed by now. The amplitude increases were paralleled by improvements of performance quality (reduced error rates) and intraindividual variability of speed (individual standard deviations ISD) but not of speed itself (no change of RTs). The ERP changes were generally restricted to later components which reflect cognitive and response monitoring processes while early components reflecting sensory processes did hardly change. The EEG studies also confirm the findings of behavioral CT studies that an adaptive difficulty of the CT appears to be important for the success. These conclusions are of course preliminary and have to be strengthened in future CT studies with EEG methods.

Benefits of EEG-Based Measures in CT Studies

The crucial advantage of brain measures over mere behavioral measures is that they can reveal changes of brain structure and function underlying performance changes due to CT. Brain measures can also be used to analyze why certain individuals profit more than others from a CT. Differences in brain changes between successful and unsuccessful trainees may help to adapt the training in a way to improve the success also in the latter. In their review Brehmer et al. (2014) stated that such an approach might provide useful information for developing individualized and more specific training programs that target factors related to individual deficits or differences in performance and its neural correlates. Brain measures can be used to track those changes in time and can also disentangle the interplay of different processes and their development during CT. For example, in the Mishra et al. (2014) study the processing of irrelevant stimuli was reduced due to CT which enabled a reduction of cognitive control. Finally, brain measures can reveal strategic or compensatory mechanisms, such as an enhancement of preparation (CNV), as observed in the Dortmund training study which was paralleled by a reduction of error rates. Such results also show that brain measures are important if not indispensable to explain changes in performance measures.

In future studies this potential of brain-derived in addition to behavioral measures should be more fully exploited. In particular, brain measures should be used to investigate differential development of brain changes in the course of CT in young and older trainees in dependence of their training success and further influence factors such as pre-training performance. This could clarify why some old subjects profit more from CT than others, as also seen in the PFIFF study. It could be that successful trainees show similar brain changes than young ones or change their neural pattern toward that of young subjects or rather recruit additional processes to be successful. For example, older subjects often show a larger CNV than young subjects in difficult but not easy tasks (Wild-Wall et al. 2007). This might also happen in the course of CT, as shown in the Dortmund training study. Changes due to CT appear to develop rather early in the course of the training. Hence, in future studies the assessment of brain functions should be performed several times during the course of CT and also after the training in follow-up sessions. Such follow-up measures should consider whether the trainees continued CT in a self-directed manner or not and, if not, how long the training effects persist.

Despite their clear benefits brain measures are not easy to achieve in comparison to behavioral measures. Also it should be considered which of the measures are appropriate and affordable. MRI and EEG-based methods appear to have complementary benefits: while ERPs can reveal changes in the timing and strength of single processes, MRI-based methods have better access to subcortical structures and can reveal changes of brain networks due to training. Ideally the two method domains should be combined. This extends the suggestion of Brehmer et al. (2014) who claimed a multimodal approach by combining several imaging methods in future CT studies. We recommend using EEG-based measures in such combinations or also as single method. EEG is easy and flexible to apply, for example, in field studies, and could be easily used in factories. Moreover, they are nonintrusive and well tolerated by older subjects and do not require large changes in the psychometric test settings. The comfortable application is also well-suited to administer multiple assessments in the course of a CT which is highly useful to track CT-related changes in time. Therefore EEG-based methods should be more often applied in future CT studies.

CT Studies with Industrial Workers

Among the most important results reported in the present chapter is the successful improvement of cognitive functions in older workers with long-term repetitive work in the PFIFF project. This shows that adverse environmental influences on cognitive aging such as unchallenging work can be counteracted by CT.

In the PFIFF project (as well as in the "Dortmund training study") CT was administered in groups by an experienced trainer and care was taken to provide individually adapted difficulty and difficulty increase of the trained tasks. In future CT studies different formats should be explored with respect to compliance, training success, and change of brain measures. For example, a continuously teacher-guided CT (as also suggested by Lampit et al. 2014) is expensive, so future CT studies should explore whether an initial guidance with subsequent self-administered CT yields similar effects as a full teacher-guided CT. Also, the optimum format of CT is unknown, and future CT studies should investigate the impact of features like adaptivity and feedback. Usually CT is rather short, and the 3-month duration in the PFIFF study (or 4 months in the "Dortmund training study") might have been too long and may have even resulted in weariness. Future studies with workers should compare CT formats of different length and assess brain and behavioral changes several times. Such studies should also include younger in addition to older workers, and differences in their pre-training cognitive states should be considered.

In particular, future CT studies with workers should focus on the influence of work-related factors that likely affect cognition (such as type of work, control about work process, and social relations) on the success of the CT. For example, shift work and in particular night shifts are known to affect cognition (Ansiau et al. 2008). Hence, in future studies participants with and without night shift should be included in CT trials and different effects assessed. A simple method to enrich the working environment is rotation of work places and job assignment. Future longitudinal studies should assess whether rotation improves cognition compared to non-rotation in older workers. In such studies also the effects of rotation and CT could be compared.

One of the most intriguing questions is whether an improvement in cognitive functions due to CT in workers also transfers to work-related behavior. By now there are only very few reports on transfer of CT benefits to real life (e.g., Rose et al. 2015). In the PFIFF project the CT improved performance and brain measures in a complex test task (the memory-based switch task). As no far-transfer tasks related to the work were used, no direct conclusions can be drawn from these results regarding the employability of elderly workers. Nevertheless, the observed CT-induced improvements of cognition and performance in this task may indicate enhanced learning ability and self-confidence and thereby improve performance in novel or complex working situations. Such situations, e.g., the change of a complex machine tool under time pressure, are common in modern work environment and are a principal source of stress. In future studies it should be explored whether CT leads to better self-reported and supervisor-reported performance and reduced stress particularly in such situations.

In summary, CT appears a most promising tool for improving mental fitness and employability in older workers. Future CT studies in occupational environments should evaluate far transfer by measurements of work efficiency, individual performance at work, risk of workplace injuries, and work-related illness.

Acknowledgments We want to thank Christiane Westedt and Ludger Blanke for their invaluable organizational and technical support in the EEG-based studies. Moreover we are grateful to Ute Lobisch for improving style and legibility of the draft and for her committed work with text and literature formatting.

References

Andel, R., Crowe, M., Pedersen, N. L., Mortimer, J., Crimmins, E., Johansson, B., & Gatz, M. (2005). Complexity of work and risk of Alzheimer's disease: A population-based study of Swedish twins. *The Journals of Gerontology Series B: Psychological Sciences and Social Sciences, 60*, 251–258.

Anguera, J. A., Boccanfuso, J., Rintoul, J. L., Al-Hashimi, O., Faraji, F., Janowich, J., ... & Gazzaley, A. (2013). Video game training enhances cognitive control in older adults. *Nature, 501*, 97–101.

Ansiau, D., Wild, P., Niezborala, M., Rouch, I., & Marquie, J. C. (2008). Effects of working conditions and sleep of the previous day on cognitive performance. *Applied Ergonomics, 39*, 99–106.

Ballesteros, S., Kraft, E., Santana, S., & Tziraki, C. (2015). Maintaining older brain functionality: A targeted review. *Neuroscience and Biobehavioral Reviews, 55*, 453–477.

Bamidis, P. D., Vivas, A. B., Styliadis, C., Frantzidis, C., Klados, M., Schlee, W., ... & Papageorgiou, S. G. (2014). A review of physical and cognitive interventions in aging. *Neuroscience and Biobehavioral Reviews, 44*, 206–220.

Borness, C., Proudfoot, J., Crawford, J., & Valenzuela, M. (2013). Putting brain training to the test in the workplace: A randomized, blinded, multisite, active-controlled trial. *PLoS One, 8*, e59982.

Brehmer, Y., Kalpouzos, G., Wenger, E., & Lovden, M. (2014). Plasticity of brain and cognition in older adults. *Psychological Research, 78*, 790–802.

Cummins, T. D., & Finnigan, S. (2007). Theta power is reduced in healthy cognitive aging. *International Journal of Psychophysiology, 66*, 10–17.

Finkel, D., Andel, R., Gatz, M., & Pedersen, N. L. (2009). The role of occupational complexity in trajectories of cognitive aging before and after retirement. *Psychology and Aging, 24*, 563–573.

Gaál, Z. A., & Czigler, I. (2018). Task-switching training and transfer: Age-related effects on late ERP components. *Journal of Psychophysiology, 32*, 106–130.

Gajewski, P. D., & Falkenstein, M. (2012). Training-induced improvement of response selection and error detection in aging assessed by task switching: Effects of cognitive, physical, and relaxation training. *Frontiers in Human Neuroscience, 6*, 130.

Gajewski, P. D., Falkenstein, M. (2018). ERP and behavioral effects of physical and cognitive training on working memory in aging: A randomized controlled study. *Neural Plasticity*, ID 3454835.

Gajewski, P. D., Wipking, C., Falkenstein, M., & Gehlert, T. (2010a). *Dortmunder Altersstudie: Studie zur Förderung der Hirnleistungsfähigkeit bei Älteren (Forschungsbericht VV04)*. Berlin: Die deutschen Versicherer (GDV).

Gajewski, P. D., Wild-Wall, N., Schapkin, S. A., Erdmann, U., Freude, G., & Falkenstein, M. (2010b). Effects of aging and job demands on cognitive flexibility assessed by task switching. *Biological Psychology, 85*, 187–199.

Gajewski, P. D., Freude, G., & Falkenstein, M. (2017). Cognitive training sustainably improves executive functioning in middle aged industry workers assessed by task switching: An randomized controlled ERP study. *Frontiers in Human Neuroscience, 11*, 81.

Hahn, M., Wild-Wall, N., & Falkenstein, M. (2011). Age-related differences in performance and stimulus processing in dual task situation. *Brain Research, 1414*, 66–76.

Hohnsbein, J., Falkenstein, M., & Hoormann, J. (1998). Performance differences in reaction tasks are reflected in event-related brain potentials (ERPs). *Ergonomics, 41*, 622–633.

Karbach, J., & Verhaeghen, P. (2014). Making working memory work: A meta-analysis of executive-control and working memory training in older adults. *Psychological Science, 25*, 2027–2037.

Kelly, M. E., Loughrey, D., Lawlor, B. A., Robertson, I. H., Walsh, C., & Brennan, S. (2014). The impact of cognitive training and mental stimulation on cognitive and everyday functioning of healthy older adults: A systematic review and meta-analysis. *Ageing Research Reviews, 15*, 28–43.

Küper, K., Gajewski, P. D., Frieg, C., & Falkenstein, M. (2017). A randomized controlled ERP study on the effects of multi-domain cognitive training and task difficulty on task switching performance in older adults. *Frontiers in Human Neuroscience, 11*, 184.

Lampit, A., Hallock, H., & Valenzuela, M. (2014). Computerized cognitive training in cognitively healthy older adults: A systematic review and meta-analysis of effect modifiers. *PLoS Medicine, 11*, e1001756.

Marquié, J. C., Duarte, L. R., Bessieres, P., Dalm, C., Gentil, C., & Ruidavets, J. B. (2010). Higher mental stimulation at work is associated with improved cognitive functioning in both young and older workers. *Ergonomics, 53*, 1287–1301.

Mishra, J., de Villers-Sidani, E., Merzenich, M., & Gazzaley, A. (2014). Adaptive training diminishes distractibility in aging across species. *Neuron, 84*, 1091–1103.

O'Brien, J. L., Edwards, J. D., Maxfield, N. D., Peronto, C. L., Williams, V. A., & Lister, J. J. (2013). Cognitive training and selective attention in the aging brain: An electrophysiological study. *Clinical Neurophysiology, 124*, 2198–2208.

Olfers, K. J. F., & Band, G. P. H. (2018). Game-based training of flexibility and attention improves task-switch performance: Near and far transfer of cognitive training in an EEG study. *Psychological Research, 82*, 186–202.

Onton, J., Delorme, A., & Makeig, S. (2005). Frontal midline EEG dynamics during working memory. *NeuroImage, 27*, 341–356.

Rose, N. S., Rendell, P. G., Hering, A., Kliegel, M., Bidelman, G. M., & Craik, F. I. (2015). Cognitive and neural plasticity in older adults' prospective memory following training with the Virtual Week computer game. *Frontiers in Human Neuroscience, 9*, 592.

Wild-Wall, N., Hohnsbein, J., & Falkenstein, M. (2007). Effects of ageing on cognitive task preparation as reflected by event-related potentials. *Clinical Neurophysiology, 118*, 558–569.

Wild-Wall, N., Gajewski, P. D., & Falkenstein, M. (2009). Kognitive Leistungsfähigkeit älterer Arbeitnehmer. *Zeitschrift für Gerontologie und Geriatrie, 42*, 299–304.

Wild-Wall, N., Falkenstein, M., & Gajewski, P. D. (2012). Neural correlates of changes in a visual search task due to cognitive training in seniors. *Neural Plasticity, 2012*, 529057.

Cognitive Training in Mild Cognitive Impairment

Benjamin Boller, Laura Prieto del Val, and Sylvie Belleville

Contents

Abstract Alzheimer's disease is characterised by a slow progression and by an extensive prodromal phase during which symptoms are dormant or very mild. The term mild cognitive impairment (MCI) has been used to refer to older adults who do not meet the criteria for dementia but who present cognitive complaints and whose cognitive abilities do not fall within the expected range given their age and education. Longitudinal studies have found that many persons with MCI will later meet these criteria and are thus in the pre-dementia phase of Alzheimer's disease. The potential impact of cognitive training could be remarkable, and these individuals make for ideal candidates for training as they retain the ability to acquire new skills. This chapter describes some of the studies that have measured the efficacy of cognitive training in MCI. One of the goals is to provide guidelines regarding the approach that may be most appropriate for persons with MCI based on cognitive outcomes, subjective out-

B. Boller (✉)
Research Center, Institut universitaire de gériatrie de Montréal and Psychology Department, Université du Québec à Trois-Rivières, Trois-Rivières, QC, Canada
e-mail: benjamin.boller@uqtr.ca

L. P. del Val
Department of Physiology, Anatomy and Cellular Biology, Pablo de Olavide University, Sevilla, Spain

S. Belleville
Research Center, Institut universitaire de gériatrie de Montréal and Psychology Department, Université de Montréal, Montreal, QC, Canada

T. Strobach, J. Karbach (eds.), *Cognitive Training*,
https://doi.org/10.1007/978-3-030-39292-5_26

comes, well-being, and outcomes of everyday life. It also describes some of the results obtained through brain imaging and discusses neuroscience-based models of training. Neuroimaging studies have demonstrated the presence of training-induced neural changes in individuals with MCI. These changes indicate that the integrity of the compensatory and restorative neural mechanisms may be relatively preserved in this population. According to the INTERACTIVE model, the neural response to training is not only modulated by the severity of the disease but also by the training modalities and personal factors such as expertise and level of cognitive reserve.

Introduction

Dementia is diagnosed when acquired cognitive impairment significantly affects the autonomy of the individual. Although dementia can have many causes, Alzheimer's disease (AD) is recognised as the most common aetiology in older adults. The cognitive changes that characterise AD are progressive and the disease evolves over up to 20 years before patients meet criteria for dementia. During this extensive prodromal phase, symptoms are dormant or very mild. The term mild cognitive impairment (MCI) has been used to refer to individuals who may be in a pre-dementia phase of AD and who have an elevated likelihood of progressing to the disease. The presence of a subjective complaint, which indicates that the individual is aware of their cognitive changes, is a main characteristic of MCI. For this reason, and because the ability to learn new skills and strategies is preserved in this population, persons with MCI are particularly well suited to benefit from cognitive stimulation, which could significantly improve their quality of life. This chapter is a qualitative review of the studies measuring the impact of cognitive training in persons with MCI. Section "Mild Cognitive Impairment as a Target for Cognitive Training" will introduce the concept of MCI and the reasons why this phase is believed to be appropriate for cognitive training. Section "Memory Training" will present studies on memory training, section "Training of Attentional Control" studies on attentional or executive training, and section "Training Imbedded in Real Life: Virtual Reality and Leisure Approaches to Cognitive Training" will present strategies to promote generalisation of the acquired skills in everyday life. Finally, studies relying on neuroimaging will be presented, followed by models of the training-induced brain changes.

Mild Cognitive Impairment as a Target for Cognitive Training

MCI represents a cognitive decline that is greater than what is considered normal based on the individual's age and educational level, but that is not significant enough to limit independence in daily life activities and meet criteria for dementia. Though the original MCI concept required the presence of memory difficulties, its current definition includes impairment in non-memory domains and the possibility that MCI may progress to neurodegenerative diseases other than AD (Albert et al. 2011). A person with MCI can be categorised based on whether one or more cognitive domains are affected (i.e. single vs. multiple domain MCI) and whether they are amnestic (a-MCI) or non-amnestic MCI. The a-MCI subtype has received considerable attention since it is the subtype that most likely represents prodromal AD.

Appropriate models of cognitive training for MCI should rely on an understanding of which functions are impaired and which are preserved. Cognitive functions have been greatly researched in MCI and a pattern characteristic of MCI symptomatology is emerging (Belleville et al. 2008). Episodic memory, which is the ability to encode and retrieve new information that is embedded in a spatio-temporal context, appears to be the cognitive component that is the most impaired. Working memory (see Könen et al., this volume), the ability to manipulate maintained information, is also impaired in a-MCI, whereas short-term memory and implicit memory seem to be preserved. Executive functions (see Karbach and Kray, this volume), on the other hand, including response inhibition, switching, cognitive flexibility, and abstract thinking, seem to diminish in MCI.

Many factors make MCI a suitable target population for cognitive training: (i) given that pharmacological treatment such as cholinesterase inhibitors (ChEIs) has not been successful in MCI (Petersen et al. 2005), non-pharmacological treatment may be an appropriate and risk-free alternative to improve cognitive functions; (ii) individuals with MCI maintain a certain degree of cognitive plasticity that allows them to learn and apply new strategies; (iii) symptomatic treatment would produce the maximum benefit when applied at the earliest time point of the AD process; (iv) observational studies indicate that cognitive stimulation can have an impact on cognitive decline and dementia; (v) a cognitively stimulating lifestyle has found to be among the most important protective factors against dementia (Barnes and Yaffe 2011).

One major support for cognitive stimulation is that it protects against age-related cognitive decline and dementia. Education, learning new things, or enjoying a challenging job are mentally invigorating and represent life course models of mental stimulation. There is growing evidence that differences in cognitive lifestyles affect age-related cognitive decline and resistance to neurodegenerative diseases. Most of the evidence comes from observational studies examining the association between different lifestyle factors and cognitive decline or dementia. Barnes and Yaffe (2011) indicated that cognitive inactivity, most often measured with level of formal education, was associated with a 59% increased risk of developing AD and was estimated to account for about 19% of AD cases worldwide. The authors estimated that reducing the prevalence of low education attainment by 10% would reduce the incidence of AD by about 534,000 cases. Thus, observational studies indicate that cognitive stimulation across the lifespan determines differences in the risk for age-related neurodegenerative diseases and that reducing cognitive inactivity has the potential to substantively affect the prevalence of cognitive impairment.

Memory Training

Episodic memory is the most severely affected cognitive function and the main complaint in MCI. Thus, cognitive training as a way of promoting the maintenance and improvement of episodic memory in older adults with MCI has attracted major attention. Memory training programs typically focus on teaching strategies to encourage richer encoding or to facilitate retrieval (see also Wenger et al., this volume) and they rely on aspects of memory that are relatively well preserved in MCI such as semantic knowledge, visual imagery, or implicit retrieval. A large number of mnemonic strategies and procedures have been used including errorless learning, spaced retrieval, mind mapping, cueing, semantic organisation and elaboration, mental imagery, and the method of loci. Some are quite effortful and demand strong metacognitive abilities such as the method of loci, which requires the individual to produce an interactive image between items he/she is to learn and a series of loci in a familiar environment. Other procedures rely on more automatic memory systems such as space retrieval, where information is recalled multiple times at increasingly longer intervals. Most studies employ a combination of mnemonic strategies so as to provide patients with a broad set of tools. Most programs comprise a face-to-face intervention in which a therapist teaches these strategies and provides guidance and practice on either an individual or small-group basis.

Several studies have found that these strategies improve proximal memory measures, whether they are tested with immediate or delayed free recall of words (Belleville et al. 2006; Olchik et al. 2013), recognition (Herrera et al. 2012), event-related prospective memory (Tappen and Hain 2014), or face-name associations (Belleville et al. 2006). Some of them show that an active control comparison group (Herrera et al. 2012; Olchik et al. 2013; Tappen and Hain 2014) benefited less than the group receiving memory training, suggesting that performance gains are not entirely attributable to non-specific stimulation. Subjective memory seems to also benefit from memory training when the program introduces the notion that older adults can cope with memory problems or when cognitive restructuring of memory-related beliefs is provided (Belleville et al. 2006; Rapp et al. 2002). Targeting these components in MCI is relevant, as it can contribute to increasing self-efficacy – the perception that individuals have control over their memory – and can reduce MCI-related anxiety and depression. Overall, these studies indicate that memory interventions are promising and can increase memory performance in persons with MCI. They also suggest that the benefits can generalise to non-cognitive domains.

Some of these studies have imbedded memory training within broader multimodal interventions to maximise the cognitive training effect (Belleville et al. 2006; Kinsella et al. 2009; Schmitter-Edgecombe and Dyck 2014). Belleville and collaborators (2006) developed a multifactorial approach to be used with healthy older adults and persons with MCI (*Méthode d'entraînement pour une mémoire optimale*, MEMO). The program teaches different mnemotechniques (e.g. method of loci, face-name association, interactive imagery, text hierarchisation, semantic elaboration) and includes training on attention and visual imagery abilities. It also provides general psychoeducational information on cognitive aging and lifestyle factors and includes a number of features to promote self-efficacy and generalisation. Belleville et al. (2006, 2018) found improvement on objective episodic memory. Results from a randomised, controlled, single-blind trial using the MEMO program in persons with MCI showed improvement on episodic memory and on strategy use in everyday life and these gains were maintained 6 months following the intervention (Belleville et al. 2018). Kinsella et al. (2009) reported a multifactorial intervention that involved memory strategies, lifestyle, education, and psychotherapeutic techniques and that included family partners. They showed improvement on everyday memory, suggesting a generalisation of the effect to broader domains and contexts. Schmitter-Edgecombe and Dyck (2014) reported similar results with a program that involved care partners and comprised an educational workshop, multifamily memory strategy training, and problem-solving sessions. The involvement of family partners may facilitate the transfer of learned strategies to everyday functioning by providing support and feedback to their relatives with MCI.

Multimodal computerised training programs have also shown interesting results when applied to individuals with MCI. These programs are designed to target a general population of brain-damaged patients and typically include exercises for a wide range of cognitive functions (e.g. attention, perception, language, gnosias, calculation) in addition to memory. Rozzini (2007) reported that treatment with ChEIs alone did not reduce memory impairment in MCI subjects, but that combining computerised cognitive training with ChEIs resulted in significant memory improvements. Whether computerised training is as effective as face-to-face training has not yet been directly tested. Notably, however, Gaitán et al. (2013) tested the efficacy of multimodal computerised training with MCI persons who already received conventional face-to-face cognitive training and found that it did not produce further memory improvement. There is no strong evidence thus far that multimodal computerised training leads to a significant transfer to complex or daily activities.

Despite the positive effects of memory training described above, some randomised-controlled studies have reported negative findings (Unverzagt et al. 2007; Vidovich et al. 2015). For instance, Unverzagt et al. (2007) found no benefit from memory training in a memory-impaired subgroup from the ACTIVE cohort. Vidovich and collaborators (2015) reported improvement on attentional control and quality of life following memory training in MCI but no improvement on primary cognitive outcomes. The lack of systematic improvement makes it difficult to determine whether cognitive training interventions are able to affect a broad set of memory-related activities. A range of factors could explain the negative findings; for instance, it may be due to the fact that the selected outcome is insufficiently sensitive to change or is not sensitive to the processes improved by the intervention. Furthermore, the training format may also be an issue. Thus, there is a need for more studies aiming to disentangle the characteristics of an effective memory training program in MCI and its impact on complex memory-related activities. There are interesting avenues researchers could take: one may be to provide interventions that include additional cognitive or non-cognitive components; another would be to involve family partners in the intervention program.

Training of Attentional Control

Attentional control and executive functions are highly involved in everyday life and executive impairment is predictive of disability in older adults. Surprisingly, very few studies have focused on exercising these comprehensive abilities. Yet, there is evidence that training can improve attentional control in older adults (Karbach and Kray, this volume). For instance, Strobach et al. (2015) found that hybrid dual-task training, i.e. training with blocks that contain both dual-task and single-task trials, improved coordination skills. The authors also found that the effect was still present when tested with slightly different tasks, suggesting a near transfer of improved coordination skills. Divided attention capacities can be trained using variable priority training as opposed to fixed priority training. In both cases, participants practice divided attention tasks but in the variable priority training, individuals are also asked to prioritise one task over the other and to vary their attentional priority across different blocks of practice. Many authors reported that variable priority training is more effective in improving dual tasking than fixed priority (Bier et al. 2014; Gagnon and Belleville 2011; Kramer et al. 1995; Lee et al. 2012; Voss et al. 2012; Zendel et al. 2016) perhaps because it allows individuals to practice top-down regulatory control and hence increases self-control capacities over attention (Bier et al. 2014). Gagnon and Belleville (2012) compared the efficacy of variable and fixed priority training in persons with MCI who experience difficulties with executive control and found that variable priority increased dual-tasking capacities when compared to fixed priority training. These results suggest that training attention with programs that promote self-monitoring and metacognition can increase dual-tasking abilities in persons with MCI. Some evidence of training efficacy on attentional control were also found from a 5-week multi-domain training that combined cognitive training with elements from cognitive rehabilitation and stimulation. Trained MCI individuals exhibited a reduced decline of attention on a 2-year follow-up compared to MCI individuals who received a control-non-specific educational program (Vidovich et al. 2015). However, more studies are needed as only a few studies have focused on training attentional control in MCI.

Training Imbedded in Real Life: Virtual Reality and Leisure Approaches to Cognitive Training

Ultimately, the goal of cognitive training is to ensure that it results in significant changes in patients' lives (Taatgen, this volume). Traditional training programs are extremely variable in their ability to show far- or even near-transfer effects. Complex cognitively stimulating activities such as volunteer work, learning new languages, or engaging in interesting hobbies have the potential to meet these requirements. These activities involve learning a range of cognitive challenges that are of increasing complexities. They promote continuous learning, are pleasurable, and hopefully promote engagement, motivation, and transfer to everyday life, particularly in those who may not feel comfortable with academic activities. They are also multimodal by nature, as they involve social interactions and require that older adults explore new environments and be physically active. Interestingly, observational studies have identified these types of activities as being protective against cognitive decline and dementia. Programs based on similar activities have been shown to promote cognition in older adults. For instance, the SYNAPSE project (Park et al. 2014), which involves photography and/or learning how to quilt, was found to improve memory when compared to a placebo condition. In the Baltimore Experience Corps study (Carlson et al. 2008), in which older adults tutored elementary school pupils, improvement was found in cognition, health, and well-being. Within the Canadian Consortium on Neurodegeneration in Aging program, the ENGAGE program (Belleville et al. 2019) combined formal memory and attentional training strategies with leisure activities (Spanish learning or music lessons) and assessed whether it improves cognitive, psychosocial, and brain variables in persons with subjective cognitive decline (SCD), i.e. individuals who worry about their cognition but who are not cognitively impaired according to conventional neuropsychological tests. Because they are rooted in the community and are enjoyable, it is expected that cognitive programs that are embedded in real life such as ENGAGE, SYNAPSE, or Experience Corps will have more enduring effects, that their efficacy will transfer more readily to everyday life, and that it will be easier to offer them largely.

Developments in technology can also contribute to introducing interventions into real-life settings and promote transfer. Virtual reality (VR), for instance, allows the creation of three-dimensional, computer-generated, interactive environments. VR reproduces daily life situations into near-realistic environments that simulate the impression of being there, and a few studies have used VR to potentiate cognitive training effects in persons at risk of AD. For instance, Man and collaborators (2012) used a virtual environment that simulated a home setting and a convenience store to train the memory of individuals with MCI. VR training involved memorising virtual objects and retrieving them within the virtual environment with a range of presentation modalities, distractors, and levels of complexity, and its efficacy was compared with a face-to-face memory training condition. The results showed greater memory performance after having trained in the VR condition but better subjective memory following the face-to-face condition. This suggests that while the memory of individuals with MCI may benefit from the enhancing effects of being trained in a virtual environment, traditional approaches may be more appropriate for addressing self-efficacy and metacognition. VR can also be used to measure transfer of cognitive training effects to activities of daily living (Shuchat et al. 2012). For instance, Bier, Ouellet, and Belleville (2018) found evidence of transfer effects in healthy older participants in close to real-life environments using a "virtual car ride." Other studies focusing on the development of an immersive VR task called the "virtual shop" are very promising to assess gains from training in situations close to everyday life (Corriveau-Lecavalier et al. 2018; Ouellet et al. 2018). Results from these studies showed that the "virtual shop" was found to be a feasible and valid measure of everyday memory in older adults. Thus, recent advances in the field of virtual reality provide new opportunities to enhance the ecological validity of cognitive training and to assess real-life cognition in MCI individuals.

The Effect of Training on Brain Structure and Function

Brain imaging can establish the neural mechanisms by which training enhances cognitive functioning and indicate the training-induced neural changes (Guye et al., Wenger and Kühn, this volume). It can show whether the intervention modified specialised regions, i.e. regions that are normally involved in the task, or activated alternative brain regions, i.e. regions that are not normally active during the task and that are newly engaged. Brain imaging can also indicate whether the intervention focused on improving the function or brain region impaired (restorative effect) or relied on the intact functions and network (compensatory effect).

The few studies that have explored neural activity changes following cognitive training in MCI suggest that it can have both compensatory and restorative effects. Belleville et al. (2011) reported that strategic memory training increased brain activation in regions involved in memory encoding before training and induced new activations in regions that were not active prior to training in individuals with MCI. Interestingly, the differences between memory encoding-related brain patterns in MCI compared to healthy older controls were attenuated after training, suggesting that some restoration took place. Furthermore, the performance improvement was correlated with a newly activated region, the right parietal area, which was normalised in MCI. These results suggest that strategic cognitive training facilitates the recruitment of an intact alternative network to compensate the impaired primary network but can also contribute to meaningful restoration. Hampstead (2012) found increased activation almost exclusively in specialised regions after associative memory training in MCI individuals. They reported increased activation during both encoding and retrieval in hippocampal regions that were less activated compared to healthy older controls before training. These results show that associative memory training has a restorative effect on the primary network. Similarly, Förster (2011) showed that a multimodal intervention reduced decline in brain glucose metabolism in MCI and early AD, suggesting that it had an effect on neuronal injury.

Cognitive training was also found to have an effect on the structure of the brain in prodromal AD. Engvig et al. (2014) reported increased grey matter volume in regions encompassing the episodic memory network following strategic associative memory training in individuals with SCD. Interestingly, the strongest volume differences were found in the right prefrontal cortex, which is activated during contextual monitoring and episodic retrieval. Thus compensatory mechanisms may mediate training-related structural adaptation. Despite no significant hippocampal volume changes, there was a significant correlation between volume change and post-training memory improvement suggesting that individual differences may modulate the extent of the structural hippocampal restoration in SCD individuals. No study has looked at the effect of cognitive training on beta amyloid (βA) deposits, which is one of the main neuropathologies associated with AD. Showing that cognitive training reduces βA deposition would be of tremendous consequences and may not be that far-fetched, as observational studies have reported that a cognitively stimulating lifestyle is associated with lower levels of βA deposits in older adults (Landau et al. 2012).

The Contribution of Brain Imaging to Models of Training

Models of brain changes associated with aging are interesting to interpret the effect of cognitive training on the brain. For instance, according to the CRUNCH model (Reuter-Lorenz and Cappell 2008), compensation in older adults is supported by both increased activation of specialised brain regions and strategic recruitment of alternative regions. Interestingly, results from neuroimaging studies suggest that individual differences such as educational attainment or cognitive level of job can modulate the effect of age on brain structure and function. For instance, higher education has been associated with less reduced brain volume in older adults (Boller et al. 2017; Solé-Padullés et al. 2009). These results are consistent with the STAC-r model which proposes that individual differences in life course events can modify neural resources and compensatory capacities (Reuter-Lorenz and Park 2014). Studies reporting training-induced brain changes show that the regions modified by training generally reflect the purported active ingredient of the intervention. Cognitive training that is strategic and that targets preserved cognitive capacities in MCI increases activation in preserved brain regions, which is indicative of compensation. In contrast, cognitive training approaches that rely on adaptive learning or repeated practice are more likely to reduce activation in specialised regions. Additionally, a range of individual factors, the genetic potential for brain plasticity, and educational background may facilitate reliance on alternative networks or structural remodelling. The location and severity of structural impairment in brain-damaged individuals may also influence the success of a compensatory vs. a restorative approach, as restoration may be impossible when structural damage is too severe, for example. Thus, the INTERACTIVE model (Belleville et al. 2014b) proposes that characteristics of subjects (i.e. cognitive reserve, severity of the disease) and training modalities (i.e. format, target) modulate the type of neural changes induced by cognitive training.

Conclusion and Future Directions

Whether cognitive training and stimulation provided later in life can be used as protective tools against cognitive decline is a major research question. Observational studies have shown an effect of early life (education) and whole-life (profession, hobbies) cognitive stimulation on age-related cognitive decline, AD, and dementia. Compensatory neuroplasticity processes are particularly active during the silent phase of AD (Clément and Belleville 2010) and could be increased to postpone the cognitive decline that leads to the more severe symptoms that define dementia. Although many studies have revealed encouraging findings when using cognitive and brain markers, researchers and clinicians still need to address numerous important questions. First, we need to gain a better understanding of the critical period during which training or stimulation should be provided. The pathological cascade leading to AD, which probably starts many years prior to the diagnosis of dementia, suggests that these programs are likely to have their highest effect when provided early during the MCI phase or perhaps prior to that stage. However, demonstrating that early training has a long-lasting effect will certainly be very challenging if the outcome is clinical, and there will therefore be a need to adapt the method to those challenges. Furthermore, the efficacy of cognitive training might benefit from better characterising those with MCI that will convert to dementia. Several studies have shown that cognitive tests can distinguish those who will later progress to dementia from those who will remain stable (e.g. Belleville et al. 2014a, 2017). Additionally, it will be critical to document the effect of individual differences on cognitive training efficacy (see also Karbach and Kray, Katz et al., this volume). For instance, younger age and higher level of education were associated with larger training gain when individuals with MCI were trained with a strategic memory training program (Belleville et al. 2006). One other critical question is whether the brain processes promoted in late-life cognitive training are the same as those that underlie differences in cognitive reserve or cognitive resilience. The findings that training increases brain activation in alternative compensatory brain networks are consistent with the notion that cognitive reserve reflects more flexible brain networks. Finally, one other major issue that needs to be addressed is the notion of transfer, as cognitive training is intended to have an effect beyond the laboratory or task that is being trained (Könen and Auerswald, Schmiedek, Taatgen, this volume). It appears that older adults may be less prone to generalise learned strategies than younger adults, and whether MCI poses limits to the generalisation of learning is an important question that will need to be resolved.

Acknowledgements SB holds a Canada Research Chair on Neuroscience of Aging and Brain Plasticity and is supported by the Canadian Institute for Health Research, the Natural Sciences and Engineering Research Council of Canada, and the Fonds du Québec Recherche-Santé. BB receives a postdoctoral fellowship from the Fondation Lemaire and Fondation Institut de Gériatrie de Montréal. LP receives a doctoral fellowship from the Spanish Ministry of Economy and Competitiveness. We would like to thank Christel Cornelis and Gabrielle Ciquier for editorial assistance.

References

Albert, M. S., DeKosky, S. T., Dickson, D., Dubois, B., Feldman, H. H., Fox, N. C., ... & Snyder, P. J. (2011). The diagnosis of mild cognitive impairment due to Alzheimer's disease: Recommendations from the National Institute on Aging-Alzheimer's Association workgroups on diagnostic guidelines for Alzheimer's disease. *Alzheimer's & Dementia, 7*, 270–279.

Barnes, D. E., & Yaffe, K. (2011). The projected effect of risk factor reduction on Alzheimer's disease prevalence. *The Lancet Neurology, 10*, 819–828.

Belleville, S., Gilbert, B., Fontaine, F., Gagnon, L., Menard, E., & Gauthier, S. (2006). Improvement of episodic memory in persons with mild cognitive impairment and healthy older adults: Evidence from a cognitive intervention program. *Dementia and Geriatric Cognitive Disorders, 22*, 486–499.

Belleville, S., Ménard, M. C., de Boysson, C., & Sylvain-Roy, S. (2008). Characterizing memory deficits in mild cognitive impairment. In W. Sossin, J. C. Lacaille, V. Castellucci, & S. Belleville (Eds.), *Progress in brain research, essence of memory* (pp. 365–375). Amsterdam: Elsevier.

Belleville, S., Clement, F., Mellah, S., Gilbert, B., Fontaine, F., & Gauthier, S. (2011). Training-related brain plasticity in subjects at risk of developing Alzheimer's disease. *Brain, 134*, 1623–1634.

Belleville, S., Fouquet, C., Duchesne, S., Collins, D. L., & Hudon, C. (2014a). Detecting early preclinical Alzheimer's disease via cognition, neuropsychiatry, and neuroimaging: Qualitative review and recommendations for testing. *Journal of Alzheimer's Disease, 42*, S375–S382.

Belleville, S., Mellah, S., de Boysson, C., Demonet, J. F., & Bier, B. (2014b). The pattern and loci of training-induced brain changes in healthy older adults are predicted by the nature of the intervention. *PLoS One, 9*, e102710.

Belleville, S., Fouquet, C., Hudon, C., Zomahoun, H. T. V., & Croteau, J. (2017). Neuropsychological measures that predict progression from mild cognitive impairment to Alzheimer's type dementia in older adults: A systematic review and meta-analysis. *Neuropsychology Review, 27*, 328–353.

Belleville, S., Hudon, C., Bier, N., Brodeur, C., Gilbert, B., Grenier, S., ... & Gauthier, S. (2018). MEMO+: Efficacy, durability and effect of cognitive training and psychosocial intervention in individuals with mild cognitive impairment. *Journal of the American Geriatrics Society, 66*, 655–663.

Belleville, S., Moussard, A., Ansaldo, A. I., Belchior, P., Bherer, L., Bier, N., ... & Jokel, R. (2019). Rationale and protocol of the ENGAGE study: A double-blind randomized controlled preference trial using a comprehensive cohort design to measure the effect of a cognitive and leisure-based intervention in older adults with a memory complaint. *Trial, 20*, 282.

Bier, B., de Boysson, C., & Belleville, S. (2014). Identifying training modalities to improve multitasking in older adults. *Age, 36*, 9688.

Bier, B., Ouellet, E., & Belleville, S. (2018). Computerized attentional training and transfer with virtual reality: Effect of age and training type. *Journal of Neuropsychology, 32*, 597–614.

Boller, B., Mellah, S., Ducharme-Laliberté, G., & Belleville, S. (2017). Relationships between years of education, regional grey matter volumes, and working memory related brain activity in healthy older adults. *Brain Imaging and Behavior, 11*, 304–317.

Carlson, M. C., Saczynski, J. S., Rebok, G. W., Seeman, T., Glass, T. A., McGill, S., ... & Fried, L. P. (2008). Exploring the effects of an "everyday" activity program on executive function and memory in older adults: Experience Corps®. *The Gerontologist, 48*, 793–801.

Clement, F., & Belleville, S. (2010). Compensation and disease severity on the memory-related activations in mild cognitive impairment. *Biological Psychiatry, 68*, 894–902.

Corriveau Lecavalier, N., Ouellet, E., Boller, B., & Belleville, S. (2018). Use of immersive virtual reality to assess episodic memory: A validation study in older adults. *Neuropsychological Rehabilitation*, 1–19.

Engvig, A., Fjell, A. M., Westlye, L. T., Skaane, N. V., Dale, A. M., Holland, D., ... & Walhovd, K. B. (2014). Effects of cognitive training on gray matter volumes in memory clinic patients with subjective memory impairment. *Journal of Alzheimer's Disease, 41*, 779–791.

Förster, S., Buschert, V. C., Teipel, S. J., Friese, U., Buchholz, H. G., Drzezga, A., ... & Buerger, K (2011). Effects of a 6-month cognitive intervention on brain metabolism in patients with amnestic MCI and mild Alzheimer's disease. *Journal of Alzheimer's Disease, 26*, 337–348.

Gagnon, L. G., & Belleville, S. (2011). Working memory in mild cognitive impairment and Alzheimer's disease: Contribution of forgetting and predictive value of complex span tasks. *Neuropsychology, 25*, 226–236.

Gagnon, L. G., & Belleville, S. (2012). Training of attentional control in mild cognitive impairment with executive deficits: Results from a double-blind randomised controlled study. *Neuropsychological Rehabilitation, 22*, 809–835.

Gaitán, A., Garolera, M., Cerulla, N., Chico, G., Rodriguez-Querol, M., & Canela-Soler, J. (2013). Efficacy of an adjunctive computer-based cognitive training program in amnestic mild cognitive impairment and Alzheimer's disease: A single-blind, randomized clinical trial. *International Journal of Geriatric Psychiatry, 28*, 91–99.

Hampstead, B. M., Stringer, A. Y., Stilla, R. F., Giddens, M., & Sathian, K. (2012). Mnemonic strategy training partially restores hippocampal activity in patients with mild cognitive impairment. *Hippocampus, 22*, 1652–1658.

Herrera, C., Chambon, C., Michel, B. F., Paban, V., & Alescio-Lautier, B. (2012). Positive effects of computer-based cognitive training in adults with mild cognitive impairment. *Neuropsychologia, 50*, 1871–1881.

Kinsella, G. J., Mullaly, E., Rand, E., Ong, B., Burton, C., Price, S., ... & Storey, E. (2009). Early intervention for mild cognitive impairment: a randomised controlled trial. *Journal of Neurology, Neurosurgery, and Psychiatry, 80*, 730–736.

Kramer, A. F., Larish, J. F., & Strayer, D. L. (1995). Training for attentional control in dual task settings: A comparison of young and old adults. *Journal of Experimental Psychology: Applied, 1*, 50–76.

Landau, S. M., Marks, S. M., Mormino, E. C., Rabinovici, G. D., Oh, H., O'Neil, J. P., ... & Jagust, W. J. (2012). Association of lifetime cognitive engagement and low beta-amyloid deposition. *Archives of Neurology, 69*, 623–629.

Lee, H., Voss, M. W., Prakash, R. S., Boot, W. R., Vo, L. T., Basak, C., ... & Kramer, A. F. (2012). Videogame training strategy-induced change in brain function during a complex visuomotor task. *Behavioural Brain Research, 232*, 348–357.

Man, D. W., Chung, J. C., & Lee, G. Y. (2012). Evaluation of a virtual reality-based memory training programme for Hong Kong Chinese older adults with questionable dementia: A pilot study. *International Journal of Geriatric Psychiatry, 27*, 513–520.

Olchik, M. R., Farina, J., Steibel, N., Teixeira, A. R., & Yassuda, M. S. (2013). Memory training (MT) in mild cognitive impairment (MCI) generates change in cognitive performance. *Archives of Gerontology and Geriatrics, 56*, 442–447.

Ouellet, E., Boller, B., Corriveau-Lecavalier, N., Cloutier, S., & Belleville, S. (2018). The virtual shop: A new immersive virtual reality environment and scenario for the assessment of everyday memory. *Journal of Neuroscience Methods, 303*, 126–135.

Park, D. C., Lodi-Smith, J., Drew, L., Haber, S., Hebrank, A., Bischof, G. N., & Aamodt, W. (2014). The impact of sustained engagement on cognitive function in older adults: The Synapse project. *Psychological Science, 25*, 103–112.

Petersen, R. C., Thomas, R. G., Grundman, M., Bennett, D., Doody, R., Ferris, S., ... & Pfeiffer, E. (2005). Vitamin E and donepezil for the treatment of mild cognitive impairment. *New England Journal of Medicine, 352*, 2379–2388.

Rapp, S., Brenes, G., & Marsh, A. P. (2002). Memory enhancement training for older adults with mild cognitive impairment: A preliminary study. *Aging & Mental Health, 6*, 5–11.

Reuter-Lorenz, P. A., & Cappell, K. A. (2008). Neurocognitive aging and the compensation hypothesis. *Current Directions in Psychological Science, 17*, 177–182.

Reuter-Lorenz, P. A., & Park, D. C. (2014). How does it STAC up? Revisiting the scaffolding theory of aging and cognition. *Neuropsychology Review, 24*, 355–370.

Rozzini, L., Costardi, D., Chilovi, B. V., Franzoni, S., Trabucchi, M., & Padovani, A. (2007). Efficacy of cognitive rehabilitation in patients with mild cognitive impairment treated with cholinesterase inhibitors. *International Journal of Geriatric Psychiatry, 22*, 356–360.

Schmitter-Edgecombe, M., & Dyck, D. G. (2014). Cognitive rehabilitation multi-family group intervention for individuals with mild cognitive impairment and their care-partners. *Journal of the International Neuropsychological Society, 20*, 897–908.

Shuchat, J., Ouellet, É., Moffat, N., & Belleville, S. (2012). Opportunities for virtual reality in cognitive training with persons with mild cognitive impairment or Alzheimer's disease. *Non-Pharmacological Therapies in Dementia, 3*, 35.

Solé-Padullés, C., Bartrés-Faz, D., Junqué, C., Vendrell, P., Rami, L., Clemente, I. C., ... & Barrios, M. (2009). Brain structure and function related to cognitive reserve variables in normal aging, mild cognitive impairment and Alzheimer's disease. *Neurobiology of Aging, 30*, 1114–1124.

Strobach, T., Frensch, P., Müller, H., & Schubert, T. (2015). Evidence for the acquisition of dual-task coordination skills in older adults. *Acta Psychologica, 160*, 104–116.

Tappen, R. M., & Hain, D. (2014). The effect of in-home cognitive training on functional performance of individuals with mild cognitive impairment and early-stage Alzheimer's disease. *Research in Gerontological Nursing, 7*, 14–24.

Unverzagt, F. W., Kasten, L., Johnson, K. E., Rebok, G. W., Marsiske, M., Koepke, K. M., ... & Rexroth, D. F. (2007). Effect of memory impairment on training outcomes in ACTIVE. *Journal of the International Neuropsychological Society, 13*, 953–960.

Vidovich, M. R., Lautenschlager, N. T., Flicker, L., Clare, L., McCaul, K., & Almeida, O. P. (2015). The PACE study: A randomized clinical trial of cognitive activity strategy training for older people with mild cognitive impairment. *The American Journal of Geriatric Psychiatry, 23*, 360–372.

Voss, M. W., Prakash, R. S., Erickson, K. I., Boot, W. R., Basak, C., Neider, M. B., ... & Kramer, A. F. (2012). Effects of training strategies implemented in a complex videogame on functional connectivity of attentional networks. *Neuroimage, 59*, 138–148.

Zendel, B. R., de Boysson, C., Mellah, S., Demonet, J. F., & Belleville, S. (2016). The impact of attentional training on event-related potentials in older adults. *Neurobiology of Aging, 47*, 10–22.

Part VI
Outlook

The Future of Cognitive Training

Lorenza S. Colzato and Bernhard Hommel

Contents

Abstract Cognitive training is becoming increasingly popular as a topic of scientific research. We discuss possible as well as necessary future developments in this area. Among other things, we emphasize the need to develop more specific, mechanistic theories to guide cognitive training programs, discuss the combination of cognitive training with other cognitive enhancement techniques, and consider the opportunities provided by virtual reality and gamification. We suggest that cognitive training programs should take individual differences more into consideration and discuss the societal and ethical background for the use of such programs.

L. S. Colzato (✉)
Cognitive Neurophysiology, Department of Child and Adolescent Psychiatry, Faculty of Medicine, TU Dresden, Dresden, Germany

Department of Cognitive Psychology, Institute of Cognitive Neuroscience, Faculty of Psychology, Ruhr University Bochum, Bochum, Germany
e-mail: lorenza.colzato@uniklinikum-dresden.de

B. Hommel
Cognitive Psychology Unit & Leiden Institute for Brain and Cognition, Leiden University, Leiden, The Netherlands

T. Strobach, J. Karbach (eds.), *Cognitive Training*,
https://doi.org/10.1007/978-3-030-39292-5_27

This chapter concludes the broad overview of cognitive training activities that this book aims to provide. Where will these activities lead us? What are the upcoming challenges? It is these future-directed questions that we would like to address in this final chapter. We will do so by mixing informed guesses about to-be-expected trends, problems, and challenges in the near future, with our wish list of developments that we would like to see without being able to judge how realistic our wishes are at this point. Among other things, we explain why more specific, mechanistic theories will be necessary to guide the development of successful cognitive training programs, how cognitive training might benefit from combining them with other cognitive enhancement techniques, and how virtual reality and gamification could be used to support the efficiency of cognitive training. We also emphasize the importance of considering individual differences and discuss the societal and ethical implications of enhancement programs.

Need for Theory

There are only few areas where Kurt Lewin's claim that "nothing is as practical as a good theory" does not apply, but hardly any to which it applies more than to the area of cognitive training (see Cochrane and Green, this volume). That people get better when they repeat doing the same thing over and over again is an insight that has been with academic psychology for more than 150 years. And yet, we still see many approaches to cognitive training that do not seem to go much beyond this general insight. The typical punishment for such theoretically parsimonious approaches is the lack of any interesting transfer from the actually trained cognitive ability to any other cognitive task or skill, which should not be surprising. To reach interesting levels of transfer requires rather good ideas about the mechanisms underlying the cognitive functions one aims to improve. But we still do not see too many of them.

For instance, theorizing about cognitive control—a particularly important cognitive function worth enhancing in many subpopulations—still does not go beyond distinguishing some general, vaguely described factors (like updating, shifting, and inhibition: Miyake et al. 2000) and related brain areas, while specific models about what these factors and areas are really doing and exactly how they operate are lacking. Consider task switching, which plays an important role in many training programs. How exactly do people switch from one task to another? What do we really know about this process and the cognitive codes it operates on, after it has been addressed in hundreds and hundreds of studies? What exactly is a task set? How is it generated from instructions? Can they become stored and retrieved? As long as we have no interesting, mechanistically detailed answers to questions of that sort, it is difficult to see how training programs can generate far transfer in systematic, generalizable ways. Generating more interesting answers is likely to require more collaboration between researchers with more theoretical and researchers with more practical skills and interests. Creating such collaborations will require flexible funding schemes and substantial resources.

Enhancing Cognitive Training

From a more practical perspective, it would seem promising to combine methods suitable for cognitive enhancement. Indeed, there is preliminary evidence that cognitive training programs can be successfully enhanced by boosting performance outcomes in various ways.

First, there is increasing evidence that cognitive training may benefit from the combination with pharmacological interventions. In particular, interventions acting on the dopaminergic system seem ideal to enhance learning in cognitive training given the role of dopamine in associative learning (Schultz et al. 1997) and executive functioning (Colzato et al. 2010, 2014). Indeed, the combined administration of L-Dopa and D-amphetamine has been found to boost language learning in healthy humans (Breitenstein et al. 2004; Knecht et al. 2004). More recently, Gilleen et al. (2014) sought to enhance performance on cognitive tasks (working memory [WM], verbal learning, and learning a new language) in healthy participants by combining cognitive training with the cognitive enhancing drug modafinil. While memory and verbal learning was unaffected, new language learning was significantly enhanced through the combination, which is at least encouraging.

Second, there is some evidence that cognitive training benefits from the combination with brain stimulation by means of *transcranial direct current stimulation* (tDCS; see Byrne et al., this volume). tDCS is a noninvasive brain stimulation technique that involves passing a constant direct electrical current through the cerebral cortex (via electrodes placed upon the scalp) flowing from the positively charged anode to the negatively charged cathode (Nitsche and Paulus 2011). This technique has developed into a promising tool to boost human cognition (Kuo and Nitsche 2012). Very recently, Richmond et al. (2014) suggested that tDCS might support WM training. Participants engaged in an adaptive WM training regime for ten sessions, concurrent with either active or sham stimulation of dorsolateral prefrontal cortex. Before and after training, a battery of tests tapping domains known to relate to WM abilities was administered. tDCS was shown to enhance learning in the verbal part of the cognitive training and to enhance near transfer to other untrained WM tasks. We emphasize that this study did not include a follow-up session and needs to be replicated and generalized to other cognitive domains. And yet, it does provide preliminary evidence that tDCS might enhance cognitive training and support far transfer. In support of these results, Au et al. (2016) also found tDCS to enhance working memory training performance. Notably, this enhancing effect was still observed several months after the training regime (over 7 days). Noteworthy, the effect was stronger when tDCS was spaced over the weekend compared to daily training, which indicates that spacing the sessions is critical to consolidate the efficacy of the cognitive training. A more recent follow-up study to assess later performance showed that the enhancing effects persisted even 1 year after the training intervention (Katz et al. 2017). These findings suggest that tDCS might be effective in supplementing cognitive training.

Third, a number of findings suggest that cognitive training may benefit from a combination with neurofeedback. Neurofeedback is a technique that teaches participants to control their own brain activity by providing systematic feedback about internal states (Sherlin et al. 2011), such as neural oscillations and slow cortical potentials assessed by means of electroencephalography (EEG; Birbaumer et al. 2009). The modulation of neural oscillations through EEG-neurofeedback has been shown to enhance different cognitive functions as a function of the frequencies of neural activity (see Gruzelier 2014 for a recent review). For instance, upregulating the upper alpha band improves mental rotation (e.g., Hanslmayr et al. 2005; Zoefel et al. 2011), upregulating gamma-band activity enhances episodic retrieval (Keizer et al. 2010), and upregulating the mu-rhythm supports declarative learning (Hoedlmoser et al. 2008). Enriquez-Geppert et al. (2014) have investigated the modulation of frontal-midline theta oscillations by neurofeedback and its putative role for executive functioning. Before beginning and after completing an individualized, eight-session gap-spaced intervention, tasks tapping executive functions were administered while measuring the EEG. Compared to a pseudo-neurofeedback group, the group receiving active neurofeedback training showed better performance in WM updating and cognitive flexibility. The idea that learning to increase frontal-midline theta amplitudes facilitates executive functions is captivating and opens the possibility to use neurofeedback to boost the efficiency of cognitive training. Unfortunately, a recent study (Gordon et al. 2019) with a large sample size ($N = 140$) showed no effect of neurofeedback (alpha amplitude) and working memory-combined training on executive functions. Hence, future studies are mandatory to develop training protocols for the optimal combination of neurofeedback and cognitive training.

Fourth, research on human-machine interfaces increasingly points to an interesting role of haptic feedback, as provided by means of somatosensory information (vibration) delivered through a user interface. Training with haptic feedback has been found to reliably support the acquisition of knowledge in chemistry (Bivall et al. 2011) and physics (Han and Black 2011), as well as object manipulation (Stepp et al. 2012). Even though it is not yet clear whether such learning improvements transfer to other tasks, the incorporation of haptic feedback in cognitive training programs represents an interesting avenue for the future.

These are just a few examples for how cognitive training techniques can be enhanced by techniques that have been shown to support learning, but progress in technology is likely to generate more and more options in the near future. While many of them are interesting indeed, their novelty brings a number of risks with it. For instance, new developments have made it possible to produce tDCS-based tools for the use in daily life. While that provides interesting opportunities for research (e.g., in freeing participants from daily visits in the lab), official tests and guidelines for the safe personal use of such devices are lacking. As pointed out by Jwa (2015), given that tDCS is currently not covered by the existing regulatory framework, there are potential risks of misusing this device, in particular as its long-term effects on the brain have not been fully investigated and understood. A recent initiative supported by several research institutes and scientists calls for a more critical and active role of the scientific community in evaluating the sometimes far-reaching, sweeping claims from the brain training industry with regard to the impact of their products on cognitive performance (Max Planck Institute on Human Development, Stanford Center on Longevity 2014).

Recently, colleagues and us (Steenbergen et al. 2016) took this recommendation to heart and tested whether and to what degree the commercial tDCS headset *foc.us* improves cognitive performance, as advertised in the media. We used a single-blind, sham-controlled, within-subject design to investigate the effect of online and offline *foc.us* tDCS—applied over the prefrontal cortex in healthy young volunteers—on WM updating. In contrast to the previous positive findings with CE-certified laboratory tDCS, active stimulation with foc.us led to a significant *decrease* in WM updating. This observation reinforces the view that the scientific community can and presumably should play a crucial role of in helping to create regulations and official guidelines for the future incorporation of cognitive and neuro-technologies in cognitive training.

Virtual and Augmented Reality

The use of virtual and augmented reality (VAR) has become popular in several areas of cognitive and clinical psychology, where it, for instance, is used to treat phobia (Juan et al. 2005; see also Cochrane and Green, this volume). These kinds of uses could also be seen as enhancing techniques, similar to those discussed in the previous section. Indeed, VAR techniques can serve to visualize instructions and provide more realistic feedback about the achievements of trainees, and they have, for instance, been successfully employed to enhance attention (e.g., Cho et al. 2002), memory (e.g., Optale et al. 2010), or sensorimotor skills (e.g., Adamovich, Fluet, Tunik and Merians 2009).

However, we think that VAR techniques are particularly well-suited to address an aspect of cognitive training that has remained underdeveloped so far: the possibility of embodied cognition. The embodied cognition approach is not particularly homogeneous and theoretically straightforward (for a discussion, see Wilson 2002), but the general idea is that cognition emerges from concrete sensorimotor interactions with one's environment, which assigns an important role to one's body. This fits with older ideomotor considerations about the emergence of cognition through action (Hommel 2015), which, for instance, have motivated the development of the theory of event coding (Hommel et al. 2001). It remains to be seen whether and in which sense the idea of embodiment increases our insight into basic cognitive functions and control processes, but if it does, we will need more realistic experimental designs and training conditions. For these purposes, VAR seems ideal.

For instance, cognitive aging is not unlikely to be associated if not facilitated by motivational decline that is produced by changes in self-perception. As elaborated elsewhere (Hommel and Kibele 2016), the retired elderly is likely to perceive herself as someone who is no longer productive. Given that most jobs allow people to exert impact on the real world, this impression is based on a real fact—retirement does mean losing this impact. To the degree that the outcome of self-perception affects motivation, this would be likely to undermine the motivation of the retired individual. This in turn would make it difficult both to maintain one's cognitive abilities and to compensate for age-related cognitive decline by means of training. VAR could help to prevent and counteract vicious cycles of this sort by turning the self-perception into a more active one. Along similar lines, VAR techniques have been successfully used to alter people's (healthy or pathological) body perception and attitudes toward their bodies by means of real-time feedback about their walking patterns (e.g., Tajadura-Jiménez et al. 2015) or increase empathy with others by virtually enacting a perspective switch (Bertrand et al. 2018).

Gamification

The widespread popularity of smart phones has led to a real explosion of "apps" to enhance cognitive functioning, ranging from simple alerts reminding the elderly to take his pill to theoretically guided programs to systematically improve specific cognitive functions, such as spatial imagination. Industry and funding agencies have taken notice of the many opportunities these techniques can open, and the current European research agenda (Horizon 2020) has various calls to promote gamification. Obviously, this is likely to strengthen this trend further in the near future, but we think that the full potential of gamification is not always appreciated. Turning psychological experiments and training procedures into apps is certainly handy for both researchers and users, especially as it allows to integrate training programs better with real-life circumstances. Even more importantly, however, gamification will make cognitive training programs more acceptable and increase the motivation to get through with them. Laboratory work on the impact of cognitive training is typically based on data collected from paid or otherwise compensated participants, which reduces the risk of dropout even with extensive training and not-so-exciting tasks. To make it to real-life circumstances, however, the format of cognitive training will need to change dramatically, so to convince individuals to participate. Like physical exercise, it can take a while before cognitive training produces benefits that are recognizable for the trainee. Continuous, fine-graded feedback helps to overcome that problem but only if improvements are visible enough to keep the trainee motivated. Especially training with more preventive aims, for which immediate benefits may not be visible at all, motivation remains an issue. Gamification can help to tackle that issue by making the process more fun and providing additional, benefit-independent reward.

Individual Tailoring

Most cognitive training programs have a one-size-fits-all design and assume that everyone benefits from the intervention more or less the same way and to more or less the same degree (see Cochrane and Green, Karbach and Kray, Katz et al., this volume). There are several reasons suggesting that this is unlikely to be true. In fact, we suggest that the efficiency of cognitive training and the successful transfer to untrained tasks will often be modulated by interindividual differences, including preexisting neurodevelopmental factors and differences with a genetic basis (see deVries and Geurts, Karbach and Kray, Katz et al., Könen and Auerswald, this volume). Accordingly, only training programs that are tailored to individual abilities, skills, and needs are likely to succeed.

In particular, we believe that substantial parts of the current controversy about the benefit of the regular use of cognitive training are due to the failure to consider individual differences. For instance, while Schmiedek et al. (2010) found positive transfer of cognitive training both in young and older adults, Owen et al. (2010) famously reported about a failure to find transfer in 11,430 participants trained online over a period of 6 weeks. The participants of Owen et al. were trained on cognitive tasks developed to improve reasoning, memory, planning, visuo-spatial skills, and attention. Participants improved in every single task, as one would expect, but the benefit did not generalize to any untrained tasks. The authors conclude that this provides "no evidence to support the widely held belief that the regular use of computerized brain trainers improves general cognitive functioning in healthy participants beyond those tasks that are actually being trained" (Owen et al. 2010, p. 777).

While we do not question the importance of such large-scale studies, we consider arguments based on mean findings in not further differentiated populations problematic, especially if individual improvements are not taken into account as well. The reason why this is important is that the functions relating psychological functions (and/or their neural underpinnings) to performance are often not linear. For instance, brainstorming-like creativity is assumed to rely on mood and on (presumably striatal) dopamine, but there is evidence that a medium (i.e., not the highest) dopamine level produces the best performance (Akbari Chermahini and Hommel 2010). Given the evidence that inducing positive mood increases the dopamine level, this suggests that individuals with a low dopamine level get better while those with a medium dopamine level do not or even get worse—which is indeed what has been observed (Akbari Chermahini and Hommel 2012).

Along the same lines, we also considered that successful transfer of game-based cognitive improvements to untrained tasks might be modulated by the genetic variability related to the catechol-O-methyltransferase (COMT)—an enzyme responsible for the degradation of dopamine (Colzato et al. 2014). Participants were genotyped for the COMT Val^{158}Met polymorphism and trained on playing "Half-Life 2," a first-person shooter game that has been shown to improve cognitive flexibility. Pre-training (baseline) and post-training measures of cognitive flexibility were acquired by means of a task-switching paradigm. As predicted, Val/Val homozygous individuals (i.e., individuals with a beneficial genetic predisposition for cognitive flexibility) showed larger beneficial transfer effects than Met/-carriers, supporting the possibility that genetic predisposition modulates transfer effects and that cognitive training promotes cognitive flexibility in individuals with a suitable genetic predisposition. Even if this study needs to be replicated with a larger sample size, we view it as proof of principle that highlights the importance of considering individual differences. Considering these differences and assessing how they interact with different training regimes will allow for the development of personalized, individually tailored training programs. Not only will these programs be more effective, but they also will be much more motivating for participants (as unnecessary failures due to person-method mismatches can be avoided) and more cost-efficient. This in turn will make the implementation of such programs more likely even in times of sparse budgets. In view of the rapid aging of European societies, the number of potential beneficiaries of such an individualized approach is dramatically increasing, and the societal need for maximizing the human cognitive potential in the elderly will grow further as the economic situation will require extensions of the working lifetime.

Societal Context

Research on and the application of cognitive training depends on the societal context, which affects the respective funding budgets and acceptability. Accordingly, it is important to consider which direction societal developments related to these issues are taking. Economically, the interest in cognitive training is mainly driven by the increasing costs of the welfare system, especially with regard to the increasing age of citizens in Western societies. Cognitive training can help, so one version of the idea, to delay cognitive decline in the elderly, which would extend the time people can live autonomously and, thus, reduce the welfare costs for the time thereafter. Along the same lines, training children could speed up the education of healthy individuals and reduce the risk of behavioral deviance and pathology, again with considerable savings for welfare and education systems (see deVries and Geurts, Johann and Karbach, Rueda et al., Schaeffner et al., Thompson and Steinbeis, this volume). But there is also a more ideological reason for the increased interest in cognitive training. Both Eastern and Western societies are continuously driven toward more individualism, which emphasizes the existence and often also the importance of individual differences over commonalities and collectivistic values. These tendencies go hand in hand with ideological developments in public opinion and within political parties, which in many countries have gravitated toward more neoliberal, individualism-heavy positions over the last 15 years or so. Among other things, this has involved a rather systematic deconstruction of the welfare system and established the view of the individual as an architect of his or her own life.

Research on cognitive training has benefited from both aspects of this trend. The economic problems of the welfare system have boosted the interest in procedures and activities that make welfare societally more affordable, and the ideological turn toward individualism provides a natural breeding ground for the public interest in procedures and activities that help to express and to further develop individual needs and interests. We do not expect that the economic problems will disappear soon, but it is possible that the ideological development leads to a swing back. To the degree that it will, the opposition and ethical objections to cognitive training programs may increase substantially.

Ethical Challenges

Like any psychological intervention, cognitive training raises all sorts of ethical issues (Bostrom and Sandberg 2009). In the following, we would like to emphasize two of them, as we suspect that they are likely to dominate future discussions. The first issue has to do with the "naturalness" of the intervention. Encouraging people to take considerable active efforts to change their mind and brain, as we would hope for effective training, must be considered unnatural, in the sense that it is likely to create a situation that without these efforts would not exist. While this is the very point of any sort of training, some people take issue with that. For instance, it has been considered that methods of cognitive enhancement may disrespect dignity and human nature, augment inauthenticity and cheating behavior, and may encourage an uncontrolled striving for excellence and perfection (Habermas 2003; Kass 2002). Such considerations are not far-fetched, as witnessed by the increasing use of cognitive enhancing drugs, such as modafinil and Ritalin, by students to boost their academic performance (Colzato and Arntz 2017; Colzato and Mourits 2017). Soon, universities may opt to prohibit drug use altogether or to tolerate it in some situations (exams). The same reasoning is also applicable to commercial brain stimulation devices, which are available on the Internet without any restrictions.

A second, somewhat related issue is that the availability of cognitive training techniques creates or at least increases a tension between two widely shared ethical principles: individual freedom and equality. While effective cognitive training programs can be taken to support the expression of the former (assuming that the "unnaturalness" objection can be overcome), it may conflict with the latter. Societies and upward mobility in particular rely increasingly on competition, which emphasizes individual performance and abilities. Cognitive training is likely to create "positional benefits" by improving one's social and economic status as compared to others. While this may be considered an acceptable individual choice, it may have repercussions for general public expectations and criteria. Once a number of individuals have demonstrated that it is possible to improve one's cognitive abilities, public pressure on other individuals could arise to improve their abilities as well. The existence of effective cognitive training programs could thus create or increase the pressure of always being "at the top," to work harder, longer, and more intensively, which in the end may exacerbate the problems one was intending to solve. In other words, the mere possibility to enhance one's cognitive abilities could increase social competition. Worse, as the probability to benefit from cognitive training may differ between individuals, the availability of training programs may contribute to the emergence and increase the size of societal gaps (cf., Bostrom and Sandberg 2009).

Counterarguments exist for both of these ethical issues. For one, any kind of psychological intervention and any kind of training must be considered equally unnatural as cognitive training. Accordingly, if one finds psychologically guided education and physical exercises of athletes acceptable, it is difficult to see in which sense cognitive training falls into an ethically different category. For instance, while objections to cognitive enhancement by means of particular diets or food supplements (Colzato et al. 2013) have not been put forward so far, the impact of cognitive-enhancing drugs and neurotechnologies, such as tDCS and neurofeedback, rest basically on the same cognitive and neural mechanisms. Obviously, this raises the question why social acceptance might be more widespread for the former than for the latter.

For another, cognitive training could well be used as a way of reducing, rather than increasing, societal/social inequalities by allowing all, and not just the economically privileged individuals, to fully explore and exploit their cognitive potential. This would not eliminate competition but create more equal terms (Savulescu 2009). Moreover, it is important to consider that the widespread use of cognitive training and the associated cognitive benefits might have rather dramatic social benefits. Indeed, some studies estimate that augmenting the average IQ of the world population by no more than 3% would reduce poverty rates by 25% (Schwartz 1994) and result in an annual economic gain of US $165–195 billion and 1.2–1.5% GDP (Salkever 1995).

Challenges for the Future: CRISPR/Cas9

We are currently facing a fast technology-driven revolution in molecular biology. Via genetic engineering we are able, at very low cost, to quickly and efficiently edit a large number of genes, a phenomenon called genome editing. The revolutionary technique CRISPR/Cas9 has been discovered in 2012 (Fineran and Charpentier 2012) and since then has been spread and used in research labs around the world. This technique allows researchers to make precise changes to the DNA in any cell or organism. It is based on a bacterial immune system that allows bacteria to fight viral infections via a programmable enzyme called Cas9, which can be programmed with little bits of RNA that allow the Cas9 protein to find a piece of DNA inside of the cell and cut it. That is, the Cas9 acts as a molecular scissor that can cut the two strands of DNA at a specific location in the genome, so that bits of DNA can then be added or removed. Once the DNA is cut, the cell's natural repair mechanisms take place and work to introduce mutations or other changes to the genome.

This technology will be used to correct mutations that cause genetic diseases like muscular dystrophy. Further, it is possible to apply genetic changes also in germline cells and early embryos via so-called gene drivers, altered genetic material that can be transferred to the offspring of a certain species, for example, in order to prevent a disease from spreading. If it will be possible to contain a disease, will we also be able to edit our genome to enhance individual working memory (or other higher-order cognitive processes) so that we will not need cognitive training after all? Popular scientific magazines and social media constantly speculate about the idea that CRISPR/Cas9 might help to create a physical and cognitive "Übermensch." Is this really true? At this point the answer is no because this technology acts on single genes whereas complex cognitive functions such as intelligence and working memory rely on multiples genes and cannot be simultaneously targeted by CRISPR/Cas9. Still, we cannot exclude that in the future new genome editing technologies might achieve this goal. When that happens, it will be important to ethically assess and regulate possible applications of genome editing by law.

Conclusion

Taken altogether, the future of cognitive training will heavily depend on theoretical, technological, and societal developments. For some of these developments, cognitive researchers are solely responsible, while they can only contribute to others. As we have tried to emphasize, cognitive training is not just one more psychological intervention but it touches important societal and ethical issues. Accordingly, as suggested by Colzato (2018), it would be wise if researchers actively engaged in public discussion of these issues to bring in the necessary expertise, so as to make sure that both risks and promises of cognitive training are realistically assessed.

References

Adamovich, S. V., Fluet, G. G., Tunik, E., & Merians, A. S. (2009). Sensorimotor training in virtual reality: A review. *Neuro Rehabilitation, 25*, 29–44.

Akbari Chermahini, S., & Hommel, B. (2010). The (b)link between creativity and dopamine: Spontaneous eye blink rates predict and dissociate divergent and convergent thinking. *Cognition, 115*, 458–465.

Akbari Chermahini, S., & Hommel, B. (2012). More creative through positive mood? Not everyone! *Frontiers in Human Neuroscience, 6*, 319.

Au, J., Katz, B., Buschkuehl, M., Bunarjo, K., Senger, T., Zabel, C., ... & Jonides, J. (2016). Enhancing working memory training with transcranial direct current stimulation. *Journal of Cognitive Neuroscience, 28*, 1419–1432.

Bertrand, P., Guegan, J., Robieux, L., McCall, C., & Zenasni, F. (2018). Learning empathy through virtual reality: Multiple strategies for training empathy-related abilities using body ownership illusions in embodied virtual reality. *Frontiers in AI and Robotics, 5*, 28.

Birbaumer, N., Ramos Murguialday, A., Weber, C., & Montoya, P. (2009). Neurofeedback and brain-computer interface clinical applications. *International Review of Neurobiology, 86*, 107–117.

Bivall, P., Ainsworth, S., & Tibell, L. A. (2011). Do haptic representations help complex molecular learning? *Science Education, 95*, 700–719.

Bostrom, N., & Sandberg, A. (2009). Cognitive enhancement: Methods, ethics, regulatory challenges. *Science and Engineering Ethics, 15*, 311–341.

Breitenstein, C., Wailke, S., Bushuven, S., Kamping, S., Zwitserlood, P., Ringelstein, E. B., & Knecht, S. (2004). D-amphetamine boosts language learning independent of its cardiovascular and motor arousing effects. *Neuropsychopharmacology, 29*, 1704–1714.

Cho, B. H., Ku, J., Jang, D. P., Kim, S., Lee, Y. H., Kim, I. Y., Lee, J. H., & Kim, S. I. (2002). The effect of virtual reality cognitive training for attention enhancement. *Cyberpsychology & Behavior, 5*, 129–137.

Colzato, L. S. (2018). Responsible cognitive enhancement: Neuroethical considerations. *Journal of Cognitive Enhancement, 2*, 331–334.

Colzato, L. S., & Arntz, F. E. (2017). Ritalin. In L. S. Colzato (Ed.), *Theory-driven approaches to cognitive enhancement*. New York: Springer.

Colzato, L. S., & Mourits, R. (2017). Modafinil. In L. S. Colzato (Ed.), *Theory-driven approaches to cognitive enhancement*. New York: Springer.

Colzato, L. S., Waszak, F., Nieuwenhuis, S., Posthuma, D., & Hommel, B. (2010). The flexible mind is associated with the Catechol-O-methyltransferase (COMT) Val158Met polymorphism: Evidence for a role of dopamine in the control of task switching. *Neuropsychologia, 48*, 2764–2768.

Colzato, L. S., Jongkees, B., Sellaro, R., & Hommel, B. (2013). Working memory reloaded: Tyrosine repletes updating in the N-Back task. *Frontiers in Behavioral Neuroscience, 7*, 700.

Colzato, L. S., van den Wildenberg, W., & Hommel, B. (2014). Cognitive control and the COMT Val158Met polymorphism: Genetic modulation of videogame training and transfer to task-switching efficiency. *Psychological Research, 78*, 670–678.

Enriquez-Geppert, S., Huster, R. J., Figge, C., & Herrmann, C. S. (2014). Self-regulation of frontal-midline theta facilitates memory updating and mental set shifting. *Frontiers in Behavioral Neuroscience, 8*, 420.

Fineran, P. C., & Charpentier, E. (2012). Memory of viral infections by CRISPR-Cas adaptive immune systems: Acquisition of new information. *Virology, 434*, 202–209.

Gilleen, J., Michalopoulou, P. G., Reichenberg, A., Drake, R., Wykes, T., Lewis, S. W., & Kapur, S. (2014). Modafinil combined with cognitive training is associated with improved learning in healthy volunteers–A randomised controlled trial. *European Neuropsychopharmacology, 24*, 529–539.

Gordon, S., Todder, D., Deutsch, I., Garbi, D., Alkobi, O., Shriki, O., ... & Meiran, N. (2019). Effects of neurofeedback and working memory-combined training on executive functions in healthy young adults. *Psychological Research, 84*, 1586–1609.

Gruzelier, J. H. (2014). EEG-neurofeedback for optimising performance. I: A review of cognitive and affective outcome in healthy participants. *Neuroscience & Biobehavioral Reviews, 44*, 124–141.

Habermas, J. (2003). *The future of human nature*. Cambridge: Polity Press.

Han, I., & Black, J. B. (2011). Incorporating haptic feedback in simulation for learning physics. *Computers & Education, 57*, 2281–2290.

Hanslmayr, S., Sauseng, P., Doppelmayr, M., Schabus, M., & Klimesch, W. (2005). Increasing individual upper alpha power by neurofeedback improves cognitive performance in human subjects. *Applied Psychophysiology and Biofeedback, 30*, 1–10.

Hoedlmoser, K., Pecherstorfer, T., Gruber, G., Anderer, P., Doppelmayr, M., Klimesch, W., & Schabus, M. (2008). Instrumental conditioning of human sensorimotor rhythm (12–15 Hz) and its impact on sleep as well as declarative learning. *Sleep, 31*, 1401.

Hommel, B. (2015). The theory of event coding (TEC) as embodied-cognition framework. *Frontiers in Psychology, 6*, 1318.

Hommel, B., & Kibele, A. (2016). Down with retirement: Implications of embodied cognition for healthy aging. *Frontiers in Psychology, 7*, 1184.

Hommel, B., Müsseler, J., Aschersleben, G., & Prinz, W. (2001). The theory of event coding (TEC): A framework for perception and action planning. *Behavioral and Brain Sciences, 24*, 849–878.

Horizon (2020) EU framework programme for research and innovation. http://ec.europa.eu/programmes/horizon2020/en/h2020-section/societal-challenges.

Juan, M. C., Alcaniz, M., Monserrat, C., Botella, C., Banos, R. M., & Guerrero, B. (2005). Using augmented reality to treat phobias. *Computer Graphics and Applications, IEEE, 25*, 31–37.

Jwa, A. (2015). Early adopters of the magical thinking cap: a study on do-it-yourself (DIY) transcranial direct current stimulation (tDCS) user community. *Journal of Law and the Biosciences, 2*, 292–335.

Kass, L. (2002). *Life, liberty, and defense of dignity: The challenge for bioethics*. San Francisco: Encounter Books.

Katz, B., Au, J., Buschkuehl, M., Abagis, T., Zabel, C., Jaeggi, S. M., & Jonides, J. (2017). Individual differences and long-term consequences of tDCS-augmented cognitive training. *Journal of Cognitive Neuroscience, 29*, 1498–1508.

Keizer, A. W., Verment, R., & Hommel, B. (2010). Enhancing cognitive control through neurofeedback: A role of gamma-band activity in managing episodic retrieval. *NeuroImage, 49*, 3404–3413.

Knecht, S., Breitenstein, C., Bushuven, S., Wailke, S., Kamping, S., Floel, A., Zwitserlood, P., & Ringelstein, E. B. (2004). Levodopa: Faster and better word learning in normal humans. *Annals of Neurology, 56*, 20–26.

Kuo, M. F., & Nitsche, M. A. (2012). Effects of transcranial electrical stimulation on cognition. *Clinical EEG and Neuroscience, 43*, 192–199.

Max Planck Institute on Human Development, Stanford Center on Longevity. A Consensus on the Brain Training Industry from the Scientific Community. 2014 Oct 20 [Cited 23 Oct 2019] In: *Stanford Centre on Longevity blog* [Internet]. Available: http://longevity3.stanford.edu/blog/2014/10/15/the-consensus-on-the-brain-training-industry-from-the-scientific-community-2/

Miyake, A., Friedman, N. P., Emerson, M. J., Witzki, A. H., Howerter, A., & Wager, T. D. (2000). The unity and diversity of executive functions and their contributions to complex "frontal lobe" tasks: A latent variable analysis. *Cognitive Psychology, 41*, 49–100.

Nitsche, M. A., & Paulus, W. (2011). Transcranial direct current stimulation–Update 2011. *Restorative Neurology and Neuroscience, 29*, 463–492.

Optale, G., Urgesi, C., Busato, V., Marin, S., Piron, L., Priftis, K., ... & Bordin, A. (2010). Controlling memory impairment in elderly adults using virtual reality memory training: A randomized controlled pilot study. *Neurorehabilitation and Neural Repair, 24*, 348–357.

Owen, A. M., Hampshire, A., Grahn, J. A., Stenton, R., Dajani, S., Burns, A. S., Howard, R. J., & Ballard, C. G. (2010). Putting brain training to the test. *Nature, 465*, 775–778.

Richmond, L. L., Wolk, D., Chein, J., & Olson, I. R. (2014). Transcranial direct current stimulation enhances verbal working memory training performance over time and near transfer outcomes. *Journal of Cognitive Neuroscience, 26*, 2443–2454.

Salkever, D. S. (1995). Updated estimates of earnings benefits from reduced exposure of children to environmental lead. *Environmental Research, 70*, 1–6.

Savulescu, J. (2009). Enhancement and fairness. In P. Healey (Ed.), *Tomorrow's people: The challenges of technologies for life- extension and enhancement*. London: Earthscan.

Schmiedek, F., Lövdén, M., & Lindenberger, U. (2010). Hundred days of cognitive training enhance broad cognitive abilities in adulthood: Findings from the COGITO study. *Frontiers in Neuroscience, 2*, 27.

Schultz, W., Dayan, P., & Montague, P. R. (1997). A neural substrate of prediction and reward. *Science, 275*, 1593–1599.

Schwartz, J. (1994). Low-level lead exposure and children's IQ: A meta-analysis and search for a threshold. *Environmental Research, 65*, 42–55.

Sherlin, L. H., Arns, M., Lubar, J., Heinrich, H., Kerson, C., Strehl, U., ... & Sterman, M. B. (2011). Neurofeedback and basic learning theory: Implications for research and practice. *Journal of Neurotherapy, 15*, 292–304.

Steenbergen, L., Sellaro, R., Hommel, B., Kühn, S., & Colzato, L. S. (2016). "Unfocus" on foc. us: Commercial tDCS headset impairs working memory. *Experimental Brain Research, 234*, 637–643.

Stepp, C. E., An, Q., & Matsuoka, Y. (2012). Repeated training with augmentative vibrotactile feedback increases object manipulation performance. *PLoS One, 7*, e32743.

Tajadura-Jiménez, A., Basia, M., Deroy, O., Fairhurst, M.T., Marquardt, N., & Bianchi-Berthouze, N. (2015). As light as your footsteps: Altering walking sounds to change perceived body weight, emotional state and gait. CHI (conference proceeding).

Wilson, M. (2002). Six views of embodied cognition. *Psychonomic Bulletin & Review, 9*, 625–636.

Zoefel, B., Huster, R. J., & Herrmann, C. S. (2011). Neurofeedback training of the upper alpha frequency band in EEG improves cognitive performance. *NeuroImage, 54*, 1427–1431.

Index

T. Strobach, J. Karbach (eds.), *Cognitive Training*,
https://doi.org/10.1007/978-3-030-39292-5

The manufacturer's authorised representative in the EU is Springer Nature Customer Service Centre GmbH, Europaplatz 3, 69115 Heidelberg, Germany. If you have any concerns regarding our products, please contact ProductSafety@springernature.com

Printed and bound by CPI Group (UK) Ltd, Croydon, CR0 4YY
30/09/2026
02225331-0004